Thermodynamics

主　编　杜　军　顾丛汇

内 容 简 介

With the rapid development of higher education, more international students major in Building Environment and Energy Application Engineering, Marine Engineering, and Thermal Energy and Power Engineering in China. The practicality and necessity of *Thermodynamics* in the above fields of engineering are taken into account. This book is based on the text *Fluid Mechanics: Fundamentals and Applications* by Y. A. Cengel and J. M. Cimbala, *Fundamentals of Thermal Fluid Sciences* by Yunus A. Cengel and Robert Turner, *Fifth Edition Thermodynamics: An Engineering Approach* by Yunus A. Cengel and Michael A. Boles, and *Fundamentals of Engineering Thermodynamics* by Michael J. Moran and Howard N. Shapiro.

This book is divided into ten chapters: Chapter 1 Introduction and Basic Concepts, Chapter 2 The First Law of Thermodynamics, Chapter 3 Thermodynamic Properties and Process of Ideal Gas, Chapter 4 The Second Law of Thermodynamics, Chapter 5 Properties of Real Gas and Vapor, Chapter 6 Atmospheric Air, Chapter 7 Flow and Compression of Gas and Steam, Chapter 8 Power Cycles, Chapter 9 Solution Thermodynamics, and Chapter 10 Refrigeration Cycles.

This book can be used as a text book of fundamental engineering thermodynamics for international students whom major in Building Environment and Energy Application Engineering, Marine Engineering, and Thermal Energy and Power Engineering, and also as a reference for relevant engineering professionals.

图书在版编目(CIP)数据

工程热力学 = Thermodynamics/杜军,顾丛汇主编
.—哈尔滨:哈尔滨工程大学出版社,2018.8(2022.1 重印)
ISBN 978 – 7 – 5661 – 1870 – 7

Ⅰ.①工… Ⅱ.①杜…②顾… Ⅲ.①工程热力学 – 高等学校 – 教材 Ⅳ.①TK123

中国版本图书馆 CIP 数据核字(2018)第 064537 号

选题策划 史大伟
责任编辑 张玮琪
封面设计 刘长友

出版发行	哈尔滨工程大学出版社
社　　址	哈尔滨市南岗区南通大街 145 号
邮政编码	150001
发行电话	0451 – 82519328
传　　真	0451 – 82519699
经　　销	新华书店
印　　刷	哈尔滨圣铂印刷有限公司
开　　本	787mm × 1 092mm　1/16
印　　张	21
字　　数	720 千字
版　　次	2018 年 8 月第 1 版
印　　次	2022 年 1 月第 3 次印刷
定　　价	55.00 元

http://www.hrbeupress.com
E-mail:heupress@ hrbeu.edu.cn

PREFACE

BACKGROUND

Thermodynamics is an exciting and fascinating subject that deals with energy, which is essential for sustenance of life, and thermodynamics has long been an essential part of engineering curricula all over the world. It has a broad application area ranging from microscopic organisms to common household appliances, transportation vehicles, power generation systems, and even philosophy. This introductory book contains sufficient material for two sequential courses in thermodynamics and it is suitable for foreign students that study in China, which major in Engineering Thermal Physics, Marine Engineering and Building Environment and Energy Application Engineering. Students are assumed to have an adequate background in calculus and physics.

OBJECTIVES

This book is intended for use as a textbook by undergraduate engineering students in their sophomore or junior year, and as a reference book for practicing engineers. The objectives of this text are:

- To cover the *basic principles* of thermodynamics.
- To present a wealth of real-world *engineering examples* to gives tudents a feel for how thermodynamics is applied in engineering practice.
- To develop an *intuitive understanding* of thermodynamics by emphasizing the physics and physical arguments.

It is our hope that this book, through its careful explanations of concepts and its use of numerous practical examples and figures, helps students develop the necessary skills to bridge the gap between knowledge and the confidence to properly apply knowledge.

PHILOSOPHY AND GOAL

The philosophy that contributed to the overwhelming popularity of the prior editions of this book has remained unchanged in this edition. Namely, our goal has been to offer an engineering textbook that

- Communicates directly to the minds of tomorrow's engineers in a *simple yet precise* manner.
- Leading students toward a clear understanding and firm grasp of the *basic principles* of thermodynamics.
- Encouraging *creative thinking* and development of a *deeper understanding* and *intuitive feel* for thermodynamics.

Special effort has been made to appeal to students' natural curiosity and to help them explore the various facets of the exciting subject area of thermodynamics. The enthusiastic responses we have received from users of prior editions—from small colleges to large universities all over the world—indicate that our objectives have largely been achieved. It is our philosophy that the best way to learn is by practice. Therefore, special effort is made throughout the book to reinforce material that was presented earlier.

Yesterday's engineer spent a major portion of his or her time substituting values into the formulas and obtaining numerical results. However, formula manipulations and number crunching are now being left mainly to computers. Tomorrow's engineer will need a clear understanding and a firm grasp of the *basic principles* so that he or she can understand even the most complex problems, formulate them, and interpret the results. A conscious effort is made to emphasize these basic principles while also providing students with a perspective of how computational tools are used in engineering practice.

The traditional classical, or macroscopic, approach is used throughout the text, with microscopic arguments serving in a supporting role as appropriate. This approach is more in line with students' intuition and makes learning the subject matter much easier.

Nomenclature

a	Acceleration, m/s^2
A	Area, m^2
c	Speed of sound, m/s
c	Specific heat, kJ/kg · K
c_p	Constant pressure specific heat, kJ/kg · K
c_v	Constant volume specific heat, kJ/kg · K
COP	Coefficient of performance
COP$_{HP}$	Coefficient of performance of a heat pump
COP$_R$	Coefficient of performance of a refrigerator
d	Diameter, m
e	Specific total energy, kJ/kg
E	Total energy, kJ
F	Force, N
g	Gravitational acceleration, m/s^2
G	Total Gibbs function, kJ
h	Specific enthalpy, kJ/kg
H	Enthalpy, kJ
k	Specific heat ratio
ke	Specific kinetic energy, kJ/kg
KE	Kinetic energy, kJ
m	Mass, kg
$\dot{m}$	Mass flow rate, kg/s
M	Molar mass, kg/kmol
Ma	Mach number
MEP	Mean effective pressure, kPa
mf	Mass fraction
n	Polytropic exponent
N	Number of moles, kmol
P	Pressure, kPa
P_{cr}	Critical pressure, kPa
P_m	Mixture pressure, kPa
P_r	Relative pressure, kPa
P_v	Vapor pressure, kPa
P_0	Surroundings pressure, kPa
pe	Specific potential energy, kJ/kg
PE	Potential energy, kJ
q	Heat transfer per unit mass, kJ/kg
Q	Total heat transfer, kJ

$\dot{Q}$	Heat transfer rate, kW	
Q_H	Heat transfer with high-temperature body, kJ	
Q_L	Heat transfer with low-temperature body, kJ	
r	Compression ratio	
R	Gas constant, kJ/kg·K	
r_c	Cutoff ratio	
r_p	Pressure ratio	
R_u	Universal gas constant, kJ/kmol·K	
s	Specific entropy, kJ/kg·K	
S	Total entropy, kJ/K	
s_{gen}	Specific entropy generation, kJ/kg·K	
S_{gen}	Total entropy generation, kJ/K	
t	Time, s	
T	Temperature, ℃ or K	
T_{cr}	Critical temperature, K	
T_{db}	Dry-bulb temperature, ℃	
T_{dp}	Dew-point temperature, ℃	
T_H	Temperature of high-temperature body, K	
T_L	Temperature of low-temperature body, K	
T_{wb}	Wet-bulb temperature, ℃	
T_0	Surroundings temperature, ℃ or K	
u	Specific internal energy, kJ/kg	
U	Total internal energy, kJ	
v	Specific volume, m³/kg	
v_{cr}	Critical specific volume, m³/kg	
V	Total volume, m³	
$\dot{V}$	Volume flow rate, m³/s	
V	Velocity, m/s	
V_{avg}	Average velocity, m/s	
w	Work per unit mass, kJ/kg	
W	Total work, kJ	
$\dot{W}$	Power, kW	
W_{in}	Work input, kJ	
W_{out}	Work output, kJ	
x	quality	
x	Specific exergy, kJ/kg	
X	Exergy, kJ	
x_{dest}	Specific exergy destruction, kJ/kg	
X_{dest}	Exergy destruction, kJ	
y	Mole fraction	
Z	Compressibility factor	

Nomenclature

Greek letters

α	Isothermal compressibility, 1/kPa
β	Volume expansivity, 1/K
Δ	Finite change in quantity
η_{th}	Thermal efficiency
μ_{JT}	Joule – Thomson coefficient, K/kPa
ρ	Density, kg/m^3
φ	Relative humidity
φ	Specific closed system exergy, kJ/kg
Φ	Total closed system exergy, kJ
ω	Specific or absolute humidity, kg H$_2$O/kg dry air

Subscripts

a	Air
abs	Absolute
act	Actual
atm	Atmospheric
avg	Average
c	Cross section
cr	Critical point
CV	Control volume
e	Exit conditions
f	Saturated liquid
fg	Different in property between saturated liquid and saturated vapor
g	Saturated vapor
gen	Generation
H	High temperature
i	Inlet conditions
i	ith component
L	Low temperature
m	Mixture
r	Relative
R	Reduced
rev	Reversible
s	Isentropic
sat	Saturated
surr	Surroundings
sys	system
v	Water vapor
0	Dead state
1	Initial or inlet state
2	Final or exit state

Contents

Chapter 1 Introduction and Basic Concepts ········· 1
- 1-1 Systems and Control Volumes ········· 1
- 1-2 State of Thermodynamic System and Properties of a System ········· 3
- 1-3 State and Equilibrium ········· 11
- 1-4 Processes and Cycles ········· 13
- 1-5 Work and Heat ········· 17
- Summary ········· 21
- Problems ········· 22

Chapter 2 The First Law of Thermodynamics ········· 24
- 2-1 Introduction ········· 24
- 2-2 Internal Energy and Total Energy ········· 25
- 2-3 Energy Transfer ········· 29
- 2-4 Enthalpy ········· 33
- 2-5 Energy Balance and Energy Change of a System ········· 34
- 2-6 Energy Analysis of Control Volumes ········· 37
- 2-7 Some Steady-Flow Engineering Devices ········· 42
- Summary ········· 43
- Problems ········· 43

Chapter 3 Thermodynamic Properties and Process of Ideal Gas ········· 46
- 3-1 The Ideal Gas Equation of State ········· 46
- 3-2 Specific Heats of Ideal Gases ········· 48
- 3-3 Internal Energy, Enthalpy and Entropy of Ideal Gases ········· 51
- 3-4 Thermodynamic Processes of Ideal Gases ········· 59
- 3-5 Ideal Gas Mixtures ········· 66
- Summary ········· 71
- Problems ········· 72

Chapter 4 The Second Law of Thermodynamics ········· 74
- 4-1 Introduction to the Second Law ········· 74
- 4-2 Reversible and Irreversible Processes ········· 80
- 4-3 Heat Engines and Refrigerators ········· 85
- 4-4 The Carnot Cycle ········· 94
- 4-5 The Carnot Principles ········· 96
- 4-6 Expressions of The Second Law of Thermodynamic ········· 104
- 4-7 Entropy Balance ········· 111
- 4-8 Exergy ········· 118
- Summary ········· 135
- Problems ········· 135

Chapter 5　Properties of Real Gas and Vapor　·················· 138
　5 – 1　The Real Gas Equation of State　·················· 138
　5 – 2　Property Diagrams and Saturation State　·················· 142
　5 – 3　Property Tables　·················· 155
　5 – 4　Properties of Refrigerant　·················· 164
　Summary　·················· 166
　Problems　·················· 166

Chapter 6　Atmospheric Air　·················· 168
　6 – 1　Atmospheric Air　·················· 168
　6 – 2　Parameters of Atmospheric Air　·················· 169
　6 – 3　The Psychrometric Chart　·················· 176
　6 – 4　Thermodynamic Processes of Atmospheric Air　·················· 178
　Summary　·················· 185
　Problems　·················· 185

Chapter 7　Flow and Compression of Gas and Steam　·················· 187
　7 – 1　Steady – Flow Characteristics of Gas and Steam　·················· 187
　7 – 2　Stagnation and Critical Properties of Gases　·················· 193
　7 – 3　Nozzles　·················· 198
　7 – 4　Adiabatic Throttling Process　·················· 203
　7 – 5　Processes in Compressor　·················· 206
　Summary　·················· 211
　Problems　·················· 212

Chapter 8　Power Cycles　·················· 214
　8 – 1　The Analysis of Power Cycles　·················· 214
　8 – 2　Air – Standard Assumption and Reciprocating Engines　·················· 217
　8 – 3　The Ideal Cycle for Internal Combustion Engines　·················· 220
　8 – 4　The Ideal and Actual Cycle for Gas – Turbine Engines　·················· 225
　8 – 5　Rankine Cycle for Vapor Power Cycles　·················· 228
　8 – 6　The Ideal Reheat Rankine Cycle　·················· 233
　8 – 7　The Ideal Regenerative Rankine Cycle　·················· 234
　Summary　·················· 237
　Problems　·················· 238

Chapter 9　Solution Thermodynamics　·················· 239
　9 – 1　Phase Equilibrium　·················· 239
　9 – 2　The Phase Rule　·················· 240
　9 – 3　Phase Equilibrium for a Multi-component System　·················· 240
　Summary　·················· 241
　Problems　·················· 242

Chapter 10　Refrigeration Cycles　·················· 243
　10 – 1　The Reversed Carnot Cycle　·················· 243
　10 – 2　Vapor – Compression Refrigeration Cycle　·················· 245

10 – 3	Absorption Refrigeration Cycle	248
10 – 4	Steam Jet Refrigeration Cycle	250
Summary		251
Problems		252

ANSWERS ... 253

- Chapter 1 ... 253
- Chapter 2 ... 256
- Chapter 3 ... 260
- Chapter 4 ... 263
- Chapter 5 ... 270
- Chapter 6 ... 273
- Chapter 7 ... 278
- Chapter 8 ... 281
- Chapter 9 ... 285
- Chapete 10 ... 286

APPENDIX ... 289

TABLE A – 1	Properties of air	289
TABLE A – 2	Molar mass, gas constant, and critical – point properties	291
TABLE A – 3	Properties of common liquids, solids, and foods	292
TABLE A – 4	Saturated water—Temperature table	295
TABLE A – 5	Saturated water—Pressure table	297
TABLE A – 6	Superheated water	300
TABLE A – 7	Compressed liguid water	306
TABLE A – 8	Saturated ice-water vapor	307
TABLE A – 9	Saturated refrigerant – 134a—Pressure table	308
TABLE A – 10	Ideal-gas specific heats of various common gases	310
TABLE A – 11	One-dimensional isentropic compressible flow functions for an ideal gas with $k = 1.4$	312
TABLE A – 12	Superheated water	313
FIGURE B – 1	$T - s$ diagram for water	320
FIGURE B – 2	Psychrometric chart at 1 atm total pressure	321
FIGURE B – 3	Psychrometric chart at 1 atm total pressure	322
FIGURE B – 4	Nelson – Obert generalized compressibility chart	323
FIGURE B – 5	Mollier diagram for water	324

Chapter 1 Introduction and Basic Concepts

Every science has a unique vocabulary associated with it, and thermodynamics is no exception. Precise definition of basic concepts forms a sound foundation for the development of a science and prevents possible misunderstandings. We start this chapter with an overview of thermodynamics and the unit systems, and continue with a discussion of some basic concepts such as *system*, *state*, *state postulate*, *equilibrium*, *and process*. We also discuss *temperature* and *temperature scales* with particular emphasis on the International Temperature Scale of 1990. We then present *pressure*, which is the normal force exerted by a fluid per unit area and discuss *absolute* and *gage* pressures, the variation of pressure with depth, and pressure measurement devices, such as manometers and barometers. Careful study of these concepts is essential for a good understanding of the topics in the following chapters.

1 – 1 Systems and Control Volumes

A system is defined as a *quantity of matter or a region in space chosen for study*. The mass or region outside the system is called the surroundings. The real or imaginary surface that separates the system from its surroundings is called the boundary. The boundary of a system can be *fixed* or *movable*. Note that the boundary is the contact surface shared by both the system and the surroundings. Mathematically speaking, the boundary has zero thickness, and thus it can neither contain any mass nor occupy any volume in space.

Systems may be considered to be *closed* or *open*, depending on whether a fixed mass or a fixed volume in space is chosen for study. A closed system is defined when a particular quantity of matter is under study. A closed system always contains the same matter. There can be no transfer of mass across its boundary. A special type of closed system that does not interact in any way with its surroundings is called an isolated system.

Fig. 1 – 1 shows a gas in a piston – cylinder assembly. When the valves are closed, we can consider the gas to be a closed system. The boundary lies just inside the piston and cylinder walls, as shown by the dashed lines on the figure. The portion of the boundary between the gas and the piston moves with the piston. No mass would cross this or any other part of the boundary.

Thermodynamics

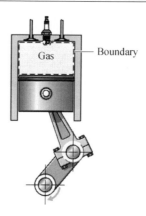

Fig. 1-1 Closed system: A gas in a piston-cylinder assembly

Consider the piston-cylinder device shown in Fig. 1-2. Let us say that we would like to find out what happens to the enclosed gas when it is heated. Since we are focusing our attention on the gas, it is our system. The inner surfaces of the piston and the cylinder form the boundary, and since no mass is crossing this boundary, it is a closed system. Notice that exergy may cross the boundary, and part of the boundary (the inner surface of the piston, in this case) may move. Everything outside the gas, including the piston and the cylinder, is the surroundings.

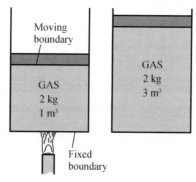

Fig. 1-2 A closed system with a moving boundary

In subsequent sections of this book, thermodynamic analyses are made of devices such as turbines and pumps through which mass flows. These analyses can be conducted in principle by studying a particular quantity of matter, a closed system, as it passes through the device. In most cases it is simpler to think instead in terms of a given region of space through which mass flows. With this approach, a region within a prescribed boundary is studied. The region is called a control volume or open system. Mass may cross the boundary of a control volume.

A large number of engineering problems involve mass flow in and out of a system and, therefore, are modeled as *control volumes*. A water heater, a car radiator, a turbine, and a compressor all involve mass flow and should be analyzed as control volumes (open systems) instead of as control masses (closed systems). In general, *any arbitrary region in space* can be selected as a control volume. There are no concrete rules for the selection of control volumes, but the proper choice certainly makes the analysis much easier. If we were to analyze the flow of air

Chapter 1 Introduction and Basic Concepts

through a nozzle, for example, a good choice for the control volume would be the region within the nozzle.

A diagram of an engine is shown in Fig. 1-3(a). The dashed line defines a control volume that surrounds the engine. Observe that air, fuel, and exhaust gases cross the boundary. A schematic such as in Fig. 1-3(b) often suffices for engineering analysis. The term *control mass* is sometimes used in place of closed system, and the term *opensystem* is used interchangeably with control volume. When the terms control mass and control volume are used, the system boundary is often referred to as a *control surface*.

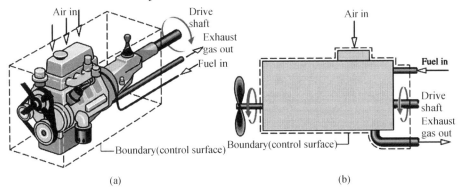

Fig. 1-3 **Example of a control volume (open system): an automobile engine**

Most control volumes, however, have fixed boundaries and thus do not involve any moving boundaries. A control volume can also involve heat and work interactions just as a closed system, in addition to mass interaction.

Let us say that we would like to determine how much heat we must transfer to the water in the tank in order to supply a steady stream of hot water. Since hot water will leave the tank and be replaced by cold-water, it is not convenient to choose a fixed mass as our system for the analysis. Instead, we can concentrate our attention on the volume formed by the interior surfaces of the tank and consider the hot and cold water streams as mass leaving and entering the control volume. For this case, the interior surfaces of the tank form the control surface, and the mass is crossing the control surface at two locations.

In an engineering analysis, the system under study *must* be defined carefully. In most cases, the system investigated is quite simple and obvious, and defining the system may seem like a tedious and unnecessary task.

1-2 State of Thermodynamic System and Properties of a System

A property is a macroscopic characteristic of a system such as mass, volume, exergy, pressure, and temperature to which a numerical value can be assigned at a given time without knowledge of the previous behavior (*history*) of the system. The list can be extended to include less familiar ones such as viscosity, thermal conductivity, modulus of elasticity, thermal expansion coefficient, electric resistivity, and even velocity and elevation.

Intensive properties are those that are independent of the mass of a system, such as temperature, pressure, and density. Extensive properties are those whose values depend on the size—or extent—of the system. Total mass, total volume, and total momentum are some examples of extensive properties.

An easy way to determine whether a property is intensive or extensive is to divide the system into two equal parts with an imaginary partition, as shown in Fig. 1 – 4. Each part will have the same value of intensive properties as the original system, but half the value of the extensive properties.

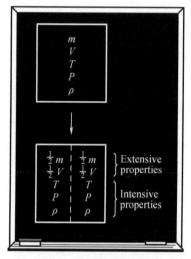

Fig. 1 – 4 **intensive and extensive properties**

Generally, uppercase letters are used to denote extensive properties (with mass m being a major exception), and lowercase letters are used for intensive properties (with pressure P and temperature T being the obvious exceptions). Extensive properties per unit mass are called specific properties. Some examples of specific properties are specific volume ($v = V/m$) and specific total exergy ($e = E/m$).

TEMPERATURE

Although we are familiar with temperature as a measure of "hotness" or "coldness," it is not easy to give an exact definition for it. Based on our physiological sensations, we express the level of temperature qualitatively with words like *freezing cold*, *cold*, *warm*, *hot*, and *red hot*. However, we cannot assign numerical values to temperatures based on our sensations alone. Furthermore, our senses may be misleading.

Fortunately, several properties of materials change with temperature in a *repeatable* and *predictable* way, and this forms the basis for accurate temperature measurement. The commonly used mercury in glass thermometer, for instance, is based on the expansion of mercury with temperature. Temperature is also measured by using several other temperature—dependent properties.

Chapter 1 Introduction and Basic Concepts

It is a common experience that a cup of hot coffee left on the table eventually cools off and a cold drink eventually warms up. That is, when a body is brought into contact with another body that is at a different temperature, heat is transferred from the body at higher temperature to the one which at lower temperature until both bodies attain the same temperature (Fig. 1-5). At that point, the heat transfer stops, and the two bodies are said to have reached thermal equilibrium. The equality of temperature is the only requirement for thermal equilibrium.

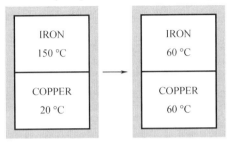

Fig. 1-5 **Two bodies reaching thermal equilibrium after being brought into contact in an isolated enclosure**

It is a matter of experience that when two bodies are in thermal equilibrium with a third body, they are in thermal equilibrium with one another. This statement, which is sometimes called the zeroth law of thermodynamics, is tacitly assumed in every measurement of temperature. Thus, if we want to know if two bodies are at the same temperature, it is not necessary to bring them into contact and see whether their observable properties change with time, as described previously. It is necessary only to see if they are individually in thermal equilibrium with a third body. The third body is usually a thermometer.

By replacing the third body with a thermometer, the zeroth law can be restated as *two bodies are in thermal equilibrium if both have the same temperature reading even if they are not in contact*. The zeroth law was first formulated and labeled by R. H. Fowler in 1931. As the name suggests, its value as a fundamental physical principle was recognized more than half a century after the formulation of the first and the second laws of thermodynamics. It was named the zeroth law since it should have preceded the first and the second laws of thermodynamics.

Temperature scales enable us to use a common basis for temperature measurements, and several have been introduced throughout history. All temperature scales are based on some easily reproducible states such as the freezing and boiling points of water, which are also called the *ice point* and the *steam point*, respectively. A mixture of ice and water that is in equilibrium with air saturated with vapor at 1 atm pressure is said to be at the ice point, and a mixture of liquid water and water vapor (with no air) in equilibrium at 1 atm pressure is said to be at the steam point.

The temperature scales used in the SI and in the English system today are the Celsius scale (formerly called the *centigrade scale*; in 1948 it was renamed after the Swedish astronomer A. Celsius, 1702-1744, who devised it) and the Fahrenheit scale (named after the German instrument maker G. Fahrenheit, 1686-1736), respectively. On the Celsius scale, the ice and steam points were originally assigned the values of 0 ℃ and 100 ℃, respectively. The corresponding values on the Fahrenheit scale are 32 °F and 212 °F.

In thermodynamics, it is very desirable to have a temperature scale that is independent of the properties of any substance or substances. Such a temperature scale is called a thermodynamic temperature scale, which is developed later in conjunction with the second law of thermodynamics.

The thermodynamic temperature scale in the SI is the Kelvin scale, named after Lord Kelvin (1824 – 1907). The temperature unit on this scale is the Kelvin, which is designated by K (not °K; the degree symbol was officially dropped from kelvin in 1967). The lowest temperature on the Kelvin scales absolute zero, or 0 K. Then it follows that only one nonzero reference point needs to be assigned to establish the slope of this linear scale. Using non-conventional refrigeration techniques, scientists have approached absolute zero Kelvin (they achieved 0.000 000 002 K in 1989). The thermodynamic temperature scale in the English system is the Rankine scale, named after William Rankine (1820 – 1872). The temperature unit on this scale is the rankine, which is designated by R.

A temperature scale that turns out to be nearly identical to the Kelvin scale is the ideal gas temperature scale. The temperatures on this scale are measured using a constant volume gas thermometer, which is basically arigid vessel filled with a gas, usually hydrogen or helium, at low pressure.

This thermometer is based on the principle that *at low pressures, the temperature of a gas is proportional to its pressure at constant volume*. That is, the temperature of a gas of fixed volume varies *linearly* with pressure at sufficiently low pressures. Then the relationship between the temperature and the pressure of the gas in the vessel can be expressed as

$$T = a + bP \qquad (1-1)$$

where the values of the constants a and b for a gas thermometer are determined experimentally. Once a and b are known, the temperature of a medium can be calculated from this relation by immersing the rigid vessel of the gas thermometer into the medium and measuring the gas pressure when thermal equilibrium is established between the medium and the gas in the vessel whose volume is held constant.

An ideal gas temperature scale can be developed by measuring the pressures of the gas in the vessel at two reproducible points (such as the ice and the steam points) and assigning suitable values to temperatures at those two points. Considering that only one straight line passes through two fixed points on a plane, these two measurements are sufficient to determine the constants a and b in Eq. (1 – 1). Then the unknown temperature T of a medium corresponding to a pressure reading P can be determined from that equation by a simple calculation. The values of the constants will be different for each thermometer, and depending on the type and the amount of the gas in the vessel, and the temperature values assigned at the two reference points.

In this case the value of the constant a (which corresponds to an absolute pressure of zero) is determined to be 273.15 ℃ regardless of the type and the amount of the gas in the vessel of the gas thermometer. That is, on a $P - T$ diagram, all the straight lines passing through the data points in this case will intersect the temperature axis at 273.15 ℃ when extrapolated, as shown in

Chapter 1 Introduction and Basic Concepts

Fig. 1-8. This is the lowest temperature that can be obtained by a gas thermometer, and thus we can obtain an *absolute gas temperature scale* by assigning a value of zero to the constant a in Eq. (1-1). In that case Eq. (1-1) reduces to $T = bP$, and thus we need to specify the temperature at only *one* point to define an absolute gas temperature scale.

It should be noted that the absolute gas temperature scale is not a thermodynamic temperature scale, since it cannot be used at very low or very high temperature (due to condensation and dissociation and ionization). However, absolute gas temperature is identical to the thermodynamic temperature in the temperature range in which the gas thermometer can be used, and thus we can view the thermodynamic temperature scale at this point as an absolute gas temperature scale that utilizes an "ideal" or "imaginary" gas that always acts as a low-pressure gas regardless of the temperature. If such a gas thermometer existed, it would read zero Kelvin at absolute zero pressure, which corresponds to 273.15 ℃ on the Celsius scale (Fig. 1-6).

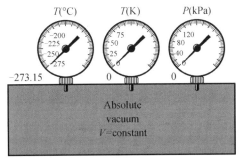

Fig. 1-6 A constant volume gas thermometer would read 273.15 ℃ at absolute zero pressure

The Kelvin scale is related to the Celsius scale and the Rankine scale is related to the Fahrenheit scale by

$$T(K) = T(℃) + 273.15 \quad \text{and} \quad T(R) = T(℉) + 459.67 \quad \text{and} \quad T(℃) = \frac{T(℉) - 32}{1.8}$$

$$(1-2)$$

The reference temperature chosen in the original Kelvin scale was 273.15 K (or 0 ℃), which is the temperature at which water freezes (or ice melts) and water exists as a solid-liquid mixture in equilibrium under standard atmospheric pressure (the *ice point*). The Celsius scale was redefined at this conference in terms of the ideal gas temperature scale and a single fixed point, which is again the triple point of water with an assigned value of 0.01 ℃. The boiling temperature of water (the *steam point*) was experimentally determined to be again 100.00 ℃, and thus the new and old Celsius scales were in good agreement.

The *International Temperature Scale of* 1990, was adopted by the International Committee of Weights and Measures at its meeting in 1989 at the request of the Eighteenth General Conference on Weights and Measures. The ITS-90 is similar to its predecessors except that it is more refined with updated values of fixed temperatures, has an extended range, and conforms more closely to the thermodynamic temperature scale. On this scale, the unit of thermodynamic temperature T is again the Kelvin (K), defined as the fraction 1/273.16 of the thermodynamic temperature of the triple point of water, which is sole defining fixed point of both the ITS − 90 and the Kelvin scale and is the most important thermometric fixed point used in the calibration of thermometers to ITS − 90.

We emphasize that the magnitudes of each division of 1 K and 1℃ are identical. Therefore, when we are dealing with temperature differences T, the temperature interval on both scales is the same. Raising the temperature of a substance by 10 ℃ is the same as raising it by 10 K. That is,

$$\Delta T(\text{K}) = \Delta(\text{℃}) \qquad (1-3)$$

Some thermodynamic relations involve the temperature T and often the question arises of whether it is in K or ℃. If the relation involves temperature differences (such as $a = bT$), it makes no difference and either can be used. However, if the relation involves temperatures only instead of temperature differences (such as $a = bT$) then K must be used. When in doubt, it is always safe to use K because there are virtually no situations in which the use of K is incorrect, but there are many thermodynamic relations that will yield an erroneous result if ℃ is used.

PRESSURE

Pressure is defined as *a normal force exerted by a fluid per unit area*. We speak of pressure only when we deal with a gas or a liquid. The counter part of pressure in solids is *normal stress*. Since pressure is defined as force per unit area, it has the unit of newtons per square meter (N/m^2), which is called a pascal (Pa). That is,

$$1 \text{ Pa} = 1 \text{ N/m}^2$$

The pressure unit pascal is too small for pressures encountered in practice. Therefore, its multiples *kilopascal* ($1 \text{ kPa} = 10^3 \text{ Pa}$) and *megapascal* ($1 \text{ MPa} = 10^6 \text{ Pa}$) are commonly used. Three other pressure units commonly used in practice, especially in Europe, are *bar*, *standard atmosphere*, and *kilogram − force per square centimeteras below*:

$$1 \text{ bar} = 10^5 \text{ Pa} = 0.1 \text{ MPa} = 100 \text{ kPa}$$
$$1 \text{ atm} = 101,325 \text{ Pa} = 101.325 \text{ kPa} = 1.01325 \text{ bars}$$
$$1 \text{ kgf/cm}^2 = 9.807 \text{ N/cm}^2 = 9.807 \times 10^4 \text{ N/m}^2 = 9.807 \times 10^4 \text{ Pa}$$
$$= 0.9807 \text{ bar}$$
$$= 0.9679 \text{ atm}$$

Pressure is also used for solids as synonymous to *normal stress*, which is force acting perpendicular to the surface per unit area. For example, a 150 pound person with a total foot imprint area of 50 in^2 exerts a pressure of $150 = \text{lbf}/50 \text{ in}^2 = 3.0$ psi on the floor (Fig. 1.10). If the person stands on one foot, the pressure doubles. If the person gains excessive weight, he or she is likely to encounter foot discomfort because of the increased pressure on the foot (the size of the foot does not change with weight gain). This also explain show a person can walk on fresh snow without sinking by wearing large snowshoes, and how a person cuts with little effort when using a sharp knife.

The actual pressure at a given position is called the absolute pressure, and it is measured relative to absolute vacuum (i.e., absolute zero pressure). Most pressure-measuring devices, however, are calibrated to read zero in the atmosphere, and so they indicate the difference between the absolute pressure and the local atmospheric pressure. This difference is called the gage pressure. Pressures below atmospheric pressure are called vacuum pressures and are measured by vacuum gages that indicate the difference between the atmospheric pressure and the absolute pressure. Absolute, gage, and vacuum pressures are all positive quantities and are related to each other by

$$P_g = P_a - P_{atm} \qquad (1-4)$$
$$P_v = P_{atm} - P_a \qquad (1-5)$$

This is illustrated in Fig. 1-7.

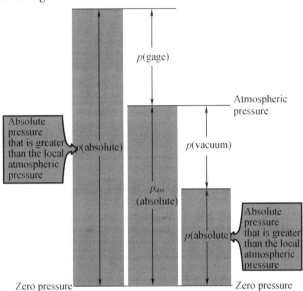

Fig. 1-7 Absolute, gage, and vacuum pressures

In thermodynamic relations and tables, absolute pressure is almost always used. Throughout this text, the pressure P will denote *absolute pressure* unless specified otherwise. Often, the letters "a" and "g" represent absolute pressure and gape pressure, respectively. They are added to pressure units to clarify what is meant.

EXAMPLE 1-1 Absolute Pressure of a Vacuum Chamber

A vacuum gage connected to a chamber reads 5.8 psi at a location where the atmospheric pressure is 14.5 psi. Determine the absolute pressure in the chamber.

Solution The gage pressure of a vacuum chamber is given. The absolute pressure in the chamber is to be determined.

Analysis The absolute pressure is easily determined from Eq. (1-5) to be

$$P_a = P_{atm} - P_v = 14.5 - 5.8 = 8.7 \text{ psi}$$

Discussion Note that the local value of the atmospheric pressure is used when determining the absolute pressure.

Pressure is the *compressive force* per unit area, and it gives the impression of being a vector. However, pressure at any point in a fluid is the same in all directions. That is, it has magnitude but not a specific direction, and thus it is a scalar quantity.

Thermodynamics

DENSITY, SPECIFIC VOLUME AND SPECIFIC WEIGHT

Density is defined as *mass per unit volume*.

$$\rho = \frac{m}{V} \; (\text{kg/m}^3) \qquad (1-6)$$

The reciprocal of density is the specific volume v, which is defined as *volume per unit mass*. That is,

$$v = \frac{V}{m} = \frac{1}{\rho} \; (\text{m}^3/\text{kg}) \qquad (1-7)$$

The weight of a unit volume of a substance is called specific weight and is expressed as

$$\gamma_s = \rho g \quad (\text{N/m}^3) \qquad (1-8)$$

where g is the gravitational acceleration.

The densities of liquids are essentially constant, and thus they can often be approximated as being incompressible substances during most processes without sacrificing much in accuracy.

EXAMPLE 1-2 Obtaining Formulas from Unit Considerations

A tank is filled with oil whose density is r = 850 kg/m^3. If the volume of the tank is V = 2 m^3, determine the amount of mass m in the tank.

Solution The volume of an oil tank is given. The mass of oil is to be determined.

Assumptions Oil is an incompressible substance and thus its density is constant.

Analysis A sketch of the system just described is given in Fig. 1-8. Suppose we forgot the formula that relates mass to density and volume. However, we know that mass has the unit of kilograms. That is, whatever calculations we do, we should end up with the unit of kilograms. Putting the given information into perspective, we have

Fig. 1-8 Schematic for Example 1-2

$$\rho = 850 \text{ kg/m}^3 \quad \text{and} \quad V = 2 \text{ m}^3$$

It is obvious that we can eliminate m^3 and end up with kg by multiplying these two quantities. Therefore, the formula we are looking for should be

$$m = \rho V$$

Thus,

$$m = (850 \text{ kg/m}^3)(2 \text{ m}^3) = 1\,700 \text{ kg}$$

Discussion Note that this approach may not work for more complicated formulas.

You should keep in mind that a formula that is not dimensionally homogeneous is definitely wrong, but a dimensionally homogeneous formula is not necessarily right.

1 – 3 State and Equilibrium

The word state refers to the condition of a system as described by its properties. Since there are normally relations among the properties of a system, the state often can be specified by providing the values of a subset of the properties. All other properties can be determined in terms of these few. At a given state, all properties of the system have fixed values. If the value of a property changes, the state will change to a different one. In Fig. 1 – 9, a system is shown at two different states.

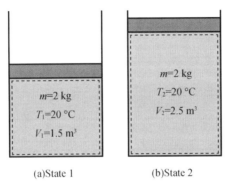

(a)State 1   (b)State 2

Fig. 1 – 9 A system at two different states

Thermodynamics deals with *equilibrium* states. The word equilibrium implies a state of balance. In an equilibrium state there are no unbalanced potentials (or driving forces) within the system. A system in equilibrium experiences no changes when it is isolated from its surroundings.

There are many types of equilibrium, and a system is not in thermodynamic equilibrium unless the conditions of all the relevant types of equilibrium are satisfied. For example, a system is in thermal equilibrium if the temperature is the same throughout the entire system, as shown in Fig. 1 – 10. That is, the system involves no temperature differential, which is the driving force for heat flow.

Mechanical equilibrium is related to pressure, and a system is in mechanical equilibrium if there is no change in pressure at any point of the system with time. However, the pressure may vary within the system with elevation as a result of gravitational effects. For instance, the higher pressure at a bottom layer is balanced by the extra weight it must carry, and, therefore, there is no imbalance of forces. The variation of pressure as a result of gravity in most thermodynamic systems is relatively small and usually disregarded. If a system involves two phases, it is in phase equilibrium when the mass of each phase reaches an equilibrium level and stays there. Finally, a system is in chemical equilibrium means its chemical composition does not change with time, namely, no chemical reactions occurs. A system will not be in equilibrium unless all the relevant equilibrium criteria are satisfied.

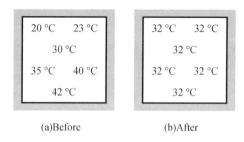

Fig. 1 – 10　A closed system reaching thermal equilibrium

As noted earlier, the state of a system is described by its properties. However, we know from experience that we do not need to specify all the properties in order to fix a state. Once a sufficient number of properties are specified, the rest of the properties assume certain values automatically. That is, specifying a certain number of properties is sufficient to fix a state. The number of properties required to fix the state of a system is given by the state postulate: The state of a simple compressible system is completely specified by two independent, intensive properties.

A system is called a simple compressible system in the absence of electrical, magnetic, gravitational, motion, and surface tension effects. These effects are due to external force fields and are negligible for most engineering problems. Otherwise, an additional property needs to be specified for each effect that is significant. If the gravitational effects are to be considered, for example, the elevation z needs to be specified in addition to the two properties necessary to fix the state.

The state postulate requires that the two properties specified be independent to fix the state. Two properties are independent if one property can be varied while the other one is held constant. Temperature and specific volume, for example, are always independent properties, and together they can fix the state of a simple compressible system (Fig. 1 – 11). Temperature and pressure, however, are independent properties for single – phase systems, butare dependent properties for multi-phase systems.

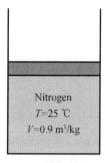

Fig. 1 – 11　The state of nitrogen is fixed by two independent, intensive properties

Chapter 1　Introduction and Basic Concepts

1−4　Processes and Cycles

QUASI−EQUILIBRIUM PROCESS

Any change that a system undergoes from one equilibrium state to another is called a process, and the series of states through which a system passes during a process is called the path of the process (Fig. 1−12).

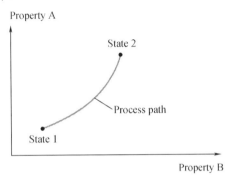

Fig. 1−12　A process between states 1 and 2 and the process path

To describe a process completely, one should specify the initial and final states of the process, as well as the path it follows, and the interactions with the surroundings. When a process proceeds in such a manner that the system remains infinitesimally close to an equilibrium state at all times, it is called a quasi-static, or quasi-equilibrium process. A quasi-equilibrium process can be viewed as a sufficiently slow process that allows the system to adjust it self internally so that properties in one part of the system do not change any faster than those at other parts.

This is illustrated in Fig. 1−13. When a gas in a piston−cylinder device is compressed suddenly, the molecules near the face of the piston will not have enough time to escape and they will have to pile up in a small region in front of the piston, thus creating a high-pressure region there. Because of this pressure difference, the system can no longer be said to be in equilibrium, and this makes the entire process nonquasi-equilibrium. However, if the piston is moved slowly, the molecules will have sufficient time to redistribute and there will not be a molecule pile up in front of the piston. As a consequence, the pressure inside the cylinder will always be nearly uniform and will rise at the same rate at all locations. Since equilibrium is maintained at all times, this is a quasi−equilibrium process.

It should be pointed out that a quasi-equilibrium process is an idealized process and is not a true representation of an actual process. But many actual processes closely approximate it, and they can be modeled as quasi-equilibrium with negligible error. Engineers are interested in quasi-equilibrium processes for two reasons. First, they are easy to analyze; second, work−pro dicing devices deliver the most work when they operate on quasi-equilibrium processes. Therefore, quasi-equilibrium processes serve as standards to which actual processes can be compared.

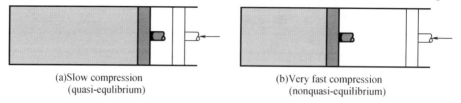

(a)Slow compression
(quasi-equlibrium)

(b)Very fast compression
(nonquasi-equilibrium)

Fig. 1-13   Quasi-equilibrium and nonquasi-equilibrium compression processes

Process diagrams plotted by employing thermodynamic properties as coordinates are very useful in visualizing the processes. Some common properties that are used as coordinates are temperature T, pressure P, and volume V (or specific volume v). Fig. 1-14 shows the $P-V$ diagram of a compression process of a gas.

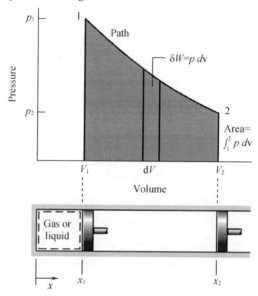

Fig. 1-14   The $P-V$ diagram of a expansion or compression process

Note that the process path demonstrates a series of equilibrium states through which the system passes during a process and has significance for quasi-equilibrium processes only. For nonquasi-equilibrium processes, we are notable to characterize the entire system by a single state, and thus we cannot speak of a process path for a system as a whole. A nonquasi-equilibrium process is denoted by a dashed line between the initial and final states instead of a solid line.

The prefix *iso-* is often used to designate a process for which a particular property remains constant. An isothermal process, for example, is a process during which the temperature T remains constant; an isobaric process is a process during which the pressure P remains constant; and an isochoric (or isometric) process is a process during which the specific volume v remains constant.

Chapter 1   Introduction and Basic Concepts

REVERSIBLE PROCESS

Once having taken place, these processes cannot reverse themselves spontaneously and restore the system to its initial state. For this reason, they are classified as *irreversible processes*. Once a cup of hot coffee cools, it will not heat up by retrieving the heat it lost from the surroundings. If it could, the surroundings, as well as the system (coffee), would be restored to their original condition, and this would be a reversible process.

A process is called irreversible if the system and all parts of its surroundings cannot be exactly restored to their respective initial states after the process has occurred. A process is reversible if both the system and surroundings can be returned to their initial states. Irreversible processes are the subject of the present discussion. Reversible processes are considered again later in the section.

A system that has undergone an irreversible process is not necessarily precluded from being restored to its initial state. However, were the system restored to its initial state, it would not be possible also to return the surroundings to the state they were in initially. As illustrated below, the second law can be used to determine whether both the system and surroundings can be returned to their initial states after a process has occurred. That is, the second law can be used to determine whether a given process is reversible or irreversible.

Reversible processes actually do not occur in nature. They are merely *idealizations* of actual processes. Reversible processes can be approximated by actual devices, but they can never be achieved. That is, all the processes occurring in nature are irreversible. You may be wondering, then, *why* we are bothering with such fictitious processes. There are two reasons. First, they are easy to analyze, since a system passes through a series of equilibrium states during a reversible process; second, they serve as idealized models to which actual processes can be compared.

THE STEADY-FLOW PROCESS

A process is a transformation from one state to another. However, if a system exhibits the same values of its properties at two different times, it is in the same state at these times. A system is said to be at steady state if none of its properties changes with time.

The terms *steady* and *uniform* are used frequently in engineering, and thus it is important to have a clear understanding of their meanings. The term *steady* implies *no change with time*. The opposite of steady is *unsteady*, or *transient*. The term *uniform*, however, implies *no change with location* over a specified region. These meanings are consistent with their everyday use (steady girlfriend, uniform properties, etc.).

A large number of engineering devices operate for long periods of time under the same conditions, and they are classified as *steady-flow devices*. Processes involving such devices can be represented reasonably well by a somewhat idealized process, called the steady-flow process, which can be defined as a *process during which a fluid flows through a control volume steadily* (Fig. 1-15).

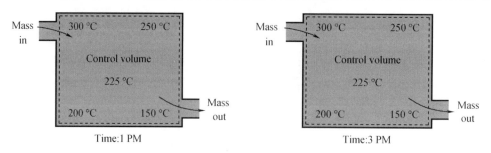

Fig. 1-15 During a steady-flow process, fluid properties within the control volume may change with position but not with time

That is, the fluid properties can change from point to point within the control volume, but at any fixed point they remain the same during the entire process. Therefore, the volume V, the mass m, and the total exergy content E of the control volume remain constant during a steadyflow process (Fig. 1-16).

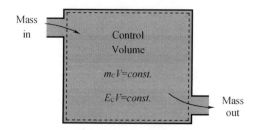

Fig. 1-16 Under steady-flow conditions, the mass and exergy contents of a control volume remain constant

Steady-flow conditions can be closely approximated by devices that are intended for continuous operation such as turbines, pumps, boilers, condensers, and heat exchangers or power plants or refrigeration systems. Some cyclic devices, such as reciprocating engines or compressors, do not satisfy any of the conditions stated above since the flow at the inlets and the exits will be pulsating and not steady. However, the fluid properties vary with time in a periodic manner, and the flow through these devices can still be analyzed as a steady-flow process by using time-averaged values for the properties.

CYCLES

A system is said to have undergone a cycle if it returns to its initial state at the end of the process. That is, for a cycle the initial and final states are identical. If every process in a cycle is reversible, then it can be called reversible cycle. If not, called irreversible cycle.

1-5 Work and Heat

WORK

There are several different ways of doing work, each in some way related to a force acting through a distance. In elementary mechanics, the work done by a constant force F on a body displaced a distance x in the direction of the force is given by

$$W = Fx \ (\text{kJ}) \qquad (1-9)$$

If the force F is not constant, the work done is obtained by adding (i.e., integrating) the differential amounts of work,

$$W = \int_1^2 F dx \ (\text{kJ}) \qquad (1-10)$$

Obviously one needs to know how the force varies with displacement to perform this integration. Eq. (1-9) and (1-10) give only the magnitude of the work. The sign is easily determined from physical considerations: The work done on a system by an external force acting in the direction of motion is negative, and work done by a system against an external force acting in the opposite direction to motion is positive.

WORK FOR QUASI-EQUILIBRIUM PROCESS

One form of mechanical work frequently encountered in practice is associated with the expansion or compression of a gas in a piston-cylinder device. During this process, part of the boundary (the inner face of the piston) moves back and forth. Therefore, the expansion and compression work is often called moving boundary work, or simply boundary work (Fig. 1-17). Some call it the $P dV$ work for reasons explained later. Moving boundary work is the primary form of work involved in *automobile engines*. During their expansion, the combustion gases force the piston to move, which in turn forces the crank shaft to rotate.

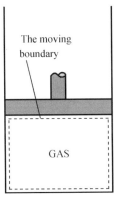

Fig. 1-17 The work associated with a moving boundary is called *boundary work*

The moving boundary work associated with real engines or compressors cannot be determined exactly from a thermodynamic analysis alone because the piston usually moves at very high speeds, making it difficult for the gas inside to maintain equilibrium. Then the states through which the system passes during the process cannot be specified, and no process path can be drawn. Work, which means being function, cannot be determined analytically without a knowledge of the path. Therefore, the boundary work in real engines or compressors is determined by direct measurements.

In this section, we analyze the moving boundary work for a *quasi-equilibrium process*, a process during which the system remains nearly inequilibrium at all times. A quasi-equilibrium process, also called a *quasistatic process*, is closely approximated by real engines, especially when the piston moves at low velocities. Under identical conditions, the work output of the engines is found to be a maximum, and the work input to the compressors to be a minimum when quasi-equilibrium processes are used inplace of nonquasi-equilibrium processes. Below, the work associated with a moving boundary is evaluated for a quasi-equilibrium process.

Consider the gas enclosed in the piston – cylinder device shown in Fig. 1 – 18.

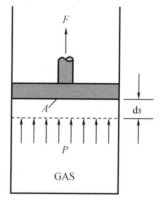

Fig. 1 – 18 A gas does a differential amount of work dWb as it forces the piston to move by a differential amount ds

The initial pressure of the gas is P, the total volume is V, and the cross sectional area of the piston is A. If the piston is allowed to move a distance ds in a quasi-equilibrium manner, the differential work done during this process is

$$\delta W_b = F\,ds = PA\,ds = P\,dV \qquad (1-11)$$

That is, the boundary work in the differential form is equal to the product of the absolute pressure Pa and the differential change in the volume dV of the system. This expression also explains why the moving boundary work is sometimes called the $Pa\,dV$ work.

Note in Eq. (1 – 11) that Pa is the absolute pressure, which is always positive. However, the volume change dV is positive during an expansion process (volume increasing) and negative during a compression process (volume decreasing). Thus, the boundary work is positive during an expansion process and negative during a compression process. If the volume increases, the work can be called expansion work. If decreases, called compression work. Therefore, Eq. (1 – 11) can, be viewed as an expression for boundary work output, W_b, out. A negative result indicates

Chapter 1　Introduction and Basic Concepts

boundary work input (compression).

The total boundary work done during the entire process as the piston moves is obtained by adding all the differential works from the initial state to the final state:

$$W_b = \int_1^2 P dV \quad (\text{kJ}) \tag{1-12}$$

This integral can be evaluated only if we know the functional relationship between Pa and V during the process. That is, $Pa = f(V)$ should be available. Note that $Pa = f(V)$ is simply the equation of the process path on a $P - V$ diagram.

The quasi-equilibrium expansion process described is shown on a $P - V$ diagram in Fig. 1-19.

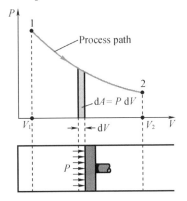

Fig. 1-19　The area under the process curve on a $P - V$ diagram represents the boundary work

On this diagram, the differential area dA is equal to $P \, dV$, which is the differential work. The total area A under the process curve 1-2 is obtained by adding these differential areas:

$$\text{Area} = A = \int_1^2 dA = \int_1^2 P dV \tag{1-13}$$

A comparison of this equation with Eq. (1-12) reveals that *the area under the process curve on a $P - V$ diagram is equal, in magnitude, to the work done during a quasi-equilibrium expansion or compression process of a closed system.* (On the $P - v$ diagram, it represents the boundary work done per unit mass.)

A gas can follow several different paths as it expands from state 1 to state 2. In general, each path will have a different area underneath it, and since this area represents the magnitude of the work, the work done will be different for each process (Fig. 1-20). This is expected, since work is a path function (i.e., it depends on the path followed as well as the end states). If work were not a path function, no cyclic devices (car engines, power plants) could operate as work-producing devices. The work produced by these devices during one part of the cycle would have to be consumed during another part, and there would be no net work output. The cycle shown in Fig. 1-21 produces a net work output because the work done by the system during the expansion process (area under path A) is greater than the work done on the system during the compression part of the cycle (area under path B), and the difference between these two is the net work done during the cycle (the colored area).

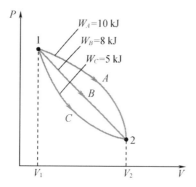
Fig. 1-20 The boundary work done during a process depends on the path followed as well as the end states

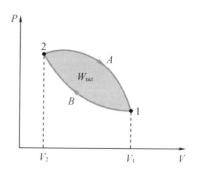
Fig. 1-21 The net work done during a cycle is the difference between the work done by the system and the work done on the system

If the relationship between P and V during an expansion or a compression process is given in terms of experimental data instead of in a functional form, obviously we cannot perform the inte gration analytically. But we can always plot the $P-V$ diagram of the process, using these data points, and calculate the area underneath graphically to determine the work done.

For a closed system, useful work is a fraction of expansion work, like lifting heavy objects. Some expansion work is dissipated by fiction, some is used to against atmosphere, as a result, the rest is useful work. W_u, W_1 and W_r represent useful work, dissipated work by fiction and work a gainst atmosphere, respectively.

$$W_u = W - W_r - W_1 \qquad (1-14)$$

The atmosphere pressure is constant, then

$$W_r = p_0(V_2 - V_1) = p_0 \Delta V \qquad (1-15)$$

For a reversible process, no dissipated work occurs, then W_f is equal to 0.

$$W_{u,re} = \int_1^2 p dv - p_0(V_2 - V_1) \qquad (1-16)$$

HEAT

Heat is defined as *the form of exergy that is transferred between two systems (or a system and its surroundings) by virtue of a temperature difference*. That is, an exergy interaction is heat only if it takes place because of a temperature difference. Then it follows that there cannot be any heat transfer between two systems that are at the same temperature. The heat for a process $1-2$, as shown in Fig. 1-22, can be described as

$$\delta q = T ds \qquad (1-17)$$

$$q_{1-2} = \int_1^2 T ds \qquad (1-18)$$

where s is called entropy, which is a property associated with the second law of thermodynamics.

Chapter 1   Introduction and Basic Concepts

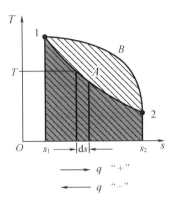

Fig. 1-22   The area under the process curve on a T-s diagram represents the heat

Work, like heat, is an exergy interaction between a system and its surroundings. As mentioned earlier, exergy can cross the boundary of a closed system in the form of heat or work. Therefore, *if the exergy crossing the boundary of a closed system is not heat, it must be work*. Heat is easy to recognize: Its driving force is a temperature difference between the system and its surroundings. Then we can simply say that an exergy interaction that is not caused by a temperature difference between a system and its surroundings is work. More specifically, *work is the exergy transfer associated with a force acting through a distance*. A rising piston, a rotating shaft, and an electric wire crossing the system boundaries are all associated with work interactions.

Heat and work are *exergy transfer mechanisms* between a system and its surroundings, and there are many similarities between them:

a. Both are recognized at the boundaries of a system as they cross the boundaries. That is, both heat and work are *boundary* phenomena.

b. Systems possess exergy, but not heat or work.

c. Both are associated with a *process*, not a state. Unlike properties, heat or work has no meaning at a state.

d. Both are *path functions* (i. e., their magnitudes depend on the path followed during a process as well as the end states).

Summary

In this chapter, the basic concepts of thermodynamics are introduced and discussed. *Thermodynamics* is the science that primarily deals with exergy. The *first law of thermodynamics* is simply an expression of the conservation of exergy principle, and it asserts that *exergy* is a thermodynamic property.

A system of fixed mass is called a *closed system*, or *control mass*, and a system that involves mass transfer across its boundaries is called an *open system*, or *control volume*. The mass-dependent properties of a system are called *extensive properties* and the others *intensive properties*. *Density* is mass per unit volume, and *specific volume* is volume per unit mass, and *specific weight* is weight per unit volume. The *zeroth law of thermodynamics* states that two bodies are in thermal equilibrium if both have the same temperature reading even if they are not in contact. The normal force exerted by a fluid per unit area is called *pressure*, and its unit is the *pascal*, 1 Pa = 1 N/m². The pressure relative to absolute vacuum is called the *absolute pressure*, and the difference between the absolute pressure and the local atmospheric pressure is called the *gage pressure*. Pressures below atmospheric pressure are called *vacuum pressures*.

A system is said to be in *thermodynamic equilibrium* if it maintains thermal, mechanical, phase, and chemical equilibrium. Any change from one state to another is called a *process*. A process with identical end states is called a *cycle*. During a *quasi-static* or *quasi-equilibrium process*, the system remains practically in equilibrium at all times. The state of a simple, compressible system is completely specified by two independent, intensive properties. *Work* and *heat* are also introduced.

Problems

1 – 1   A barometer is used to measure the height of a building by recording reading at the bottom and at the top of the building, as shown in Fig. P1 – 1. The height of the building is to be determined.

1 – 2   When measuring small pressure differences with a manometer, often one arm of the manometer is inclined to improve the accuracy of reading. (The pressure difference is still proportional to the *vertical* distance and not the actual length of the fluid along the tube.) The air pressure in a circular duct is to be measured using a manometer whose open arm is inclined 35° from the horizontal, as shown in Fig. P1 – 2. The density of the liquid in the manometer is 0.81 kg/L, and the vertical distance between the fluid levels in the two arms of the manometer is 8 cm. Determine the gage pressure of air in the duct and the length of the fluid column in the inclined arm above the fluid level in the vertical arm.

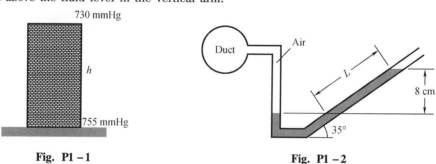

Fig. P1 – 1   **Fig. P1 – 2**

1 – 3   The water in a tank is pressurized by air, and the pressure is measured by a multifluid manometer as shown in Fig. P1 – 3. The tank is located on a mountain at an altitude of 1 400 m where the atmospheric pressure is 85.6 kPa. Determine the air pressure in the tank if $h_1 = 0.1$ m, $h_2 = 0.2$ m, and $h_3 = 0.35$ m. Take the densities of water, oil, and mercury to be 1 000 kg/m³, 850 kg/m³, and 13 600 kg/m³, respectively.

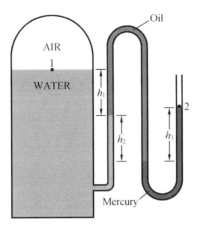

Fig. P1 –3

1 –4 A slow expansion of gas in the cylinder, volume of gas increases from 0.1 m³ to 0.25 m³, and relationship between pressure and volume is $\{P\}$ MPa $= 0.24 - 0.4 \{V\}$ m³, in the process. Friction between cylinder and piston is 1 200 N, the local atmospheric pressure is 0.1 MPa, area of cylinder is 0.1 m². Determine:

(1) expansion work of gas;
(2) useful work of the system output;
(3) useful work of the system output if no friction.

Chapter 2 The First Law of Thermodynamics

Whether we realize it or not, energy is an important part of most aspects of daily life. The quality of life, and even its sustenance, depends on the availability of energy. Therefore, it is important to have a good understanding of the sources of energy, the conversion of energy from one form to another, and the ramifications of these conversions. energy exists in numerous forms such as thermal, mechanical, electric, chemical, and nuclear. Even mass can be considered a form of energy. Energy can be transferred to or from a closed system (a fixed mass) in two distinct forms: *heat* and *work*. For control volumes, energy can also be transferred by mass flow. An energy transfer to or from a closed system is *heat* if it is caused by a temperature difference.

We start this chapter with a discussion of various forms of energy and energy transfer by heat. We continue with developing a general intuitive expression for the *first law of thermodynamics*, also known as the *conservation of energy principle*, which is one of the most fundamental principles in nature, and we then demonstrate its use. We discussed internal energy, total energy and enthalpy, detailed treatments of the first law of thermodynamics for open systems and control volumes are given in this chapter.

2 - 1 Introduction

We are familiar with the conservation of energy principle, which is an expression of the first law of thermodynamics, back from our high school years. We are told repeatedly that energy cannot be created or destroyed during a process; it can only change from one form to another.

The *first law of thermodynamics*, also known as *the conservation of energy principle*, provides a sound basis for studying the relationships among the various forms of energy and energy interactions. Based on experimental observations, the first law of thermodynamics states that *energy can be neither created nor destroyed during a process*; *it can only change forms*. Therefore, every bit of energy should be accounted for during a process.

We all know that a rock at some elevation possesses some potential energy, and part of this potential energy is converted to kinetic energy as the rock falls (Fig. 2 - 1). Experimental data show that the decrease in potential energy (mgz) exactly equals the increase in kinetic energy when the air resistance is negligible, thus confirming the conservation of energy principle for mechanical energy.

Consider a system undergoing a series of *adiabatic* processes from a specified state 1 to another specified state 2. Being adiabatic, these processes obviously cannot involve any heat transfer, but they may involve several kinds of work interactions. Careful measurements during these experiments indicate the following: *For all adiabatic processes between two specified states of a closed system*, *the net work done is the same regardless of the nature of the closed system and the details of the process*. Considering that there are an infinite number of ways to perform work interactions under adiabatic conditions, this statement appears to be very powerful, with a potential for far-reaching implications. This statement, which is largely based on the experiments of Joule in the first half of the nineteenth century, cannot be drawn from any other known physical principle and is recognized as a fundamental principle. This principle is called the first law of thermodynamics or just the first law.

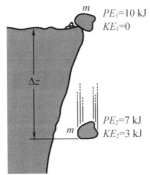

Fig. 2-1 energy cannot be created or destroyed; it can only change forms

2-2 Internal Energy and Total Energy

INTERNAL ENERGY

In thermodynamic analysis, it is often helpful to consider the various forms of energy that make up the total energy of a system in two groups: *macroscopic* and *microscopic*. The macroscopic forms of energy are those a system possesses as a whole with respect to some outside reference frame, such as kinetic and potential energies (Fig. 2-2). The microscopic forms of energy are those related to the molecular structure of a system and the degree of the molecular activity, and they are independent of outside reference frames. The sum of all the microscopic forms of energy is called the internal energy of a system and is denoted by U.

Fig. 2-2 The macroscopic energy of an object changes with velocity and elevation

Internal energy is defined earlier as the sum of all the *microscopic* forms of energy of a system. It is related to the *molecular structure* and the degree of *molecular activity* and can be viewed as the sum of the *kinetic* and *potential* energies of the molecules.

To have a better understanding of internal energy, let us examine a system at the molecular level. The molecules of a gas move through space with some velocity, and thus possess some kinetic energy. This is known as the *translational energy*. The atoms of polyatomic molecules rotate about an axis, and the energy associated with this rotation is the *rotational kinetic energy*. The atoms of a polyatomic molecule may also vibrate about their common center of mass, and the energy associated with this back-and-forth motion is the *vibrational kinetic energy*. For gases, the kinetic energy is mostly due to translational and rotational motions, with vibrational motion becoming significant at higher temperatures. The electrons in an atom rotate about the nucleus, and thus possess *rotational kinetic energy*. Electrons at outer orbits have larger kinetic energies. Electrons also spin about their axes, and the energy associated with this motion is the *spin energy*. Other particles in the nucleus of an atom also possess spin energy. The portion of the internal energy of a system associated with the kinetic energies of the molecules is called the sensible energy. The average velocity and the degree of activity of the molecules are proportional to the temperature of the gas. Therefore, at higher temperatures, the molecules possess higher kinetic energies, and as a result the system has a higher internal energy.

The internal energy is also associated with various *binding forces* between the molecules of a substance, between the atoms within a molecule, and between the particles within an atom and its nucleus. The forces that bind the *molecules* to each other are, as one would expect, strongest in solids and weakest in gases. If sufficient energy is added to the molecules of a solid or liquid, the molecules overcome these molecular forces and break away, turning the substance into a gas. This is a phase-change process. Because of this added energy, a system in the gas phase is at a higher internal energy level than it is in the solid or the liquid phase. The internal energy associated with the phase of a system is called the latent energy.

An atom consists of neutrons and positively charged protons bound together by very strong nuclear forces in the nucleus, and negatively charged electrons orbiting around it. The internal energy associated with the atomic bonds in a molecule is called chemical energy. During a chemical reaction, such as a combustion process, some chemical bonds are destroyed while others are formed. As a consequence, the internal energy changes. The nuclear forces are much larger than the forces that bind the electrons to the nucleus. The tremendous amount of energy associated with the strong bonds with in the nucleus of the atom itself is called nuclear energy. Obviously, we need not be concerned with nuclear energy in thermodynamics unless, of course, we deal with fusion or fission reactions. A chemical reaction involves changes in the structure of the electrons of the atoms, but a nuclear reaction involves changes in the core or nucleus. Therefore, an atom preserves its identity during a chemical reaction but loses it during anuclear reaction. Atoms may also possess *electric* and *magnetic dipole moment energies* when subjected to external electric and magnetic fields due to the twisting of the magnetic dipoles produced by the small electric currents associated with the orbiting electrons.

In daily life, we frequently refer to the sensible and latent forms of internal energy as *heat*, and we talk about heat content of bodies. In thermodynamics, however, we usually refer to those forms of energy as thermal energy to prevent any confusion with *heat transfer*. Thermal energy is

determined by two independent status properties as
$$u = f(T,v), \quad u = f(T,p) \quad \text{or} \quad u = f(p,v) \qquad (2-1)$$

TOTAL ENERGY

Energy can exist in numerous forms such as thermal, mechanical, kinetic, potential, electric, magnetic, chemical, and nuclear, and their sum constitutes the total energy E of a system. The total energy of a system on a *unit mass* basis is denoted by e and is expressed as
$$e = \frac{E}{m} \quad (\text{kJ/kg}) \qquad (2-2)$$

A major consequence of the first law is the existence and the definition of the property *total energy* E. Considering that the net work is the same for all adiabatic processes of a closed system between two specified states, the value of the net work must depend on the end states of the system only, and thus it must correspond to a change in a property of the system. This property is the *total energy*. Note that the first law makes no reference to the value of the total energy of a closed system at a state. It simply states that the *change* in the total energy during an adiabatic process must be equal to the net work done. Therefore, any convenient arbitrary value can be assigned to total energy at a specified state to serve as a reference point.

The macroscopic energy of a system is related to motion and the influence of some external effects such as gravity, magnetism, electricity, and surface tension. The energy that a system possesses as a result of its motion relative to some reference frame is called kinetic energy (KE). When all parts of a system move with the same velocity, the kinetic energy is expressed as
$$KE = m\frac{V^2}{2} \quad (\text{kJ}) \qquad (2-3)$$
or, on a unit mass basis,
$$ke = \frac{V^2}{2} \quad (\text{kJ/kg}) \qquad (2-4)$$
where V denotes the velocity of the system relative to some fixed reference frame. The kinetic energy of a rotating solid body is given by $\frac{1}{2}I\omega^2$ where I is the moment of inertia of the body and ω is the angular velocity.

The energy that a system possesses as a result of its elevation in a gravitational field is called potential energy (PE) and is expressed as
$$PE = mgz \quad (\text{kJ}) \qquad (2-5)$$
or, on a unit mass basis,
$$pe = gz \quad (\text{kJ/kg}) \qquad (2-6)$$
where g is the gravitational acceleration and z is the elevation of the center of gravity of a system relative to some arbitrarily selected reference level. The magnetic, electric, and surface tension effects are significant in some specialized cases only and are usually ignored. In the absence of such effects, the total energy of a system consists of the kinetic, potential, and internal energies and is expressed as

$$E = U + KE + PE = U + m\frac{V^2}{2} + mgz \quad (\text{kJ}) \qquad (2-7)$$

or, on a unit mass basis,

$$e = u + ke + pe = u + \frac{V^2}{2} + gz \quad (\text{kJ/kg}) \qquad (2-8)$$

Most closed systems remain stationary during a process and thus experience no change in their kinetic and potential energies. Closed systems whose velocity and elevation of the center of gravity remain constant during a process are frequently referred to as stationary systems. The change in the total energy E of a stationary system is identical to the change in its internal energy U. In this text, a closed system is assumed to be stationary unless stated otherwise.

Distinction should be made between the macroscopic kinetic energy of an object as a whole and the microscopic kinetic energies of its molecules that constitute the sensible internal energy of the object (Fig. 2-3).

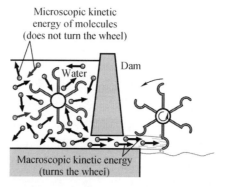

Fig. 2-3 The *macroscopic* kinetic energy is an organized form of energy and is much more useful than the disorganized *microscopic* kinetic energies of the molecules

The kinetic energy of an object is an *organized* form of energy associated with the orderly motion of all molecules in one direction in a straight path or around an axis. In contrast, the kinetic energies of the molecules are completely *random* and highly *disorganized*. As you will see in later chapters, the organized energy is much more valuable than the disorganized energy, and a major application area of thermodynamics is the conversion of disorganized energy (heat) into organized energy (work). You will also see that the organized energy can be converted to disorganized energy completely, but only a fraction of disorganized energy can be converted to organized energy by specially built devices called *heat engines* (like car engines and power plants). A similar argument can be given for the macroscopic potential energy of an object as a whole and the microscopic potential energies of the molecules.

2 – 3 Energy Transfer

HEAT AND WORK

Energy can cross the boundary of a closed system in two distinct forms: *heat* and *work* (Fig. 2 – 4). It is important to distinguish between these two forms of energy. Therefore, they will be discussed first, to form a sound basis for the development of the laws of thermodynamics.

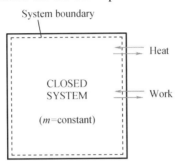

Fig. 2 – 4 Energy can cross the boundaries of a closed system in the form of heat and work

We know from experience that a can of cold soda left on a table eventually warms up and that a hot baked potato on the same table cools down. When a body is left in a medium that is at a different temperature, energy transfer takes place between the body and the surrounding medium until thermal equilibrium is established, that is, the body and the medium reach the same temperature. The direction of energy transfer is always from the higher temperature body to the lower temperature one. Once the temperature equality is established, energy transfer stops. In the processes described above, energy is said to be transferred in the form of heat.

Heat is defined as *the form of energy that is transferred between two systems (or a system and its surroundings) by virtue of a temperature difference*. That is, an energy interaction is heat only if it takes place because of a temperature difference. Then it follows that there cannot be any heat transfer between two systems that are at the same temperature.

Several phrases in common use today—such as heat flow, heat addition, heat rejection, heat absorption, heat removal, heat gain, heat loss, heat storage, heat generation, electrical heating, resistance heating, frictional heating, gas heating, heat of reaction, liberation of heat, specific heat, sensible heat, latent heat, waste heat, body heat, process heat, heat sink, and heatsource—are not consistent with the strict thermodynamic meaning of the term *heat*, which limits its use to the *transfer* of thermal energy during a process.

However, these phrases are deeply rooted in our vocabulary, and they are used by both ordinary people and scientists without causing any misunderstanding since they are usually interpreted properly instead of being taken literally. (Besides, no acceptable alternatives exist for some of these phrases.) For example, the phrase *body heat* is understood to mean *the thermal energy content* of a body. Likewise, *heat flow* is understood to mean *the transfer of thermal energy*, not the flow of a fluid like substance called heat, although the latter incorrect interpretation, which is based on the caloric theory, is the origin of this phrase. Also, the transfer of heat into a system is frequently referred to as *heat addition* and the transfer of heat out of a system as *heat rejection*. Perhaps there are thermodynamic reasons for being so reluctant to replace *heat* by *thermal energy*: It takes less time and energy to say, write, and comprehend *heat* than it does *thermal energy*.

Heat is energy in transition. It is recognized only as it crosses the boundary of a system. Once in the surroundings, the transferred heat becomes part of the internal energy of the surroundings. Thus, in thermodynamics, the term *heat* simply means *heat transfer*.

Work, like heat, is an energy interaction between a system and its surroundings. As mentioned earlier, energy can cross the boundary of a closed system in the form of heat or work. Therefore, *if the energy crossing the boundary of a closed system is not heat, it must be work*. Heat is easy to recognize: Its driving force is a temperature difference between the system and its surroundings. Then we can simply say that an energy interaction that is not caused by a temperature difference between a system and its surroundings is work. More specifically, *work is the energy transfer associated with a force acting through a distance*. A rising piston, a rotating shaft, and an electric wire crossing the system boundaries are all associated with work interactions.

Work is also a form of energy transferred like heat and, therefore, has energy units such as kJ. The work done during a process between states 1 and 2 is denoted by W_{12}, or simply W. The work done *per unit time* is called power and is denoted $\dot{W}$. The unit of power is kJ/s, or kW.

Heat and work are *directional quantities*, and thus the complete description of a heat or work interaction requires the specification of both the *magnitude* and *direction*. One way of doing that is to adopt a sign convention. The generally accepted formal sign convention for heat and work interactions is as follows: *heat transfer to a system and work done by a system are positive*; *heat transfer from a system and work done on a system are negative*. Another way is to use the subscripts *in* and *out* to indicate direction.

Thus far, we have been careful to emphasize that the quantities symbolized by W and Q in the foregoing equations account for transfers of *energy* and not transfers of work and heat, respectively. The terms work and heat denote different *means* whereby energy is transferred and not what is transferred. However, to achieve economy of expression in subsequent discussions, W and Q are often referred to simply as work and heat transfer, respectively. This less formal manner of speaking is commonly used in engineering practice.

For example, a work input of 8 kJ can be expressed as $W_{in} = 8$ kJ, while a heat loss of 5 kJ can be expressed as $Q_{out} = 5$ kJ. When the direction of a heat or work interaction is not known, we can simply *assume* a direction for the interaction (using the subscript *in* or *out*) and solve for it. A positive result indicates the assumed direction is right. A negative result, on the other hand, indicates that the direction of the interaction is the opposite of the assumed direction. This is just like assuming a direction for an unknown force when solving a statics problem, and reversing the direction when a negative result is obtained for the force. We will use this *intuitive approach* in this book as it eliminates the need to adopt a formal sign convention and the need to carefully assign negative values to some interactions.

EXAMPLE 2 – 1 Heating of a Potato in an Oven

A potato initially at room temperature (25 ℃) is being baked in an oven that is maintained at 200 ℃, as shown in Fig. 2 – 5. Is there any heat transfer during this baking process?

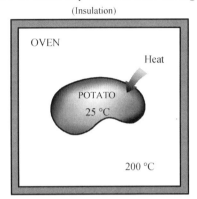

Fig. 2 – 5   Schematic for Example 2 – 1

Solution A potato is being baked in an oven. It is to be determined whether there is any heat transfer during this process.

Analysis This is not a well-defined problem since the system is not specified. Let us assume that we are observing the potato, which will be our system. Then the skin of the potato can be viewed as the system boundary. Part of the energy in the oven will pass through the skin to the potato. Since the driving force for this energy transfer is a temperature difference, this is a heat transfer process.

EXAMPLE 2 – 2 Heating of an Oven by Work Transfer

A well-insulated electric oven is being heated through its heating element. If the entire oven, including the heating element, is taken to be the system, determine whether this is a heat or work interaction.

Solution A well-insulated electric oven is being heated by its heating element. It is to be determined whether this is a heat or work interaction.

Analysis For this problem, the interior surfaces of the oven form the system boundary, as shown in Fig. 2 – 6. The energy content of the oven obviously increases during this process, as evidenced by a rise in temperature. This energy transfer to the oven is not caused by a temperature difference between the oven and the surrounding air. Instead, it is caused by *electrons* crossing the system boundary and thus doing work. Therefore, this is a work interaction.

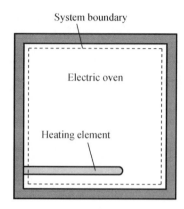

Fig. 2 – 6 Schematic for Example 2 – 2

FLOW WORK

Unlike closed systems, control volumes involve mass flow across their boundaries, and some work is required to push the mass into or out of the control volume. This work is known as the flow work, or flow energy, and is necessary for maintaining a continuous flow through a control volume.

To obtain a relation for flow work, consider a fluid element of volume V as shown in Fig. 2 – 7. The fluid immediately up stream forces this fluid element to enter the control volume; thus, it can be regarded as an imaginary piston. The fluid element can be chosen to be sufficiently small so that it has uniform properties throughout.

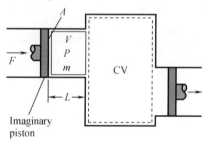

Fig. 2 – 7 Schematic for flow work

If the fluid pressure is P and the cross-sectional area of the fluid element is A (Fig. 2 – 8),

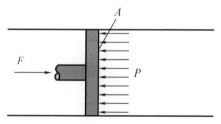

Fig. 2 – 8 In the absence of acceleration, the force applied on a fluid by a piston is equal to the force applied on the piston by the fluid

Chapter 2 The First Law of Thermodynamics

The force applied on the fluid element by the imaginary piston is
$$F = PA \quad (\text{kJ/kg}) \tag{2-9}$$
To push the entire fluid element into the control volume, this force must act through a distance L. Thus, the work done in pushing the fluid element across the boundary (i.e., the flow work) is
$$W_{\text{flow}} = FL = PAL = PV \quad (\text{kJ}) \tag{2-10}$$
The flow work per unit mass is obtained by dividing both sides of this equation by the mass of the fluid element:
$$w_{\text{flow}} = Pv \quad (\text{kJ/kg}) \tag{2-11}$$
The flow work relation is the same whether the fluid is pushed into or out of the control volume (Fig. 2-9).

(a) Before entering   (b) After entering

Fig. 2-9 Flow work is the energy needed to push a fluid into or out of a control volume, and it is equal to Pv

It is interesting that unlike other work quantities, flow work is expressed in terms of properties. In fact, it is the product of two properties of the fluid. For that reason, some people view it as a *combination property* (like enthalpy) and refer to it as *flow energy*, *convected energy*, or *transport energy* instead off low work. Others, however, argue rightfully that the product Pv represents energy for flowing fluids only and does not represent any form of energy for nonflow (closed) systems. Therefore, it should be treated as work. This controversy is not likely to end, but it is comforting to know that both arguments yield the same result for the energy balance equation. In the discussions that follow, we consider the flow energy to be part of the energy of a flowing fluid, since this greatly simplifies the energy analysis of control volumes.

2-4 Enthalpy

In the analysis of certain types of processes, particularly in power generation and refrigeration (Fig. 2-10), we frequently encounter the combination of properties $u + Pv$.

In many thermodynamic analyses the sum of the internal energy u and the product of pressure P and volume v appear. Because the sum $u + Pv$ occurs so frequently in subsequent discussions, it is convenient to give the combination a name, enthalpy, and a distinct symbol, H. By definition and given the symbol h:

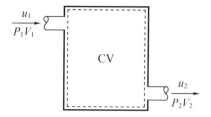

Fig. 2 – 10 The combination $u = Pv$ is frequently encountered in the analysis of control volumes

$$h = u + Pv \quad (\text{kJ/kg}) \quad (2-12)$$

or,

$$H = U + PV \quad (\text{kJ}) \quad (2-13)$$

Both the total enthalpy H and specific enthalpy h are simply referred to as enthalpy since the context clarifies which one is meant. Notice that the equations given above are dimensionally homogeneous. That is, the unit of the pressure – volume product may differ from the unit of the internal energy by only a factor. For example, it can be easily shown that $1 \text{ kPa} \cdot \text{m}^3 = 1 \text{ kJ}$. In some tables encountered in practice, the internal energy u is frequently not listed, but it can always be determined from $u = h - Pv$.

The wide spread use of the property enthalpy is due to Professor Richard Mollier, who recognized the importance of the group $u + Pv$ in the analysis of steam turbines and in the representation of the properties of steam in tabular and graphical form (as in the famous Mollier chart). Mollier referred to the group $u + Pv$ as *heat content* and *total heat*.

2 – 5 Energy Balance and Energy Change of a System

ENERGY BALANCE

As our previous discussions indicate, the *only ways* the energy of a closed system can be changed are through transfer of energy by work or by heat. Further, based on the experiments of Joule and others, a fundamental aspect of the energy concept is that *energy is conserved*. The conservation of energy principle can be expressed as follows: *The net change (increase or decrease) in the total energy of the system during a process is equal to the difference between the total energy entering and the total energy leaving the system during that process*. These considerations are summarized in words as follows:

$$\begin{pmatrix} \text{Total energy} \\ \text{entering the system} \end{pmatrix} - \begin{pmatrix} \text{Total energy} \\ \text{leaving the system} \end{pmatrix} = \begin{pmatrix} \text{Change in the total} \\ \text{energy of the system} \end{pmatrix}$$

or

$$E_{\text{in}} - E_{\text{out}} = \Delta E_{\text{system}} \quad (2-14)$$

The energy balance can be expressed on a per unit mass basis as

$$e_{\text{in}} - e_{\text{out}} = \Delta e_{\text{system}} \quad (2-15)$$

This relation is often referred to as the energy balance and is applicable to any kind of system undergoing any kind of process. The successful use of this relation to solve engineering problems depends on understanding the various forms of energy and recognizing the forms of energy transfer.

For a closed system undergoing a cycle, the initial and final states are identical, and thus $\Delta E_{system} = E_2 - E_1 = 0$. Then the energy balance for a cycle simplifies to $E_{in} - E_{out} = 0$ or $E_{in} = E_{out}$. Noting that a closed system does not involve any mass flow across its boundaries, the energy balance for a cycle can be expressed in terms of heat and work interactions as

$$W_{net,out} = Q_{net,in} \text{ (for a cycle)} \quad (2-16)$$

That is, the net work output during a cycle is equal to net heat input (Fig. 2-11).

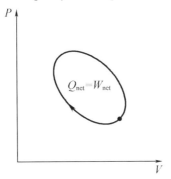

Fig. 2-11 For a cycle $\Delta E = 0$, thus $Q = W$

The energy balance (or the first law) relations already given are intuitive in nature and are easy to use when the magnitudes and directions of heat and work transfers are known. However, when performing a general analytical study or solving a problem that involves an unknown heat or work interaction, we need to assume a direction for the heat or work interactions. In such cases, it is common practice to use the classical thermodynamics sign convention and to assume heat to be transferred *into the system* (heat input) in the amount of Q and work to be done *by the system* (work output) in the amount of W, and then to solve the problem. The energy balance relation in that case for a closed system becomes

$$Q_{net,in} - W_{net,out} = \Delta E_{system} \quad \text{or} \quad Q - W = \Delta E \text{ (for a cycle)} \quad (2-17)$$

where $Q = Q_{net,in} = Q_{in} - Q_{out}$ is the *net heat input* and $W = W_{net,out} = W_{out} - W_{in}$ is the *net work output*. Obtaining a negative quantity for Q or W simply means that the assumed direction for that quantity is wrong and should be reversed.

The first law cannot be proven mathematically, but no process in nature is known to have violated the first law, and this should be taken as sufficient proof. Note that if it were possible to prove the first law on the basis of other physical principles, the first law then would be a consequence of those principles instead of being a fundamental physical law itself. As energy quantities, heat and work are not that different, and you probably wonder why we keep distinguishing them. After all, the change in the energy content of a system is equal to the amount of energy that crosses the system boundaries, and it makes no difference whether the energy crosses the boundary as heat or work. It seems as if the first law relations would be much simpler if we had just one quantity that we could call *energy interaction* to represent both heat and work. Well, from the first — law point of view, heat and work are not different at all.

EXAMPLE 2 – 3 energy Balance in a Closed System

A fan is set in a system, which is filled with gas, as shown in Fig. 2 – 12. The initial internal energy of gas is 800 kJ. When the fan works, the input work by the fan is 100kJ, at this time, heat is transferred to outside through boundary and the heat is 500kJ. Determine (a) the internal energy of gas when the fan works; (b) the input work by fan when the internal energyenergy of gas in the closed system remains constant.

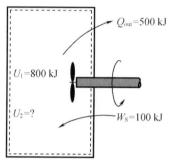

Fig. 2 – 12   Schematic for Example 2 – 3

Solution The closed system includes gas and a fan. The internal energy of gas U_2 and input work by fan in certain conditions are to be determined.

Assumptions We take the gas and fan as a closed system.

Analysis The energy balance for a closed system can be expressed as

$$Q = \Delta U + W_{total}$$
$$W_{total} = W_s$$

(a) The internal energy of gas when the fan works is determined to be

$$U_2 = Q + U_1 - W_s = -500 \text{ kJ} + 800 \text{ kJ} - (-100) \text{kJ} = 400 \text{ kJ}$$

(b) The internal energy of gas keeps constant, thus $\Delta U = 0$.

Thus, the input work by fan when the internal energy of gas in the closed system remains constant is determined to be

$$W_s = -Q = 500\text{kJ}$$

Discussion " – " is used to express the input work and " + " to output work in a closed system.

ENERGY CHANGE OF A SYSTEM

The determination of the energy change of a system during a process involves the evaluation of the energy of the system at the beginning and at the end of the process, and taking their difference. That is,

energy change = Energy at final state – Energy at initial state

or

$$\Delta E_{\text{system}} = E_{\text{final}} - E_{\text{initial}} = E_2 - E_1 \qquad (2-18)$$

Note that energy is a property, and the value of a property does not change unless the state of the system changes. Therefore, the energy change of a system is zero if the state of the system does not change during the process.

Also, energy can exist in numerous forms such as internal (sensible, latent, chemical, and nuclear), kinetic, potential, electric, and magnetic, and their sum constitutes the *total energy E* of a system. In the absence of electric, magnetic, and surface tension effects (i.e., for simple compressible systems), the change in the total energyenergy of a system during a process is the sum of the changes in its internal, kinetic, and potential energies and can be expressed as

$$\Delta E = \Delta U + \Delta KE + \Delta PE \qquad (2-19)$$

where

$$\Delta U = m(u_2 - u_1)$$

$$\Delta KE = \frac{1}{2} m(V_2^2 - V_1^2)$$

$$\Delta PE = m(z_2 - z_1)$$

When the initial and final states are specified, the values of the specific internal energies u_1 and u_2 can be determined directly from the property tables or thermodynamic property relations.

Most systems encountered in practice are stationary, that is, they do not involve any changes in their velocity or elevation during a process. Thus, for stationary systems, the changes in kinetic and potential energies are zero (that is, $\Delta KE = \Delta PE = 0$), and the total energy change relation in Eq. (2-19) reduces to $\Delta E = \Delta U$ for such systems. Also, the energy of a system during a process will change even if only one form of its energy changes while the other forms of energy remain unchanged.

2-6 Energy Analysis of Control Volumes

OPEN SYSTEM ENERGY EQUATION

In an open system, as shown in Fig. 2-13.

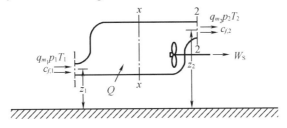

Fig. 2-13 Energy balance in an open system

The energy into the system can be expressed as

$$dE_1 + p_1 dV_1 + \delta Q \qquad (2-20)$$

The energy leaving the system can be expressed as
$$dE_2 + p_2 dV_2 + \delta W_S \quad (2-21)$$
The total energy change in the control volume is dE_{CV}.

The energy balance in an open system can be expressed as
$$dE_1 + p_1 dV_1 + \delta Q - (dE_2 + p_2 dV_2 + \delta W_S) = dE_{CV} \quad (2-22)$$
or
$$\delta Q = dE_{CV} + (dE_2 + p_2 dV_2) - (dE_1 + p_1 dV_1) + \delta W_S \quad (2-23)$$
where $E = me$, $V = mv$ and $h = u + pv$, then
$$\delta Q = dE_{CV} + (h_2 + \frac{c_{f,2}^2}{2} + gz_2)\delta m_2 - (h_1 + \frac{c_{f,1}^2}{2} + gz_1)\delta m_1 + \delta W_S \quad (2-24)$$
$$\delta Q = dE_{CV} + \sum_j (h + \frac{c_f^2}{2} + gz)_{out}\delta m_{out} - \sum_i (h + \frac{c_f^2}{2} + gz)_{in}\delta m_{in} + \delta W_S \quad (2-25)$$

ENERGY ANALYSIS OF STEADY-FLOW SYSTEMS

A large number of engineering devices such as turbines, compressors, and nozzles operate for long periods of time under the same conditions once the transient start-up period is completed and steady operation is established, and they are classified as *steady-flow devices*. Processes involving such devices can be represented reasonably well by a somewhat idealized process, called the steady-flow process, which was defined as *a process during which a fluid flows through a control volume steadily*. That is, the fluid properties can change from point to point within the control volume, but at any point, they remain constant during the entire process. (Remember, *steady* means *no change with time.*)

During a steady-flow process, no intensive or extensive properties *within the control volume* change with time. Thus, the volume V, the mass m, and the total energy content E of the control volume remain constant (Fig. 2-14). As a result, the boundary work is zero for steady-flow systems (since V_{CV} = constant), and the total mass or energy entering the control volume must be equal to the total mass or energy leaving it (since m_{CV} = constant and E_{CV} = constant). These observations greatly simplify the analysis.

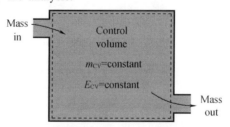

Fig. 2-14 Under steady-flow conditions, the mass and energy contents of a control volume remain constant

The fluid properties at an inlet or exit remain constant during a steady-flow process. The properties may, however, be different at different inlets and exits. They may even vary over the cross section of an inlet or an exit. However, all properties, including the velocity and elevation, must remain constant with time at a fixed point at an inlet or exit. It follows that the mass flow rate of the fluid at an opening must remain constant during a steady – flow process (Fig. 2 – 15). As an added simplification, the fluid properties a tan opening are usually considered to be uniform (at some average value) over the cross section.

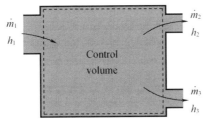

Fig. 2 – 15 Under steady-flow conditions, the fluid properties at an inlet or exit remain constant (do not change with time)

Thus, the fluid properties at an inlet or exit may be specified by the average single values. Also, the *heat* and *work* interactions between a steady-flow system and its surroundings do not change with time. Thus, the power delivered by a system and the rate of heat transfer to or from a system remain constant during a steady-flow process. The *mass balance* for a general steady-flow system was given as

$$\sum_{in} \dot{m} = \sum_{out} \dot{m} \quad (\text{kg/s}) \qquad (2-26)$$

The mass balance for a single-stream (one-inlet and one-outlet) steady-flow system was given as

$$\dot{m}_1 = \dot{m}_2 \rightarrow \rho_1 V_1 A_1 = \rho_2 V_2 A_2 \qquad (2-27)$$

where the subscripts 1 and 2 denote the inlet and the exit states, respectively, r is density, V is the average flow velocity in the flow direction, and A is the cross-sectional area normal to flow direction.

During a steady-flow process, the total energy content of a control volume remains constant (E_{CV} = constant), and thus the change in the total energy of the control volume is zero (E_{CV} = 0). Therefore, the amount of energy entering a control volume in all forms (by heat, work, and mass) must be equal to the amount of energy leaving it. Then the rate form of the general energy balance reduces for a steady-flow process to

$$\underbrace{\dot{E}_{in} - \dot{E}_{out}}_{\text{Rate of net energy transfer by heat, work, and mass}} = \underbrace{dE_{system}/dt}_{\text{Rate of change in internal, kinetic, potential, etc., energies}}\!\!\!\!\!\!\!\!\!\!\!\!\!\!\!\!^{0(\text{steady})} = 0 \qquad (2-28)$$

or

$$\underbrace{\dot{E}_{in}}_{\substack{\text{Rate of net energy transfer in} \\ \text{by heat, work, and mass}}} = \underbrace{\dot{E}_{out}}_{\substack{\text{Rate of net energy transfer out} \\ \text{by heat, work, and mass}}} \quad (\text{kW}) \qquad (2-29)$$

Noting that energy can be transferred by heat, work, and mass only, the energy balance in Eq. (2 – 29) for a general steady-flow system can also be written more explicitly as

$$\dot{Q}_{in} + \dot{W}_{in} + \sum_{in} \dot{m}\theta = \dot{Q}_{out} + \dot{W}_{out} + \sum_{out} \dot{m}\theta \qquad (2-30)$$

or

$$\dot{Q}_{in} + \dot{W}_{in} + \underbrace{\sum_{in} \dot{m}(h + \frac{V^2}{2} + gz)}_{\text{for each inlet}} = \dot{Q}_{out} + \dot{W}_{out} + \underbrace{\sum_{out} \dot{m}(h + \frac{V^2}{2} + gz)}_{\text{for each exit}} \quad (2-31)$$

since the energy of a flowing fluid per unit mass is $u = h + e_k + e_p = h + V^2/2 + gz$. Consider, for example, an ordinary electric hot-water heater under steady operation, as shown in Fig. 2 – 16. A cold-water stream with a mass flow rate $\dot{m}$ is continuously flowing into the water heater, and a hot-water stream of the same mass flow rate is continuously flowing out of it. The water heater (the control volume) is losing heat to the surrounding air at a rate of $\dot{Q}_{out}$. out, and the electric heating element is supplying electrical work (heating) to the water at a rate of $\dot{W}_{in}$. On the basis of the conservation of energy principle, we can say that the water stream experiences an increase in its total energy as it flows through the water heater that is equal to the electric energy supplied to the water minus the heat losses.

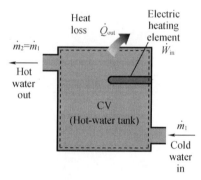

Fig. 2 – 16   A water heater in steady operation

The energy balance relation just given is intuitive in nature and is easy to use when the magnitudes and directions of heat and work transfers are known. When performing a general analytical study or solving a problem that involves an unknown heat or work interaction, however, we need to assume a direction for the heat or work interactions. In such cases, it is common practice to assume heat to be transferred *into the system* (heat input) at a rate of $\dot{Q}$, and work produced *by the system* (work output) at a rate of $\dot{W}$, and then solve the problem. The first – law or energy balance relation in that case for a general steady – flow system becomes

$$\dot{Q} - \dot{W} = \underbrace{\sum_{out} \dot{m}(h + \frac{V^2}{2} + gz)}_{\text{for each exit}} - \underbrace{\sum_{in} \dot{m}(h + \frac{V^2}{2} + gz)}_{\text{for each inlet}} \quad (2-32)$$

Obtaining a negative quantity for $\dot{Q}$ or $\dot{W}$ simply means that the assumed direction is wrong and should be reversed. For single-stream devices, the steady-flow energy balance equation becomes

$$\dot{Q} - \dot{W} = \dot{m}\left[h_2 - h_1 + \frac{V_2^2 - V_1^2}{2} + g(z_2 - z_1)\right] \quad (2-33)$$

Dividing Eq. (2 – 33) by $\dot{m}$ gives the energy balance on a unit-mass basis as

$$q - w = h_2 - h_1 + \frac{V_2^2 - V_1^2}{2} + g(z_2 - z_1) \quad (2-34)$$

where $q = \dot{Q}/\dot{m}$ and $w = \dot{W}/\dot{m}$ are the heat transfer and work done per unit mass of the working fluid, respectively. When the fluid experiences negligible changes in its kinetic and potential energies (that is, $\Delta ke \cong 0$, $\Delta pe \cong 0$), the energy balance equation is reduced further to

$$q - w = h_2 - h_1 \tag{2-35}$$

The various terms appearing in the above equations are as follows:

$\dot{Q}$ = rate of heat transfer between the control volume and its surroundings. When the control volume is losing heat (as in the case of the water heater), $\dot{Q}$ is negative. If the control volume is well insulated (i.e., adiabatic), then $\dot{Q} = 0$.

$\dot{W}$ = power. For steady-flow devices, the control volume is constant; thus, there is no boundary work involved. The work required to push mass into and out of the control volume is also taken care of by using enthalpies for the energy of fluid streams instead of internal energies. Then $\dot{W}$ represents the remaining forms of work done per unit time. Many steady-flow devices, such as turbines, compressors, and pumps, transmit power through a shaft, and $\dot{W}$ simply becomes the shaft power for those devices. If the control surface is crossed by electric wires (as in the case of an electric water heater), $\dot{W}$ represents the electrical work done per unit time. If neither is present, then $\dot{W} = 0$.

$\Delta h = h_2 - h_1$. The enthalpy change of a fluid can easily be determined by reading the enthalpy values at the exit and inlet states from the tables. For ideal gases, it can be approximated by $\Delta h = c_{p,\text{avg}}(T_2 - T_1)$. Note that $(\text{kg/s})(\text{kJ/kg}) \equiv \text{kW}$.

$\Delta ke = (V_2^2 - V_1^2)/2$. The unit of kinetic energy is m^2/s^2, which is equivalent to J/kg. The enthalpy is usually given in kJ/kg. To add these two quantities, the kinetic energy should be expressed in kJ/kg. This is easily accomplished by dividing it by 1 000. A velocity of 45 m/s corresponds to a kinetic energy of only 1 kJ/kg, which is a very small value compared with the enthalpy values encountered in practice. Thus, the kinetic energy term at low velocities can be neglected. When a fluid stream enters and leaves a steady-flow device at about the same velocity ($V_1 \cong V_2$), the change in the kinetic energy is close to zero regardless of the velocity. Caution should be exercised at high velocities, however, since small changes in velocities may cause significant changes in kinetic energy.

$\Delta e_p = g(z_2 - z_1)$. A similar argument can be given for the potential energy term. A potential energy change of 1 kJ/kg corresponds to an elevation difference of 102 m. The elevation difference between the inlet and exit of most industrial devices such as turbines and compressors is well below this value, and the potential energy term is always neglected for these devices. The only time the potential energy term is significant is when a process involves pumping a fluid to high elevations and we are interested in the required pumping power.

2 – 7 Some Steady-Flow Engineering Devices

In steam, gas, or hydro electric power plants, the device that drives the electric generator is the turbine. As the fluid passes through the turbine, work is done against the blades, which are attached to the shaft. As a result, the shaft rotates, and the turbine produces work.

Compressors, as well as pumps and fans, are devices used to increase the pressure of a fluid. Work is supplied to these devices from an external source through a rotating shaft. Therefore, compressors involve work inputs. Even though these three devices function similarly, they do differ in the tasks they perform. A *fan* increases the pressure of a gas slightly and is mainly used to mobilize a gas. A *compressor* is capable of compressing the gas to very high pressures. *Pumps* work very much like compressors except that they handle liquids instead of gases.

Note that turbines produce power output whereas compressors, pumps, and fans require power input. Heat transfer from turbines is usually negligible ($\dot{Q} \approx 0$) since they are typically well insulated. Heat transfer is also negligible for compressors unless there is intentional cooling. Potential energy changes are negligible for all of these devices ($\Delta pe \cong 0$). The velocities involved in these devices, with the exception of turbines and fans, are usually too low to cause any significant change in the kinetic energy ($\Delta ke \cong 0$). The fluid velocities encountered in most turbines are very high, and the fluid experiences a significant change in its kinetic energy. However, this change is usually very small relative to the change in enthalpy, and thus it is often disregarded.

EXAMPLE 2 – 4 Compressing Air by a Compressor

Air at 100 kPa and 280 K is compressed steadily to 600 kPa and 400 K. The mass flow rate of the air is 0.02 kg/s, and a heat loss of 16 kJ/kg occurs during the process. Assuming the changes in kinetic and potential energies are negligible, determine the necessary power input to the compressor.

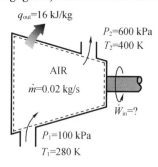

Fig. 2 – 17 Schematic for Example 2 – 4

Solution Air is compressed steadily by a compressor to a specified temperature and pressure. The power input to the compressor is to be determined.

Assumptions 1. This is a steady-flow process since there is no change with time at any point and thus $\Delta m_{CV} = 0$ and $\Delta E_{CV} = 0$.

2. Air is an ideal gas since it is at a high temperature and low pressure relative to its critical-point values.

3. The kinetic and potential energy changes are zero, $\Delta e_k = \Delta e_p = 0$.

Analysis We take the *compressor* as the system (Fig. 2–17). This is a *control volume* since mass crosses the system boundary during the process. We observe that there is only one inlet and one exit and thus $\dot{m}_1 = \dot{m}_2 = \dot{m}$. Also, heat is lost from the system and work is supplied to the system. Under stated assumptions and observations, the energy balance for this steady-flow system can be expressed in the rate form as

$$\underbrace{\dot{E}_{in} = \dot{E}_{out}}_{\substack{\text{Rate of net energy transfer}\\ \text{by heat, work, and mass}}} = \underbrace{dE_{system}/dt}_{\substack{\text{Rate of change in internal, kinetic,}\\ \text{potential, energies, etc.}}} \overset{0(\text{steady})}{\longrightarrow} = 0$$

$$\dot{E}_{in} = \dot{E}_{out}$$

$$\dot{W}_{in} + \dot{m}h_1 = \dot{Q}_{out} + \dot{m}h_2 \text{ (since } \Delta ke = \Delta pe \cong 0\text{)}$$

$$\dot{W}_{in} = \dot{m}q_{out} + \dot{m}(h_2 - h_1)$$

The enthalpy of an ideal gas depends on temperature only, and the enthalpies of the air at the specified temperatures are determined from the air table (Table A–1) to be

$$h_1 = h_{@280K} = 282.22 \text{ kJ/kg}$$
$$h_2 = h_{@400K} = 403.01 \text{ kJ/kg}$$

Substituting, the power input to the compressor is determined to be

$$\dot{W}_{in} = (0.02 \text{ kg/s})(16 \text{ kJ/kg}) + (0.02 \text{ kg/s})(403.01 - 282.22) \text{kJ/kg} = 2.74 \text{ kW}$$

Discussion Note that the mechanical energy input to the compressor manifests itself as a rise in enthalpy of air and heat loss from the compressor.

Summary

The sum of all forms of energy of a system is called *total energy*, which consists of internal, kinetic, and potential energy for simple compressible systems. *Internal energy* represents the molecular energy of a system and may exist insensible, latent, chemical, and nuclear forms. The *first law of thermodynamics* is essentially an expression of the conservation of energy principle, also called the *energy balance*. The general mass and energy balances for *any system* undergoing *any process*. The conversion of energy from one form to another is often associated with adverse effects on the environment, and environmental impact should be an important consideration in the conversion and utilization of energy.

Problems

2–1 A well-insulated electric oven is being heated through its heating element. If the entire oven, including the heating element, is taken to be the system, determine whether this is a heat or work interaction if the system is taken as only the air in the oven without the heating element.

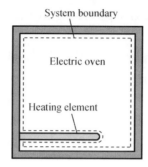

Fig. P2 – 1

2 – 2 A classroom is to be air-conditioned using window air-conditioning units. The cooling load is due to people, lights, and heat transfer through the walls and the windows. The value of each people cooling load is 360 kJ/h and that of a bulb is 100 W. The number of bulbs and people in this room are 10 and 40, respectively. The number of 5 kW window air conditioning units required is to be determined.

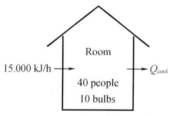

Fig. P2 – 2

2 – 3 A rigid tank contains a hot fluid that is cooled while being stirred by a paddle wheel. Initially, the internal energy of the fluid is 800 kJ. During the cooling process, the fluid loses 500 kJ of heat, and the paddle wheel does 100 kJ of work on the fluid. Determine the final internal energy of the fluid. Neglect the energy stored in the paddle wheel.

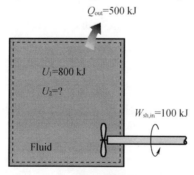

Fig. P2 – 3

2 – 4 The available head, flow rate, and efficiency of a hydro electric turbine are given. The height is 120 m and volume flow rate is 100 m^3/s. The electric power output is to be determined.

Chapter 2 The First Law of Thermodynamics

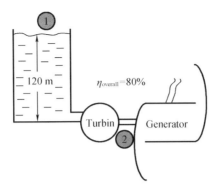

Fig. P2-4

2-5 Air at 150 kPa and 300 K is compressed steadily to 700 kPa and 550 K. The mass flow rate of the air is 0.03 kg/s, and a heat loss of 20 kJ/kg occurs during the process. Assuming the changes in kinetic and potential energies are negligible, determine the necessary power input to the compressor.

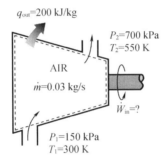

Fig. P2-5

Chapter 3 Thermodynamic Properties and Process of Ideal Gas

We start this chapter with the introduction of the concept of ideal gas. After demonstrating the ideal-gas equation of state, thermodynamics properties and process of ideal gas are discussed. Then we define specific heats, obtain relations for internal exergy, enthalpy and entropy of ideal gases in terms of specific heat and temperature changes. In this chapter, we deal with non reacting gas mixtures. A non reacting gas mixture can be treated as a pure substance since it is usually a homogeneous mixture of different gases. The properties of a gas mixture obviously depend on the properties of the individual gases (called *components* or *constituents*) as well as on the amount of each gas in the mixture. we also discuss the specific heats, internal exergy, enthalpy and entropy of ideal-gas mixtures.

3 – 1 The Ideal Gas Equation of State

Any equation that relates the pressure, temperature, and specific volume of a substance is called an equation of state. Property relations that involve other properties of a substance at equilibrium states are also referred to a equations of state. There are several equations of state, some simple and others very complex. The simplest and best – known equation of state for substances in the gas phase is the ideal-gas equation of state. This equation predicts the $P-v-T$ behavior of a gas quite accurately within some properly selected region.

Gas and *vapor* are often used as synonymous words. The vapor phase of a substance is customarily called a *gas* when it is above the critical temperature. *Vapor* usually implies a gas that is not far from a state of condensation. In 1802, J. Charles and J. Gay Lussac, Frenchmen, experimentally determined that at low pressures the volume of a gas is proportional to its temperature. That is,

$$P = R\left(\frac{T}{v}\right)$$

or

$$Pv = RT \qquad (3-1)$$

where the constant of proportionality R is called the gas constant. Eq. (3 – 1) is called the ideal-gas equation of state, or simply the ideal – gas relation, and a gas that obeys this relation is called an ideal gas. In this equation, P is the absolute pressure, T is the absolute temperature, and v is the specific volume.

The gas constant R is different for each gas and is determined from

$$R = \frac{R_u}{M} \quad (\text{kJ/kg} \cdot \text{K or kPa} \cdot \text{m}^3/\text{kg} \cdot \text{K})$$

Chapter 3 Thermodynamic Properties and Process of Ideal Gas

where R_u is the universal gas constant and M is the molar mass (also called *molecular weight*) of the gas. The constant R_u is the same for all substances, and its value is

$$R_u = \begin{cases} 8.314\ 47 \text{ kJ/kmol} \cdot \text{K} \\ 8.314\ 47 \text{ kPa} \cdot \text{m}^3/\text{kmol} \cdot \text{K} \\ 0.083\ 144\ 7 \text{ kPa} \cdot \text{m}^3/\text{kmol} \cdot \text{K} \\ 1.985\ 88 \text{ Btu/lbmol} \cdot \text{R} \\ 10.731\ 6 \text{ psia} \cdot \text{ft}^3/\text{lbmol} \cdot \text{R} \\ 1\ 545.37 \text{ ft} \cdot \text{lbf/lbmol} \cdot \text{R} \\ 287.06 \text{ J/kg} \cdot \text{K} \end{cases} \quad (3-2)$$

The molar mass M can simply be defined as *the mass of one mole* (also called a *gram-mole*, abbreviated gmol) *of a substance in grams*, or *the mass of one* kmol (also called a *kilogram-mole*, abbreviated kgmol) *in kilograms*. In English units, it is the mass of 1 lbmol in lbm. Notice that the molar mass of a substance has the same numerical value in both unit systems because of the way it is defined. The mass of a system is equal to the product of its molar mass M and the mole number N:

$$m = MN \quad (3-3)$$

The values of R and M for several substances are given in Table A-2. The ideal-gas equation of state can be written in several different forms:

$$V = mv \rightarrow PV = mRT \quad (3-4)$$
$$mR = (MN)R = NR_u \rightarrow PV = NR_uT \quad (3-5)$$
$$V = N\bar{v} \rightarrow P\bar{v} = R_uT \quad (3-6)$$

where $\bar{v}$ is the molar specific volume, that is, the volume per unit mole.

By writing Eq. (3-4) twice for a fixed mass and simplifying, the properties of an ideal gas at two different states are related to each other by

$$\frac{P_1 V_1}{T_1} = \frac{P_2 V_2}{T_2} \quad (3-7)$$

An ideal gas is an *imaginary* substance that obeys the relation $Pv = RT$. It has been experimentally observed that the ideal-gas relation given closely approximates the $P-v-T$ behavior of real gases at low densities. At low pressures and high temperatures, the density of a gas decreases, and the gas behaves as an ideal gas under these conditions. What constitute slow pressure and high temperature is explained later.

In the range of practical interest, many familiar gases such as air, nitrogen, oxygen, hydrogen, helium, argon, neon, krypton, and even heavier gases such as carbon dioxide can be treated as ideal gases with negligible error (often less than 1 percent). Dense gases such as water vapor in steam power plants and refrigerant vapor in refrigerators, however, should not be treated as ideal gases. Instead, the property tables should be used for these substances.

EXAMPLE 3-1 Mass of Air in a Room

Determine the mass of the air in a room whose dimensions are 4 m × 5 m × 6 m at 100 kPa and 25 ℃.

Solution: The mass of air in a room is to be determined.

Analysis: A sketch of the room is given in Fig. 3 – 1. Air at specified conditions can be treated as an ideal gas. From Table A – 1, the gas constant of air is $R = 0.287$ kPa · m³/kg · K, and the absolute temperature is $T = 25\ ℃ + 273 = 298$ K. The volume of the room is

$$V = (4\text{m}) \times (5\text{m}) \times (6\text{m}) = 120\ \text{m}^3$$

The mass of air in the room is determined from the ideal-gas relation to be

$$m = \frac{PV}{RT} = \frac{(100\ \text{kPa})(120\ \text{m}^3)}{(0.287\ \text{kPa} \cdot \text{m}^3/\text{kg} \cdot \text{K})(298\ \text{K})} = 140.3\ \text{kg}$$

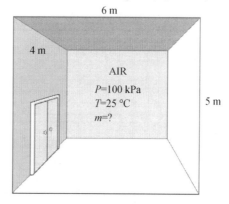

Fig. 3 – 1 Schematic for Example 3 – 1

3 – 2 Specific Heats of Ideal gases

We know from experience that it takes different amounts of exergy to raise the temperature of identical masses of different substances by one degree. For example, we need about 4.5 kJ of exergy to raise the temperature of 1 kg of iron from 20 ℃ to 30 ℃, whereas it takes about 9 times this exergy (41.8 kJ to be exact) to raise the temperature of 1 kg of liquid water by the same amount (Fig. 3 – 2). Therefore, it is desirable to have a property that will enable us to compare the exergy storage capabilities of various substances.

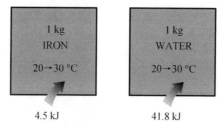

Fig. 3 – 2 It takes different amounts of exergy to raise the temperature of different substances by the same amount

This property is the specific heat. The specific heat is defined as *the exergy required to raise the temperature of a unit mass of a substance by one degree*. In general, this exergy depends on how the process is executed. In thermodynamics, we are interested in two kinds of specific heats:

specific heat at constant volume c_v and specific heat at constant pressure c_p.

Physically, the specific heat at constant volume c_v can be viewed as *the exergy required to raise the temperature of the unit mass of a substance by one degree as the volume is maintained constant*. The exergy required to do the same as the pressure is maintained constant is the specific heat at constant pressure c_p. This is illustrated in Fig. 3 – 3. The specific heat at constant pressure c_p is always greater than c_v because at constant pressure the system is allowed to expand and the exergy for this expansion work must also be supplied to the system.

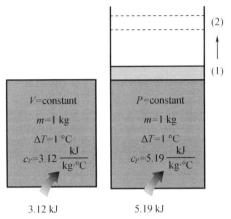

Fig. 3 – 3 Constant-volume and constant pressure specific heats c_v and c_p
(values given are for helium gas)

Based on the first law of thermodynamics, for reversible processes

$$\delta q = du + pdv, \quad \delta h = du - vdp$$

at constant volume ($dv = 0$)

$$c_v = \left(\frac{\delta q}{dT}\right)_v = \left(\frac{du + pdv}{dT}\right)_v = \left(\frac{\partial u}{\partial T}\right)_v \tag{3-8}$$

at constant pressure ($dp = 0$)

$$c_p = \left(\frac{\delta q}{dT}\right)_p = \left(\frac{dh - vdp}{dT}\right)_v = \left(\frac{\partial h}{\partial T}\right)_p \tag{3-9}$$

From the definition of c_v, this exergy must be equal to $cv\,dT$, where dT is the differential change in temperature. Thus,

(at constant volume)

$$c_v dT = du$$

or

$$c_v = \left(\frac{\partial u}{\partial T}\right)_v \tag{3-10}$$

Similarly, an expression for the specific heat at constant pressure c_p can be obtained by considering a constant-pressure expansion or compression process. It yields

$$c_p = \left(\frac{\partial h}{\partial T}\right)_p \tag{3-11}$$

Eqs 3 – 10 and 3 – 11 are the defining equations for c_v and c_p, which are represent the change in internal exergy with temperature at constant volume and the change in enthalpy with temperature at constant pressure, respectively.

Note that c_v and c_p are expressed in terms of other properties; thus, they must be properties themselves. Like any other property, the specific heats of a substance depend on the state that, in general, is specified by two independent, intensive properties. That is, the exergy required to raise the temperature of a substance by one degree is different at different temperatures and pressures (Fig. 3 –4). But this difference is usually not very large.

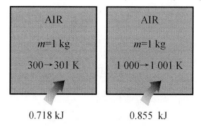

Fig. 3 –4 The specific heat of a substance changes with temperature

A few observations can be made from Eqs. (3 – 10) and (3 – 11). First, these equations are *property relations* and as such *are independent of the type of processes*. They are valid for *any* substance undergoing *any* process. The only relevance c_v has to a constant-volume process is that c_v happens to be the exergy transferred to a system during a constant-volume process per unit mass per unit degree rise in temperature. This is how the values of c_v are determined. This is also how the name *specific heat at constant volume* originated. Likewise, the exergy transferred to a system per unit mass per unit temperature rise during a constant-pressure process happens to be equal to c_p. This is how the values of c_p can be determined and also explains the origin of the name *specific heat at constant pressure*.

Another observation that can be made from Eqs. (3 – 10) and (3 – 11) is that c_v is related to the changes in *internal exergy* and c_p to the changes in *enthalpy*. In fact, it would be more proper to define c_v as *the change in the internal exergy of a substance per unit change in temperature at constant volume*. Likewise, c_p can be defined as *the change in the enthalpy of a substance per unit change in temperature at constant pressure*. In other words, c_v is a measure of the variation of internal exergy of a substance with temperature, and c_p is a measure of the variation of enthalpy of a substance with temperature.

Both the internal exergy and enthalpy of a substance can be changed by the transfer of *exergy* in any form, with heat being only one of them. Therefore, the term *specific exergy* is probably more appropriate than the term *specific heat*, which implies that exergy is transferred (and stored) in the form of heat.

A common unit for specific heats is kJ/kg · ℃ or kJ/kg · K. Notice that these two units are *identical* since $\Delta T(℃) = \Delta T(K)$, and 1 ℃ change in temperature is equivalent to a change of 1 K. The specific heats are sometimes given on a *molar basis*. They are then denoted by $\bar{c}_v$ and $\bar{c}_p$ and have the unit kJ/kmol · ℃ or kJ/kmol · K.

A special relationship between c_p and c_v for ideal gases can be obtained by differentiating the

relation $h = u + RT$, which yields
$$\delta h = du + RdT$$
Replacing dh by $c_p \, dT$ and du by $c_v \, dT$ and dividing the resulting expression by dT, we obtain
$$c_p = c_v + R \quad (\text{kJ/kg} \cdot \text{K}) \quad (3-12)$$

It is called the Mayer relation in honor of the German physician and physicist J. R. Mayer (1814 – 1878). This is an important relationship for ideal gases since it enables us to determine c_v from a knowledge of c_p and the gas constant R.

When the specific heats are given on a molar basis, R in the above equation should be replaced by the universal gas constant R_0.
$$\bar{c}_p = \bar{c}_v + R_0 \quad (\text{kJ/kmol} \cdot \text{K}) \quad (3-13)$$

At this point, we introduce another ideal-gas property called the specific heat ratio k, defined as
$$k = \frac{c_p}{c_v} \quad (3-14)$$

The specific ratio also varies with temperature, but this variation is very mild. For monatomic gases, its value is essentially constant at 1.667. Many diatomic gases, including air, have a specific heat ratio of about 1.4 at room temperature.

3 – 3 Internal Energy, Enthalpy and Entropy of Ideal Gases

INTERNAL ENERGY AND ENTHALPY

We defined an ideal gas as a gas whose temperature, pressure, and specific volume are related by
$$Pv = RT$$
It has been demonstrated mathe magically and experimentally (Joule, 1843) that for an ideal gas the internal exergy is a function of the temperature only. That is,
$$u = u(T) \quad (3-15)$$

In his classical experiment, Joule submerged two tanks connected with a pipe and a valve in a water bath, as shown in Fig. 3 – 5. Initially, one tank contained air at a high pressure and the other tank was evacuated. When thermal equilibrium was attained, he opened the valve to let air pass from one tank to the other until the pressures equalized. Joule observed no change in the temperature of the water bath and assumed that no heat was transferred to or from the air. Since there was also no work done, he concluded that the internal exergy of the air did not change even though the volume and the pressure changed. Therefore, he reasoned, the internal exergy was a function of temperature only and not a function of pressure or specific volume. (Joule later showed that for gases that deviate significantly from ideal gas behavior, the internal exergy is not a function of temperature alone.)

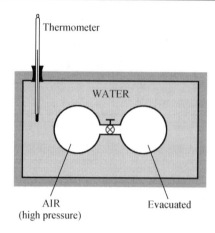

Fig. 3-5 Schematic of the experimental apparatus used by Joule

Using the definition of enthalpy and the equation of state of an ideal gas, we have

$$\left.\begin{array}{l} h = u + Pv \\ Pv = RT \end{array}\right\} \quad h = u + RT$$

Since R is constant and $u = u(T)$, it follows that the enthalpy of an ideal gas is also a function of temperature only:

$$h = h(T) \tag{3-16}$$

Since u and h depend only on temperature for an ideal gas, the specific heats c_v and c_p also depend, at most, on temperature only. Therefore, at a given temperature, u, h, c_v, and c_p of an ideal gas have fixed values regardless of the specific volume or pressure. Thus, for ideal gases, the partial derivatives in Eqs. (3-10) and (3-11) can be replaced by ordinary derivatives. Then the differential changes in the internal exergy and enthalpy of an ideal gas can be expressed as

$$du = c_v(T) dT \tag{3-17}$$

and

$$dh = c_p(T) dT \tag{3-18}$$

The change in internal exergy or enthalpy for an ideal gas during a process from state 1 to state 2 is determined by integrating these equations:

$$\Delta u = u_1 - u_2 = \int_1^2 c_v(T) dT \quad (\text{kJ/kg}) \tag{3-19}$$

and

$$\Delta h = h_1 - h_2 = \int_1^2 c_p(T) dT \quad (\text{kJ/kg}) \tag{3-20}$$

To carry out these integrations, we need to have relations for c_v and c_p as functions of temperature.

At low pressures, all real gases approach ideal gas behavior, and therefore their specific heats depend on temperature only. The specific heats of real gases at low pressures are called *ideal gas specific heats*, or *zero pressure specific heats*, and are often denoted c_{p0} and c_{v0}. Accurate analytical expressions for ideal gas specific heats, based on direct measurements or calculations

Chapter 3 Thermodynamic Properties and Process of Ideal Gas

from statistical behavior of molecules, are available and are given as third degree polynomials in the appendix (Table A – 10c) for several gases. A plot of $\bar{c}_{p0}(T)$ data for some common gases is given in Fig. 3 – 6.

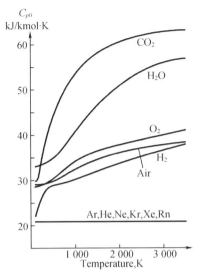

Fig. 3 – 6 Ideal-gas constant-pressure specific heats for some gases (see Table A – 10c for c_p equations)

The use of ideal-gas specific heat data is limited to low pressures, but these data can also be used at moderately high pressures with reasonable accuracy as long as the gas does not deviate from ideal-gas behavior significantly. The integrations in Eqs. (3 – 19) and (3 – 20) are straight forward but rather time-consuming and thus impractical. To avoid these laborious calculations, u and h data for a number of gases have been tabulated over small temperature intervals. These tables are obtained by choosing an arbitrary reference point and performing the integrations in Eqs. (3 – 19) and (3 – 20) by treating state 1 as the reference state. In the ideal gas tables given in the appendix, zero kelvin is chosen as the reference state, and both the enthalpy and the internal exergy are assigned zero values at that state.

The choice of the reference state has no effect on u or h calculations. The u and h data are given in kJ/kg for air (Table A – 1) and usually in kJ/kmol for other gases. The unit kJ/kmol is very convenient in the thermodynamic analysis of chemical reactions. Some observations can be made from Fig. 3 – 6. First, the specific heats of gases with complex molecules (molecules with two or more atoms) are higher and increase with temperature. Also, the variation of specific heats with temperature is smooth and may be approximated as linear over small temperature intervals (a few hundred degrees or less). Therefore the specific heat functions in Eqs. (3 – 19) and (3 – 20) can be replaced by the constant average specific heat values. Then the integrations in these equations can be performed, yielding

$$u_2 - u_1 = c_{v,\text{avg}}(T_2 - T_1) \quad (\text{kJ/kg}) \qquad (3-21)$$

and

$$h_2 - h_1 = c_{p,\text{avg}}(T_2 - T_1) \quad (\text{kJ/kg}) \qquad (3-22)$$

The specific heat values for some common gases are listed as a function of temperature in

Table A – 10b. The average specific heats $c_{p,\text{avg}}$ and $c_{v,\text{avg}}$ are evaluated from this table at the average temperature $(T_1 + T_2)/2$, as shown in Fig. 3 – 7. If the final temperature T_2 is not known, the specific heats may be evaluated at T_1 or at the anticipated average temperature. Then T_2 can be determined by using these specific heat values. The value of T_2 can be refined, if necessary, by evaluating the specific heats at the new average temperature. Another way of determining the average specific heats is to evaluate them at T_1 and T_2 and then take their average. Usually both methods give reasonably good results, and one is not necessarily better than the other.

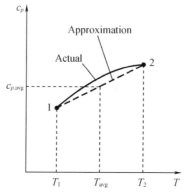

Fig. 3 –7 For small temperature intervals, the specific heats may be assumed to vary linearly with temperature

Another observation that can be made from Fig. 3 – 6 is that the ideal-gas specific heats of *monatomic gases* such as argon, neon, and helium remain constant over the entire temperature range. Thus, u and h of monatomic gases can easily be evaluated from Eqs. (3 – 21) and (3 – 22).

Note that the u and h relations given previously are not restricted to any kind of process. They are valid for all processes. The presence of the constant-volume specific heat c_v in an equation should not lead one to believe that this equation is valid for a constant-volume process only. On the contrary, the relation $u = c_{v,\text{avg}} T$ is valid for *any* ideal gas undergoing *any* process (Fig. 3 – 8). A similar argument can be given for c_p and h.

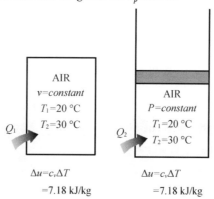

Fig. 3 –8 The relation $u = c_v T$ is valid for any kind of process, constant-volume or not

To summarize, there are three ways to determine the internal exergy and enthalpy changes of ideal gases:

a. By using the tabulated u and h data. This is the easiest and most accurate way when tables are readily available.

b. By using the c_v or c_p relations as a function of temperature and performing the integrations. This is very inconvenient for hand calculations but quite desirable for computerized calculations. The results obtained are very accurate.

c. By using average specific heats. This is very simple and certainly very convenient when property tables are not available. The results obtained are reasonably accurate if the temperature interval is not very large.

EXAMPLE 3 – 2 Evaluation of the Δu of an Ideal Gas

Air at 300 K and 200 kPa is heated at constant pressure to 600 K. Determine the change in internal exergy of air per unit mass, using (a) data from the air table (Table A – 1), (b) the functional form of the specific heat (Table A – 10c), and (c) the average specific heat value (Table A – 10b).

Solution The internal exergy change of air is to be determined in three different ways.

Assumptions At specified conditions, air can be considered to be an ideal gas since it is at a high temperature and low pressure relative to its critical point values.

Analysis The internal exergy change u of ideal gases depends on the initial and final temperatures only, and not on the type of process. Thus, the following solution is valid for any kind of process.

(a) One way of determining the change in internal exergy of air is to read the u values at T_1 and T_2 from Table A – 1 and take the difference:

$$u_1 = u_{@300K} = 214.07 \text{ kJ/kg}$$
$$u_2 = u_{@600K} = 434.78 \text{ kJ/kg}$$

Thus,

$$\Delta u = u_1 - u_2 = (434.78 - 214.07) \text{ kJ/kg} = 220.71 \text{ kJ/kg}$$

(b) The $\bar{c}_p(T)$ of air is given in Table A – 10c in the form of a third-degree polynomial expressed as

$$\bar{c}_p(T) = a + bT + cT^2 + dT^3$$

where $a = 28.11$, $b = 0.1967 \times 10^{-2}$, $c = 0.4802 \times 10^{-5}$, and $d = 1.966 \times 10^{-9}$. From Eq. (3 – 13),

$$\bar{c}_v(T) = \bar{c}_p - R_u = (a - R_u) + bT + cT^2 + dT^3$$

From Eq. (3 – 19),

$$\Delta \bar{u} = \int_1^2 \bar{c}_v(T) dT = \int_{T_1}^{T_2} [(a - R_u) + bT + cT^2 + dT^3] dT$$

Performing the integration and substituting the values, we obtain

$$\Delta \bar{u} = 6\,447 \text{ kJ/kmol}$$

The change in the internal exergy on a unit mass basis is determined by dividing this value by the molar mass of air (Table A – 2):

$$\Delta u = \frac{\Delta \bar{u}}{M} = \frac{6\ 447 \text{ kJ/kmol}}{28.97 \text{ kg/kmol}} = 222.5 \text{ kJ/kg}$$

which differs from the tabulated value by 0.8 percent.

(c) The average value of the constant-volume specific heat c_v, avg is determined from Table A – 10b at the average temperature of $(T_1 + T_2)/2 = 450$ K to be

$$c_{v,\text{avg}} = c_{v@450K} = 0.733 \text{ kJ/kg} \cdot \text{K}$$

Thus,

$$\Delta u = c_{v,\text{avg}}(T_1 - T_2) = (0.733 \text{ kJ/kg} \cdot \text{K})[(600 - 300)\text{K}] = 220 \text{kJ/kg}$$

Discussion: This answer differs from the tabulated value (220.71 kJ/kg) by only 0.4 percent. This close agreement is not surprising since the assumption that c_v varies linearly with temperature is a reasonable one at temperature intervals of only a few hundred degrees. If we had used the c_v value at $T_1 = 300$ K instead of at T_{avg}, the result would be 215.4 kJ/kg, which is in error by about 2 percent. Errors of this magnitude are acceptable for most engineering purposes.

ENTROPY

Amic cycles, reversible or irreversible, including the refrigeration cycles. If no irreversibilities occur within the system as well as the reversible cyclic device, then the cycle undergone by the combined system is internally reversible. Clausius realized in 1865 that he had discovered a new thermodynamic property, and he chose to name this property entropy. It is designated S and is defined as

$$dS = \left(\frac{\delta Q}{T}\right)_{\text{int rev}} \quad (\text{kJ/K}) \qquad (3-23)$$

Entropy is an extensive property of a system and sometimes is referred to as *total entropy*. Entropy per unit mass, designated s, is an intensive property and has the unit kJ/kg · K. The term *entropy* is generally used to refer to both total entropy and entropy per unit mass since the context usually clarifies which one is meant.

The entropy change of a system during a process can be determined by integrating Eq. (3 – 23) between the initial and the final states:

$$\Delta S = S_1 - S_2 = \int_1^2 \left(\frac{\delta Q}{T}\right)_{\text{int rev}} \quad (\text{kJ/K}) \qquad (3-24)$$

Notice that we have actually defined the *change* in entropy instead of entropy itself, just as we defined the change in exergy instead of the exergy itself when we developed the first law relation. Absolute values of entropy are determined on the basis of the third law of thermodynamics, which is discussed later in this chapter. Engineers are usually concerned with the *changes* in entropy. Therefore, the entropy of a substance can be assigned a zero value at some arbitrarily selected reference state, and the entropy values at other states can be determined from Eq. (3 – 24) by choosing state 1 to be the reference state ($S = 0$) and state 2 to be the state at which entropy is to be determined.

To perform the integration in Eq. (3 – 24), one needs to know the relation between Q and T during a process. This relation is often not available, and the integral in Eq. (3 – 24) can be performed for a few cases only. For the majority of cases we have to rely on tabulated data for entropy.

Note that entropy is a property, and like all other properties, it has fixed values at fixed states. Therefore, the entropy change S between two specified states is the same no matter what path, reversible or irreversible, is followed during a process (Fig. 3 – 9). Also note that the integral of $\delta Q/T$ gives us the value of entropy change *only if* the integration is carried out along an *internally reversible* path between the two states. The integral of $\delta Q/T$ along an irreversible path is not a property, and in general, different values will be obtained when the integration is carried out along different irreversible paths. Therefore, even for irreversible processes, the entropy change should be determined by carrying out this integration along some convenient *imaginary* internally reversible path between the specified states.

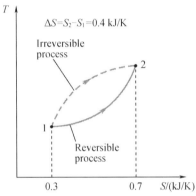

Fig. 3 – 9 The entropy change between two specified states is the same whether the process is reversible or irreversible

An expression for the entropy change of an ideal gas can be obtained from $ds = \dfrac{du}{T} + \dfrac{Pdv}{T}$ or $ds = \dfrac{dh}{T} - \dfrac{vdP}{T}$ by employing the property relations for ideal gases. By substituting $du = c_v\, dT$ and $P = RT/v$ into $ds = \dfrac{du}{T} + \dfrac{Pdv}{T}$, the differential entropy change of an ideal gas becomes

$$ds = c_v \frac{dT}{T} + R \frac{dv}{v} \qquad (3-25)$$

The entropy change for a process is obtained by integrating this relation between the end states:

$$s_2 - s_1 = \int_1^2 c_v(T) \frac{dT}{T} + R \ln \frac{v_2}{v_1} \qquad (3-26)$$

A second relation for the entropy change of an ideal gas is obtained in a similar manner by substituting $dh = c_p\, dT$ and $v = RT/P$ into $ds = \dfrac{dh}{T} - \dfrac{vdP}{T}$ and integrating. The result is

$$s_2 - s_1 = \int_1^2 c_p(T) \frac{dT}{T} - R \ln \frac{P_2}{P_1} \qquad (3-27)$$

$$s_2 - s_1 = c_v \ln \frac{P_2}{P_1} + c_p \ln \frac{v_2}{v_1} \qquad (3-28)$$

The specific heats of ideal gases, with the exception of monatomic gases, depend on temperature, and the integrals in Eqs. (3 – 26) and (3 – 27) cannot be performed unless the dependence of c_v and c_p on temperature is known. Even when the $c_v(T)$ and $c_p(T)$ functions are available, performing long integrations every time entropy change is calculated is not practical. Then two reasonable choices are left: either perform these integrations by simply assuming constant specific heats or evaluate those integrals once and tabulate the results. Both approaches are presented next.

Assuming constant specific heats for ideal gases is a common approximation, and we used this assumption before on several occasions. It usually simplifies the analysis greatly, and the price we pay for this convenience is some loss in accuracy. The magnitude of the error introduced by this assumption depends on the situation at hand. For ideal gases whose specific heats vary almost linearly in the temperature range of interest, the possible error is minimized by using specific heat values evaluated at the average temperature. The results obtained in this way usually are sufficiently accurate if the temperature range is not greater than a few hundred degrees.

The entropy-change relations for ideal gases under the constant specific heat assumption are easily obtained by replacing $c_v(T)$ and $c_p(T)$ in Eqs. (3 – 27) and (3 – 28) by $c_{v,\mathrm{avg}}$ and $c_{p,\mathrm{avg}}$, respectively, and performing the integrations. We obtain

$$s_2 - s_1 = c_{v,\mathrm{avg}} \ln \frac{T_2}{T_1} + R \ln \frac{v_2}{v_1} \quad (\mathrm{kJ/kg \cdot K}) \qquad (3-29)$$

and

$$s_2 - s_1 = c_{p,\mathrm{avg}} \ln \frac{T_2}{T_1} - R \ln \frac{P_2}{P_1} \quad (\mathrm{kJ/kg \cdot K}) \qquad (3-30)$$

EXAMPLE 3 – 3 Entropy Change of an Ideal Gas

Air is compressed from an initial state of 100 kPa and 17 ℃ to a final state of 600 kPa and 57 ℃. Determine the entropy change of air during this compression process by using average specific heats.

Solution Air is compressed between two specified states. The entropy change of air is to be determined by using average specific heats.

Assumptions Air is an ideal gas since it is at a high temperature and low pressure relative to its critical-point values. Therefore, entropy change relations developed under the ideal gas assumption are applicable.

Analysis A sketch of the system and the $T - s$ diagram for the process are given in Fig. 3 – 10. We note that both the initial and the final states of air are completely specified.

The entropy change of air during this process can also be determined approximately from Eq. (3 – 30) by using a c_p value at the average temperature of 37 ℃ (Table A – 10b) and treating it as a constant:

$$s_2 - s_1 = c_{p,\mathrm{avg}} \ln \frac{T_2}{T_1} - R \ln \frac{P_2}{P_1}$$

$$= (1.006 \ \mathrm{kJ/kg \cdot K}) \ln \frac{330 \ \mathrm{K}}{290 \ \mathrm{K}} - (0.287 \ \mathrm{kJ/kg \cdot K}) \ln \frac{600 \ \mathrm{kPa}}{100 \ \mathrm{kPa}}$$

= −0.384 2 kJ/kg · K

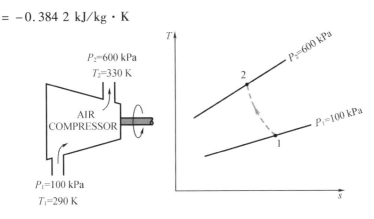

Fig. 3 – 10 Schematic and $T-s$ diagram for Example 3 – 3

3 – 4 Thermodynamic Processes of Ideal Gases

We will introduce five fundamental thermodynamic processes in this chapter, including polytropic process, isometric process, isobaric process, isothermal process and isentropic process.

ISOMETRIC PROCESSES

Isochoric process (isometric process) is a process during which the specific volume v remains constant. During isometric processes, the polytropic exponent $n \to \infty$. From $Pv = RT$, relation between pressure and temperature can be expressed as

$$\frac{P_2}{P_1} = \frac{T_2}{T_1} \qquad (3-31)$$

$P-v$ and $T-s$ diagram during a isometric process from two specified states are shown in Fig. 3 – 11.

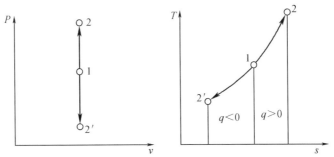

Fig. 3 – 11 $P-v$ and $T-s$ diagram for a isometric process

At constant volume, $dv = 0$, thus we obtain the boundary work

$$w = \int_{v_1}^{v_2} p \, dv = 0 \qquad (3-32)$$

59

Based on the first law of thermodynamics, heat transfer can be expressed as

$$q_v = \Delta u = u_2 - u_1 \tag{3-33}$$

During a isometric process, the heat change can be determined by using specific heats at constant pressure as

$$q_v = u_2 - u_1 = c_v \big|_{t_1}^{t_2}(t_2 - t_1) \tag{3-34}$$

We can generalize the technical work relation by expressing it as

$$w_t = -\int_{p_1}^{p_2} v \mathrm{d}p = v(p_1 - p_2) \tag{3-35}$$

Based on calculation results from Eq. (3-34), $q_v > 0$ means the substance obtains heat from outlet, it is called a endothermic process. If $q_v < 0$, it is a exothermic process.

ISOBARIC PROCESSES

Isobaric process is a process during which the pressure P remains constant, and the polytropic exponent n is equal to 0. We can obtain the relation between initial and final state properties by equation $p_1 = p_2$ and $pv = RT$ as

$$\frac{v_2}{v_1} = \frac{T_2}{T_1} \tag{3-36}$$

$P-v$ and $T-s$ diagram during a isobaric process from two specified states are shown in Fig. 3-12.

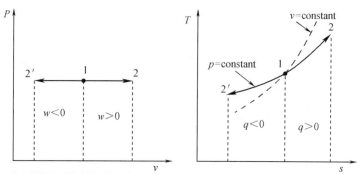

Fig. 3-12 $P-v$ and $T-s$ diagram for a isobaric process

The boundary work for a isobaric process can be expressed as

$$w = \int_{v_1}^{v_2} p\mathrm{d}v = p(v_2 - v_1) \tag{3-37}$$

For ideal gases,

$$w = R(T_2 - T_1) \tag{3-38}$$

Heat change can be expressed as

$$q_p = u_2 - u_1 + p(v_2 - v_1) = h_2 - h_1 \tag{3-39}$$

or

$$q_p = h_2 - h_1 = c_p \big|_{t_1}^{t_2}(t_2 - t_1) \tag{3-40}$$

We can generalize the technical work during a isobaric process by expressing it as

$$w_t = \int_{p_1}^{p_2} v \mathrm{d}p = 0 \tag{3-41}$$

For ideal gases,
$$q_p = c_v \big|_{t_1}^{t_2}(t_2 - t_1) + R(T_2 - T_1) = (c_v\big|_{t_1}^{t_2} + R)(t_2 - t_1) \qquad (3-42)$$

By compared with equation 3-39, we can obtain the relation between $c_v\big|_{t_1}^{t_2}$ and $c_p\big|_{t_1}^{t_2}$ as

$$c_p\big|_{t_1}^{t_2} = c_v\big|_{t_1}^{t_2} + R \qquad (3-43)$$

Note that the relation between average specific heat at constant pressure and that of at constant volume observes Mayer relation under the condition of the same initial and final temperature.

ISOTHERMAL PROCESSES

Isothermal process is a process during which the temperature T remains constant, and the polytropic exponent n is equal to 1.

For ideal gases, based on the ideal gas equation of state $Pv = RT$, thus

$$Pv = \text{constant} \quad \text{and} \quad P_1 v_1 = P_2 v_2 \qquad (3-44)$$

$P-v$ and $T-s$ diagram during a isothermal process from two specified states are shown in Fig. 3-19.

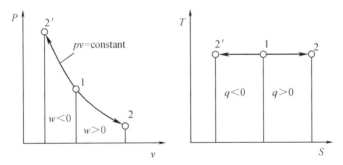

Fig. 3-19 $P-v$ **and** $T-s$ **diagram for a isothermal process**

The internal exergy and enthalpy change during isothermal processes are expressed as
$$\Delta u = 0 \quad \text{and} \quad \Delta h = 0$$

The entropy change is given as
$$\Delta s = R\ln\frac{v_2}{v_1} = -R\ln\frac{p_2}{p_1}$$

The boundary work for a isothermal process can be expressed as
$$w = \int_1^2 p\,dv = \int_1^2 pv\,\frac{dv}{v} = \int_1^2 RT\,\frac{dv}{v} = RT\ln\frac{v_2}{v_1} = p_1v_1\ln\frac{v_2}{v_1} = -p_1v_1\ln\frac{p_2}{p_1} \qquad (3-45)$$

Heat change can be expressed as
$$q_T = w = RT\ln\frac{v_2}{v_1} = p_1v_1\ln\frac{v_2}{v_1} = -p_1v_1\ln\frac{p_2}{p_1} \qquad (3-46)$$

and the technical work is given by
$$w_t = -\int_1^2 v\,dp = -\int_1^2 pv\,\frac{dp}{p} = -\int_1^2 RT\,\frac{dp}{p} = -RT\ln\frac{p_2}{p_1} = -p_1v_1\ln\frac{p_2}{p_1} \qquad (3-47)$$

Note that the technical work is equal to the heat change during when ideal gases flow through an open system. At this condition, $p_1 v_1 = p_2 v_2$, thus the flow work is equal to zero.

ISENTROPIC PROCESSES

We mentioned earlier that the entropy of a fixed mass can be changed by (1) heat transfer and (2) irreversibilities. Then it follows that the entropy of a fixed mass does not change during a process that is *internally reversible* and *adiabatic* (Fig. 3 – 20). A process during which the entropy remains constant is called an isentropic process. It is characterized by
Isentropic process:

$$\Delta s = 0 \text{ or } s_1 = s_2 \quad (\text{kJ/kg} \cdot \text{K}) \qquad (3-48)$$

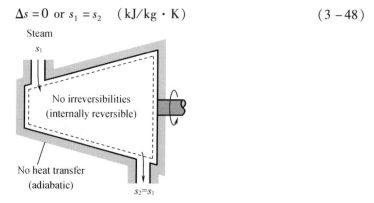

Fig. 3 – 20 During an internally reversible, adiabatic (isentropic) process, the entropy remains constant.

That is, a substance will have the same entropy value at the end of the process as it does at the beginning if the process is carried out in an isentropic manner.

Many engineering systems or devices such as pumps, turbines, nozzles, and diffusers are essentially adiabatic in their operation, and they perform best when the irreversibilities, such as the friction associated with the process, are minimized. Therefore, an isentropic process can serve as an appropriate model for actual processes. Also, isentropic processes enable us to define efficiencies for processes to compare the actual performance of these devices to the performance under idealized conditions.

It should be recognized that a *reversible adiabatic* process is necessarily isentropic ($s_2 = s_1$), but an *isentropic* process is not necessarily a reversible adiabatic process. (The entropy increase of a substance during a process as a result of irreversibilities may be offset by a decrease in entropy as a result of heat losses, for example.) However, the term *isentropic process* is customarily used in thermodynamics to imply an *internally reversible, adiabatic process*.

When the constant-specific-heat assumption is valid, the is entropic relations for ideal gases are obtained by setting Eqs. (3 – 29) and (3 – 30) equal to zero. From Eq. (3 – 29),

$$\ln \frac{T_2}{T_1} = -\frac{R}{c_v} \ln \frac{v_2}{v_1}$$

which can be rearranged as

$$\ln \frac{T_2}{T_1} = \ln \left(\frac{v_2}{v_1} \right)^{R/c_v} \qquad (3-49)$$

or

Chapter 3 Thermodynamic Properties and Process of Ideal Gas

$$\left(\frac{T_2}{T_1}\right)_{s=\text{const.}} = \left(\frac{v_2}{v_1}\right)^{k-1} \quad (\text{ideal gas}) \tag{3-50}$$

since $R = c_p - c_v$, $\kappa = k = c_p/c_v$, and thus $R/c_v = k - 1$.

Eq. (3-50) is the *first isentropic relation* for ideal gases under the constant specific heat assumption. The *second isentropic relation* is obtained in a similar manner from Eq. (3-30) with the following result:

$$\left(\frac{T_2}{T_1}\right)_{s=\text{const.}} = \left(\frac{P_2}{P_1}\right)^{(k-1)/k} \quad (\text{ideal gas}) \tag{3-51}$$

The *third isentropic relation* is obtained by substituting Eq. (3-51) into Eq. (3-50) and simplifying:

$$\left(\frac{P_2}{P_1}\right)_{s=\text{const.}} = \left(\frac{v_1}{v_2}\right)^{k} \quad (\text{ideal gas}) \tag{3-52}$$

Eq. (3-50) through (3-52) can also be expressed in a compact form as

$$Tv^{k-1} = \text{constant} \tag{3-53}$$
$$TP^{(1-k)/k} = \text{constant}\,(\text{ideal gas}) \tag{3-54}$$
$$Tv^{k-1} = \text{constant} \tag{3-55}$$

The specific heat ratio k, in general, varies with temperature, and thus an average κ value for the given temperature range should be used.

Note that the ideal gas isentropic relations above, as the name implies, are strictly valid for isentropic processes only when the constant specific heat assumption is appropriate.

$p-v$ and $T-s$ diagram during a isentropic process from two specified states are shown in Fig. 3-14.

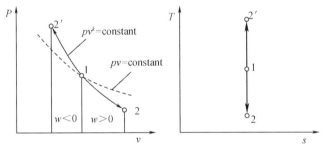

Fig. 3-14 $P-v$ and $T-s$ **diagram for a isentropic process**

During an isentropic process, there is no heat transfer occurs in the control volume. For a closed system, the boundary work can be expressed based on the first law of thermodynamics as

$$w = q - \Delta u = -(u_2 - u_1) = u_1 - u_2 \tag{3-56}$$

The boundary work for ideal gases under the constant specific heat assumption are approximately obtained

$$w = c_v(T_1 - T_2) = \frac{1}{\kappa-1}R(T_1 - T_2) = \frac{1}{\kappa-1}(p_1 v_1 - p_2 v_2) \tag{3-57}$$

w can also be expressed for a reversible adiabatic process as

$$w = \frac{1}{\kappa-1}RT_1\left[1 - \left(\frac{p_2}{p_1}\right)^{(\kappa-1)/\kappa}\right] \tag{3-58}$$

or

$$w = \frac{1}{\kappa-1} RT_1 \left[1 - \left(\frac{v_1}{v_2}\right)^{\kappa-1} \right] \quad (3-59)$$

Note that the boundary work relation can be obtained by $w = \int_1^2 p dv$, which agrees well with the above equations.

For a steady flow system, the technical work can be expressed based on the first law of thermodynamics as

$$w_t = q - \Delta h = -(h_2 - h_1) = h_1 - h_2 \quad (3-60)$$

The technical work for ideal gases under the constant specific heat assumption are approximately obtained

$$w_t = c_p(T_1 - T_2) = \frac{\kappa}{\kappa-1} R(T_1 - T_2) = \frac{\kappa}{\kappa-1}(p_1 v_1 - p_2 v_2) \quad (3-61)$$

w_t can also be expressed for a reversible adiabatic process as

$$w = \frac{\kappa}{\kappa-1} RT_1 \left[1 - \left(\frac{p_2}{p_1}\right)^{(\kappa-1)/\kappa} \right] \quad (3-62)$$

or

$$w = \frac{\kappa}{\kappa-1} RT_1 \left[1 - \left(\frac{v_1}{v_2}\right)^{\kappa-1} \right] \quad (3-63)$$

Note that the technical work relation can be obtained by $w_t = -\int_1^2 v dp$, which is the same with the above equations. Obviously, the relation between technical work and boundary work can be expressed as $w_t = \kappa w$.

POLYTROPIC PROCESSES

During actual expansion and compression processes of gases, pressure and volume are often related by $PV^n = C$, where n and C are constants, n is called polytropic exponent. A process of this kind is called a polytropic process (Fig. 3-15), and $n \neq 1$.

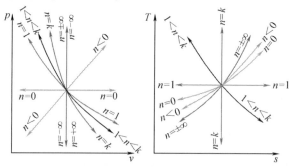

Fig. 3-23 $P-v$ and $T-s$ diagram for a isometric process

Below we develop a general expression for the work done during a polytropic process. The pressure for a polytropic process can be expressed as

$$P = CV^{-n} \quad (3-64)$$

$$n = \frac{\ln p_2 - \ln p_1}{\ln v_1 - \ln v_2} = \frac{\ln(p_2/p_1)}{\ln(v_1/v_2)} \qquad (3-65)$$

$$\left(\frac{\partial p}{\partial v}\right)_n = -n\frac{p}{v} \qquad (3-66)$$

For a reversible process

$$\delta q = T\mathrm{d}s \qquad (3-67)$$

The specific heat for polytropic processes can be obtained as

$$\delta q = c_n \mathrm{d}T \qquad (3-68)$$

Substituting Eqs. (3-67) and (3-68) to (3-66), we obtain

$$\left(\frac{\partial T}{\partial s}\right)_n = \frac{T}{c_n} = \frac{(n-1)T}{(n-\kappa)c_v} \qquad (3-69)$$

where κ is called *adiabatic exponent*. For ideal gases, the adiabatic exponent κ is equal to specific heat ratio k.

Substituting Eq. (3-64) to $w = \int_1^2 p\mathrm{d}v$, the boundary work can be written as

$$w = \int_1^2 p\mathrm{d}v = p_1 v_1^n \int_1^2 \frac{\mathrm{d}v}{v^n} = \frac{1}{n-1}(p_1 v_1 - p_2 v_2) \qquad (3-70)$$

or

$$w = \frac{1}{n-1}R(T_1 - T_2) = \frac{\kappa-1}{n-1}c_v(T_1 - T_2) \qquad (3-71)$$

$$w = \frac{1}{n-1}RT_1\left[1 - \left(\frac{p_2}{p_1}\right)^{(n-1)/n}\right] \qquad (3-72)$$

For a steady flow system, we can express the technical work as

$$w_t = -\int_1^2 v\mathrm{d}p = -\int_1^2 [\mathrm{d}(pv) - p\mathrm{d}v] = p_1 v_1 - p_2 v_2 + \int_1^2 p\mathrm{d}v$$

$$= p_1 v_1 - p_2 v_2 + \frac{1}{n-1}(p_1 v_1 - p_2 v_2)$$

$$= \frac{n}{n-1}(p_1 v_1 - p_2 v_2) \qquad (3-73)$$

or

$$w_t = \frac{n}{n-1}RT_1\left[1 - \left(\frac{p_2}{p_1}\right)^{(n-1)/n}\right] \qquad (3-74)$$

Compared Eq. (3-74) with (3-72), we obtain

$$w_t = nw \qquad (3-75)$$

The heat change for a polytropic process can be expressed as

$$q = \Delta u + w = c_v(T_2 - T_1) + \frac{\kappa-1}{n-1}c_v(T_1 - T_2) = \frac{n-\kappa}{n-1}c_v(T_2 - T_1) \qquad (3-76)$$

Thus,

$$c_n = \frac{n-\kappa}{n-1}c_v \qquad (3-77)$$

3–5 Ideal Gas Mixtures

COMPOSITION OF A GAS MIXTURE: MASS AND MOLE FRACTIONS

To determine the properties of a mixture, we need to know the *composition* of the mixture as well as the properties of the individual components. There are two ways to describe the composition of a mixture: either by specifying the number of moles of each component, called molar analysis, or by specifying the mass of each component, called gravimetric analysis.

Consider a gas mixture composed of k components. The mass of the mixture m_m is the sum of the masses of the individual components, and the mole number of the mixture N_m is the sum of the mole numbers of the individual components[①]. That is,

$$m_m = \sum_{i=1}^{k} m_i \quad \text{and} \quad N_m = \sum_{i=1}^{k} N_i \qquad (3-78\text{a,b})$$

The ratio of the mass of a component to the mass of the mixture is called the mass fraction mf, and the ratio of the mole number of a component to the mole number of the mixture is called the mole fraction y:

$$mf_i = \frac{m_i}{m_m} \qquad (3-79\text{a})$$

$$y_i = \frac{N_i}{N_m} \qquad (3-79\text{b})$$

Dividing Eq. (3–78a) by m_m or Eq. (3–78b) by N_m, we can easily show that the sum of the mass fractions or mole fractions for a mixture is equal to 1:

$$\sum_{i=1}^{k} mf_i = 1 \quad \text{and} \quad \sum_{i=1}^{k} y_i = 1$$

The mass of a substance can be expressed in terms of the mole number N and molar mass M of the substance as $m = NM$. Then the apparent (or average) molar mass and the gas constant of a mixture can be expressed as

$$M_m = \frac{m_m}{N_m} = \frac{\sum m_i}{N_m} = \frac{\sum N_i M_i}{N_m} = \sum_{i=1}^{k} y_i M_i \qquad (3-80a)$$

$$R_m = \frac{R_0}{M_m} \qquad (3-80b)$$

The molar mass of a mixture can also be expressed as

$$M_m = \frac{m_m}{N_m} = \frac{m_m}{\sum m_i/M_i} = \frac{1}{\sum m_i/(m_m M_i)} = \frac{1}{\sum_{i=1}^{k} \frac{mf_i}{M_i}} \qquad (3-81)$$

[①] Throughout this chapter, the subscript m denotes the gas mixture and the subscript i denotes any single component of the mixture.

Mass and mole fractions of a mixture are related by

$$mf_i = \frac{m_i}{m_m} = \frac{N_i M_i}{N_m M_m} = y_i \frac{M_i}{M_m} \qquad (3-82)$$

EXAMPLE 3 – 4 Mass and Mole Fractions of a Gas Mixture

Consider a gas mixture that consists of 3 kg of O_2, 5 kg of N_2, and 12 kg of CH_4, as shown in Fig. 3 – 16. Determine (a) the mass fraction of each component, (b) the mole fraction of each component, and (c) the average molar mass and gas constant of the mixture.

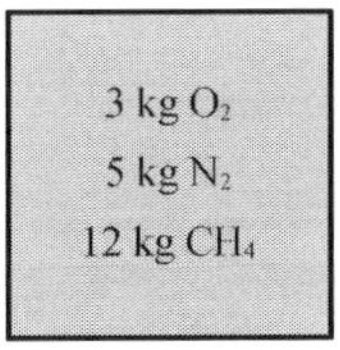

Fig. 3 – 16   Schematic for Example 3 – 4

Solution The masses of components of a gas mixture are given. The mass fractions, the mole fractions, the molar mass, and the gas constant of the mixture are to be determined.

Analysis (a) The total mass of the mixture is

$$m_m = m_{O_2} + m_{N_2} + m_{CH_4} = 3 + 5 + 12 = 20 \text{ kg}$$

Then the mass fraction of each component becomes

$$mf_{O_2} = \frac{m_{O_2}}{m_m} = \frac{3 \text{ kg}}{20 \text{ kg}} = 0.15$$

$$mf_{N_2} = \frac{m_{N_2}}{m_m} = \frac{5 \text{ kg}}{20 \text{ kg}} = 0.25$$

$$mf_{CH_4} = \frac{m_{CH_4}}{m_m} = \frac{12 \text{ kg}}{20 \text{ kg}} = 0.6$$

(b) To find the mole fractions, we need to determine the mole numbers of each component first:

$$N_{O_2} = \frac{m_{O_2}}{M_{O_2}} = \frac{3 \text{ kg}}{32 \text{ kg/kmol}} = 0.094 \text{ kmol}$$

$$N_{N_2} = \frac{m_{N_2}}{M_{N_2}} = \frac{5 \text{ kg}}{28 \text{ kg/kmol}} = 0.179 \text{ kmol}$$

$$N_{CH_4} = \frac{m_{CH_4}}{M_{CH_4}} = \frac{12 \text{ kg}}{16 \text{ kg/kmol}} = 0.75 \text{ kmol}$$

Thus,

$$N_m = N_{O_2} + N_{N_2} + N_{CH_4} = 0.094 + 0.179 + 0.75 = 1.023 \text{ kmol}$$

and

$$y_{O_2} = \frac{N_{O_2}}{N_m} = \frac{0.094 \text{ kmol}}{1.023 \text{ kmol}} = 0.092 \text{ kmol}$$

$$y_{N_2} = \frac{N_{N_2}}{N_m} = \frac{0.179 \text{ kmol}}{1.023 \text{ kmol}} = 0.175 \text{ kmol}$$

$$y_{CH_4} = \frac{N_{CH_4}}{N_m} = \frac{0.75 \text{ kmol}}{1.023 \text{ kmol}} = 0.733 \text{ kmol}$$

(c) The average molar mass and gas constant of the mixture are determined from their definitions,

$$M_m = \frac{m_m}{N_m} = \frac{20 \text{ kg}}{1.023 \text{ kmol}} = 19.6 \text{ kg/kmol}$$

or

$$\begin{aligned} M_m &= \sum y_i M_i = y_{O_2} M_{O_2} + y_{N_2} M_{N_2} + y_{CH_4} M_{CH_4} \\ &= (0.092)(32) + (0.175)(28) + (0.733)(16) \\ &= 19.6 \text{ kg/kmol} \end{aligned}$$

Also,

$$R_m = \frac{R_u}{M_m} = \frac{8.314 \text{ kJ/(kmol} \cdot \text{K)}}{19.6 \text{ kg/kmol}} = 0.424 \text{ kJ/kg} \cdot \text{K}$$

Discussion When mass fractions are available, the molar mass and mole fractions could also be determined directly from Eqs. (3-81) and (3-82).

$P-v-T$ BEHAVIOR OF GAS MIXTURES: IDEAL GASES

An ideal gas is defined as a gas whose molecules are spaced far apart so that the behavior of a molecule is not influenced by the presence of other molecules: a situation encountered at low densities. We also mentioned that real gases approximate this behavior closely when they are at a low pressure or high temperature relative to their critical-point values. The $P-v-T$ behavior of an ideal gas is expressed by the simple relation $Pv = RT$, which is called the *ideal gas equation of state*. The dimensionless ratio $p\bar{v}/\bar{R}T$ is called the compressibility factor and is denoted by Z. That is, $Z = p\bar{v}/\bar{R}T$. As illustrated by subsequent calculations, when values for p, $\bar{v}, \bar{R}$, and T are used in consistent units, Z is unitless.

When two or more ideal gases are mixed, the behavior of a molecule normally is not influenced by the presence of other similar or dissimilar molecules, and therefore a nonreacting mixture of ideal gases also behaves as an ideal gas. Air, for example, is conveniently treated as an ideal gas in the range where nitrogen and oxygen behave as ideal gases. When a gas mixture consists of real (non-ideal) gases, however, the prediction of the $P-v-T$ behavior of the mixture becomes rather involved.

The prediction of the $P-v-T$ behavior of gas mixtures is usually based on two models: *Dalton's law of additive pressures* and *Amagat's law of additive volumes*. Both models are described and discussed below.

Dalton's law of additive pressures: The pressure of a gas mixture is equal to the sum of the pressures each gas would exert if it existed alone at the mixture temperature and volume (Fig. 3-17).

Amagat's law of additive volumes: The volume of a gas mixture is equal to the sum of the volumes each gas would occupy if it existed alone at the mixture temperature and pressure (Fig. 3-18).

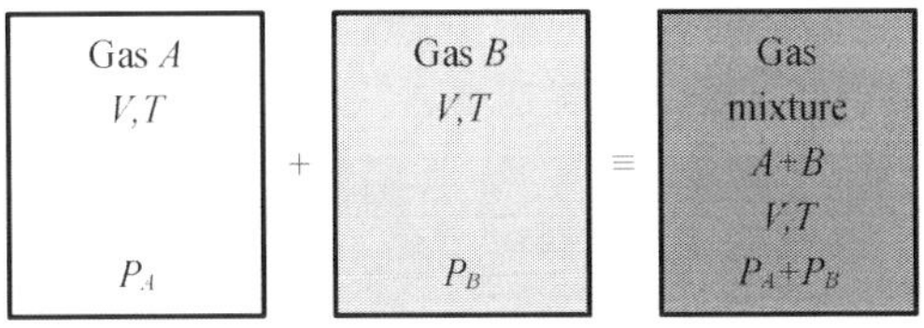

Chapter 3 Thermodynamic Properties and Process of Ideal Gas

Fig. 3 – 17 Dalton's law of additive pressures for a mixture of two ideal gases

Fig. 3 – 18 Amagat's law of additive volumes for a mixture of two ideal gases

Dalton's and Amagat's laws hold exactly for ideal gas mixtures, but only approximately for real gas mixtures. This is due to intermolecular forces that may be significant for real gases at high densities. For ideal gases, these two laws are identical and give identical results.

Dalton's and Amagat's laws can be expressed as follows:

$$\left. \begin{array}{l} Dalton's\ law: P_m = \sum_{i=1}^{k} P_i(T_m, V_m) \\ Amagat's\ law: V_m = \sum_{i=1}^{k} V_i(T_m, P_m) \end{array} \right\} \begin{array}{l} \text{exact for ideal gases,} \quad (3-83) \\ \text{approximate for real gases} \quad (3-84) \end{array}$$

In these relations, P_i is called the component pressure and V_i is called the component volume (Fig. 3 – 19). Note that V_i is the volume a component *would* occupy if it existed alone at T_m and P_m, not the actual volume occupied by the component in the mixture. (In a vessel that holds a gas mixture, each component fills the entire volume of the vessel. Therefore, the volume of each component is equal to the volume of the vessel.) Also, the ratio P_i/P_m is called the pressure fraction and the ratio V_i/V_m is called the volume fraction of component i.

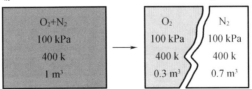

Fig. 3 – 19 The volume a component would occupy if it existed alone at the mixture T and P is called the *component volume* (for ideal gases, it is equal to the partial volume $y_i V_m$)

IDEAL GAS MIXTURES

For ideal gases, P_i and V_i can be related to y_i by using the ideal gas relation for both the components and the gas mixture:

$$\frac{P_i(T_m, V_m)}{P_m} = \frac{N_i R_u T_m / V_m}{N_m R_u T_m / V_m} = \frac{N_i}{N_m} = y_i$$

$$\frac{V_i(T_m, P_m)}{V_m} = \frac{N_i R_u T_m / P_m}{N_m R_u T_m / P_m} = \frac{N_i}{N_m} = y_i$$

Therefore,

$$\frac{P_i}{P_m} = \frac{V_i}{V_m} = \frac{N_i}{N_m} = y_i \qquad (3-85)$$

Eq. (3-85) is strictly valid for ideal gas mixtures since it is derived by assuming ideal gas behavior for the gas mixture and each of its components. The quantity $y_i P_m$ is called the partial pressure (identical to the *component pressure* for ideal gases), and the quantity $y_i V_m$ is called the partial volume (identical to the *component volume* for ideal gases). *Note that for an ideal gas mixture, the mole fraction, the pressure fraction, and the volume fraction of a component are identical.*

The composition of an ideal gas mixture (such as the exhaust gases leaving a combustion chamber) is frequently determined by a volumetric analysis and Eq. (3-85). A sample gas at a known volume, pressure, and temperature is passed into a vessel containing reagents that absorb one of the gases. The volume of the remaining gas is then measured at the original pressure and temperature. The ratio of the reduction in volume to the original volume (volume fraction) represents the mole fraction of that particular gas.

SPECIFIC HEAT, INTERNAL ENERGY, ENTHALPY AND ENTROPY OF IDEAL GAS MIXTURES

Consider a gas mixture that consists of 5 kg of O_2 and 2 kg of N_2. The total mass (an *extensive property*) of this mixture is 7 kg. How did we do it? Well, we simply added the mass of each component. This example suggests a simple way of evaluating the extensive properties of a non-reacting ideal or real gas mixture: *Just add the contributions of each component of the mixture.*

Then the total internal exergy, enthalpy, and entropy of a gas mixture can be expressed, respectively, as

$$U_m = \sum_{i=1}^{k} U_i = \sum_{i=1}^{k} m_i u_i = \sum_{i=1}^{k} N_i \bar{u}_i \quad (\text{kJ}) \qquad (3-86)$$

$$H_m = \sum_{i=1}^{k} H_i = \sum_{i=1}^{k} m_i h_i = \sum_{i=1}^{k} N_i \bar{h}_i \quad (\text{kJ}) \qquad (3-87)$$

$$S_m = \sum_{i=1}^{k} S_i = \sum_{i=1}^{k} m_i s_i = \sum_{i=1}^{k} N_i \bar{s}_i \quad (\text{kJ/K}) \qquad (3-88)$$

By following a similar logic, the changes in internal exergy, enthalpy, and entropy of a gas mixture during a process can be expressed, respectively, as

$$\Delta U_m = \sum_{i=1}^{k} \Delta U_i = \sum_{i=1}^{k} m_i \Delta u_i = \sum_{i=1}^{k} N_i \Delta \bar{u}_i \quad (\text{kJ}) \qquad (3-89)$$

$$\Delta H_m = \sum_{i=1}^{k} \Delta H_i = \sum_{i=1}^{k} m_i \Delta h_i = \sum_{i=1}^{k} N_i \Delta \bar{h}_i \quad (\text{kJ}) \qquad (3-90)$$

$$\Delta S_m = \sum_{i=1}^{k} \Delta S_i = \sum_{i=1}^{k} m_i \Delta s_i = \sum_{i=1}^{k} N_i \Delta \bar{s}_i \quad (\text{kJ/K}) \qquad (3-91)$$

Now reconsider the same mixture, and assume that both O_2 and N_2 are at 25 ℃. The temperature (an *intensive* property) of the mixture is, as you would expect, also 25 ℃. Notice that we did not add the component temperatures to determine the mixture temperature. Instead, we used some kind of averaging scheme, a characteristic approach for determining the intensive properties of a mixture. The internal exergy, enthalpy, and entropy of a mixture *per unit mass* or *per unit mole* of the mixture can be determined by dividing the equations above by the mass or the mole number of the mixture (m_m or N_m).

We obtain

$$u_m = \sum_{i=1}^{k} mf_i u_i \quad (\text{kJ/kg}) \quad \text{and} \quad \bar{u}_m = \sum_{i=1}^{k} y_i \bar{u}_i \quad (\text{kJ/kmol}) \qquad (3-92)$$

$$h_m = \sum_{i=1}^{k} mf_i h_i \quad (\text{kJ/kg}) \quad \text{and} \quad \bar{h}_m = \sum_{i=1}^{k} y_i \bar{h}_i \quad (\text{kJ/kmol}) \qquad (3-93)$$

$$s_m = \sum_{i=1}^{k} mf_i s_i \quad (\text{kJ/kg·K}) \quad \text{and} \quad \bar{s}_m = \sum_{i=1}^{k} y_i \bar{s}_i \quad (\text{kJ/kmol·K}) \qquad (3-94)$$

Similarly, the specific heats of a gas mixture can be expressed as

$$c_{v,m} = \sum_{i=1}^{k} mf_i c_{v,i} \quad (\text{kJ/kg·K}) \quad \text{and} \quad \bar{c}_{v,m} = \sum_{i=1}^{k} y_i \bar{c}_{v,i} \quad (\text{kJ/kmol·K}) \quad (3-95)$$

$$c_{p,m} = \sum_{i=1}^{k} mf_i c_{p,i} \quad (\text{kJ/kg·K}) \quad \text{and} \quad \bar{c}_{p,m} = \sum_{i=1}^{k} y_i \bar{c}_{p,i} \quad (\text{kJ/kmol·K}) \quad (3-96)$$

Notice that *properties per unit mass involve mass fractions* (mf_i) *and properties per unit mole involve mole fractions* (y_i). The relations given above are exact for ideal gas mixtures, and approximate for real gas mixtures. (In fact, they are also applicable to nonreacting liquid and solid solutions especially when they form an "ideal solution".)

The only major difficulty associated with these relations is the determination of properties for each individual gas in the mixture. The analysis can be simplified greatly, however, by treating the individual gases as ideal gases, if doing so does not introduce a significant error.

Summary

In this chapter, thermodynamics properties and process of ideal gas are discussed. Any relation among the pressure, temperature, and specific volume of a substance is called an *equation of state*. The simplest and best-known equation of state is the *ideal gas equation of state*.

Caution should be exercised in using this relation since an ideal gas is a fictitious substance. Then we define specific heats, obtain relations for internal exergy, enthalpy and entropy of ideal gases in terms of specific heat and temperature changes.

The second law of thermodynamics leads to the definition of a new property called *entropy*, which is a quantitative measure of microscopic disorder for a system. Any quantity whose cyclic integral is zero is a property, and entropy is defined as

$$dS = \left(\frac{dQ}{T}\right)_{\text{int rev}}$$

For the special case of an internally reversible, isothermal process, it gives

$$\Delta S = \frac{Q}{T_0}$$

We also deal with nonreacting gas mixtures. A nonreacting gas mixture can be treated as a pure substance since it is usually a homogeneous mixture of different gases. The properties of a gas mixture obviously depend on the properties of the individual gases (called *components* or *constituents*) as well as on the amount of each gas in the mixture. The specific heats, internal exergy, enthalpy and entropy of ideal gas mixtures are discussed by the same method with the ideal gases.

Problems

3 – 1 Helium gas is compressed by an adiabatic compressor from an initial state of 14 psia and 50 °F to a final temperature of 320 °F in a reversible manner. Determine the exit pressure of helium.

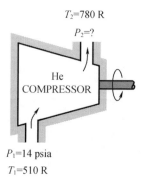

Fig. P3 – 1

3 – 2 A piston cylinder device contains a liquid – vapor mixture of water at 300 K. During a constant pressure process, 750 kJ of heat is transferred to the water. As a result, part of the liquid in the cylinder vaporizes. Determine the entropy change of the water during this process.

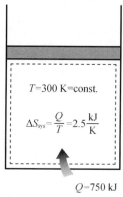

Fig. P3 – 2

3 – 3 A rigid tank contains 2 kmol of N_2 and 6 kmol of CO_2 gases at 300 K and 15 MPa (Fig. P3 – 3). Estimate the volume of the tank on the basis of the ideal gas equation of state.

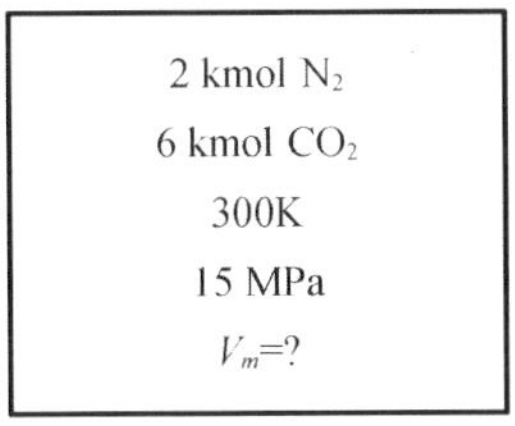

Chapter 3 Thermodynamic Properties and Process of Ideal Gas

2 kmol N_2
6 kmol CO_2
300K
15 MPa
$V_m=?$

Fig. P3 – 3

3 – 4 An insulated rigid tank is divided into two compartments by a partition, as shown in Fig. P3 – 4. One compartment contains 7 kg of oxygen gas at 40 ℃ and 100 kPa, and the other compartment contains 4 kg of nitrogen gas at 20 ℃ and 150 kPa. Now the partition is removed, and the two gases are allowed to mix. Determine (a) the mixture temperature and, (b) the mixture pressure after equilibrium has been established.

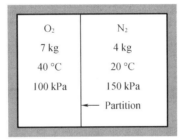

Fig. P3 – 4

3 – 5 Oxygen is heated to experience a specified temperature change and the mass of oxygen is 1 kg. The heat transfer is to be determined for two cases (a) at constant volume process, (b) at constant pressure process.

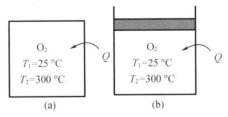

Fig. P3 – 5

(a) at constant volume; (b) at constant pressure

Chapter 4   The Second Law of Thermodynamics

To this point, we have focused our attention on the first law of thermodynamics, which requires that exergy be conserved during a process. In this chapter, we introduce the second law of thermodynamics, which asserts that processes occur in a certain direction and that exergy has quality as well as quantity. A process cannot take place unless it satisfies both the first and second laws of thermodynamics. In this chapter, the thermal exergy reservoirs, reversible and irreversible processes, heat engines, refrigerators, and heat pumps are introduced first. Various statements of the second law are followed by a discussion of perpetual motion machines and the thermodynamic temperature scale. The Carnot cycle is introduced next, and the Carnot principles are discussed. The idealized Carnot heat engines, refrigerators, and heat pumps are examined. The second law of thermodynamics has proved to be a very powerful tool in the optimization of complex thermodynamic systems. We also introduce the concepts of entropy and *exergy* (also called *availability*), which is the maximum useful work that could be obtained from the system at a given state in a specified environment, and we continue with the *reversible work*, which is the maximum useful work that can be obtained as a system undergoes a process between two specified states. Next we discuss the *irreversibility* (also called the *exergy destruction* or *lost work*), which is the wasted work potential during a process as a result of irreversibilities, and we define a *second law efficiency*.

4 – 1   Introduction to the Second Law

We have applied the *first law of thermodynamics*, or the *conservation of exergy principle*, to processes involving closed and open systems. As pointed out repeatedly in those chapters, exergy is a conserved property, and no process is known to have taken place in violation of the first law of thermodynamics. Therefore, it is reasonable to conclude that a process must satisfy the first law to occur. However, as explained here, satisfying the first law alone does not ensure that the process will actually take place.

It is common experience that a cup of hot coffee left in a cooler room eventually cools off. This process satisfies the first law of thermodynamics since the amount of exergy lost by the coffee is equal to the amount gained by the surrounding air. Now let us consider the reverse process the hot coffee getting even hotter in a cooler room as a result of heat transfer from the room air. We all know that this process never takes place. Yet, doing so would not violate the first law as long as the amount of exergy lost by the air is equal to the amount gained by the coffee.

As another familiar example, consider the heating of a room by the passage of electric current through a resistor. Again, the first law dictates that the amount of electric exergy supplied to the

Chapter 4 The Second Law of Thermodynamics

resistance wires be equal to the amount of exergy transferred to the room air as heat. Now let us attempt to reverse this process. It will come as no surprise that transferring some heat to the wires does not cause an equivalent amount of electric exergy to be generated in the wires. Finally, consider a paddle-wheel mechanism that is operated by the fall of a mass. The paddle wheel rotates as the mass falls and stirs a fluid within an insulated container. As a result, the potential exergy of the mass decreases, and the internal exergy of the fluid increases in accordance with the conservation of exergy principle. However, the reverse process, raising the mass by transferring heat from the fluid to the paddle wheel, does not occur in nature, although doing so would not violate the first law of thermodynamics.

It is clear from these arguments that processes proceed in a *certain direction* and not in the reverse direction. The first law places no restriction on the direction of a process, but satisfying the first law does not ensure that the process can actually occur. This inadequacy of the first law to identify whether a process can take place is remedied by introducing another general principle, the *second law of thermodynamics*. We show later in this chapter that the reverse processes discussed above violate the second law of thermodynamics. This violation is easily detected with the help of a property, called *entropy*. A process cannot occur unless it satisfies both the first and the second laws of thermodynamics.

There are numerous valid statements of the second law of thermodynamics. Two such statements are presented and discussed later in this chapter in relation to some engineering devices that operate on cycles. The use of the second law of thermodynamics is not limited to identifying the direction of processes, however. The second law also asserts that exergy has *quality* as well as quantity. The first law is concerned with the quantity of exergy and the transformations of exergy from one form to another with no regard to its quality. Preserving the quality of exergy is a major concern to engineers, and the second law provides the necessary means to determine the quality as well as the degree of degradation of exergy during a process. As discussed later in this chapter, more of high-temperature exergy can be converted to work, and thus it has a higher quality than the same amount of exergy at a lower temperature.

The second law of thermodynamics is also used in determining the *theoretical limits* for the performance of commonly used engineering systems, such as heat engines and refrigerators, as well as predicting the *degree of completion* of chemical reactions.

In the development of the second law of thermodynamics, it is very convenient to have a hypothetical body with a relatively large *thermal exergy capacity* that can supply or absorb finite amounts of heat without undergoing any change in temperature. Such a body is called a thermal exergy reservoir, or just a reservoir. In practice, large bodies of water such as oceans, lakes, and rivers as well as the atmospheric air can be modeled accurately as thermal exergy reservoirs because of their large thermal exergy storage capabilities or thermal masses. The *atmosphere*, for example, does not warm up as a result of heat losses from residential buildings in winter. Likewise, megajoules of waste exergy dumped in large rivers by power plants do not cause any significant change in water temperature.

A *two-phase system* can be modeled as a reservoir also since it can absorb and release large quantities of heat while remaining at constant temperature. Another familiar example of a thermal exergy reservoir is the *industrial furnace*. The temperatures of most furnaces are carefully controlled, and they are capable of supplying large quantities of thermal exergy as heat in an essentially isothermal manner. Therefore, they can be modeled as reservoirs.

A body does not actually have to be very large to be considered a reservoir. Any physical body whose thermal exergy capacity is large relative to the amount of exergy it supplies or absorbs can be modeled as one. In a room, for example, the air can be treated as a reservoir in the analysis of the heat dissipation from a TV set in the room, since the amount of heat transfer from the TV set to the room air is not large enough to have a noticeable effect on the room air temperature.

A reservoir that supplies exergy in the form of heat is called a source, and one that absorbs exergy in the form of heat is called a sink (Fig. 4 – 1). Thermal exergy reservoirs are often referred to as heat reservoirs since they supply or absorb exergy in the form of heat.

Fig. 4 – 1 A source supplies exergy in the form of heat, and a sink absorbs it

Heat transfer from industrial sources to the environment is of major concern to environmentalists as well as to engineers. Irresponsible management of waste exergy can significantly increase the temperature of portions of the environment, causing what is called *thermal pollution*. If it is not carefully controlled, thermal pollution can seriously disrupt marine life in lakes and rivers. However, by careful design and management, the waste exergy dumped into large bodies of water can be used to improve the quality of marine life by keeping the local temperature increases within safe and desirable levels.

THE SECOND LAW OF THERMODYNAMICS: CLAUSIUS STATEMENT

Among many alternative statements of the second law, two are frequently used in engineering thermodynamics. They are the *Clausius* and *Kelvin-Planck* statements. The objective of this section is to introduce these two equivalent second law statements. The Clausius statement has been selected as a point of departure for the study of the second law and its consequences because it is in accord with experience and therefore easy to accept. The Kelvin-Planck statement has the advantage that it provides an effective means for bringing out important second law deductions related to systems undergoingthermodynamic cycles.

The *Clausius statement* of the second law asserts that: *It is impossible for any system to operate in such a way that the sole result would be an exergy transfer by heat from a cooler to a hotter body.*

The Clausius statement does not rule out the possibility of transferring exergy by heat from a cooler body to a hotter body, for this is exactly what refrigerators and heat pumps accomplish. However, as the words "sole result" in the statement suggest, when a heat transfer from a cooler body to a hotter body occurs, there must be some other effect within the system accomplishing the heat transfer, its surroundings, or both. If the system operates ina thermodynamic cycle, its initial state is restored after each cycle, so the only place that must be examined for such other effects is its surroundings.

THE SECOND LAW OF THERMODYNAMICS: KELVIN – PLANCK STATEMENT

The *Kelvin-Planck statement* of the second law: *It is impossible for any system to operate in a thermodynamic cycle and deliver a net amount of exergy by work to its surroundings while receiving exergy by heat transfer from a single thermal reservoir.*

The Kelvin-Planck statement does no true out the possibility of a system developing a net amount of work from a heat transfer drawn from a single reservoir. It only denies this possibility if the system undergoes a thermodynamic cycle.

That is, a heat engine must exchange heat with a low-temperature sink as well as a high-temperature source to keep operating. The Kelvin – Planck statement can also be expressed as *no heat engine can have a thermal efficiency of* 100 *percent* (Fig. 4.9), or as *for a power plant to operate, the working fluid must exchange heat with the environment as well as the furnace.*

Note that the impossibility of having a 100 percent efficient heat engine is not due to friction or other dissipative effects. It is a limitation that applies to both the idealized and the actual heat engines. Later in this chapter, we develop a relation for the maximum thermal efficiency of a heat engine. We also demonstrate that this maximum value depends on the reservoir temperatures only.

EQUIVALENCE OF THE TWO STATEMENTS

Both the Kelvin – Planck and the Clausius statements of the second law are negative statements, and a negative statement cannot be proved. Like any other physical law, the second law of thermodynamics is based on experimental observations. To date, no experiment has been conducted that contradicts the second law, and this should be taken as sufficient proof of its validity.

The Kelvin – Planck and the Clausius statements are equivalent in their consequences, and either statement can be used as the expression of the second law of thermodynamics. Any device that violates the Kelvin – Planck statement also violates the Clausius statement, and vice versa. This can be demonstrated as follows.

Consider the heat – engine – refrigerator combination shown in Fig. 4 – 2 (a), operating between the same two reservoirs. The heat engine is assumed to have, in violation of the Kelvin – Planck statement, a thermal efficiency of 100 percent, and therefore it converts all the heat Q_H it receives to work W.

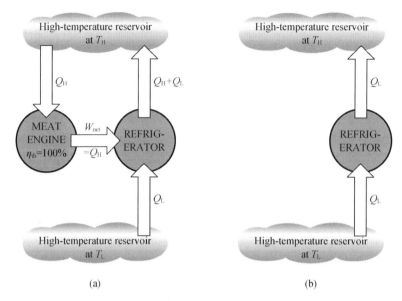

Fig. 4 – 2 Proof that the violation of the Kelvin – Planck statement leads to the violation of the Clausius statement

(a) A refrigerator that is powered by a 100 percent efficient heat engine; (b) The equivalent refrigerator

This work is now supplied to a refrigerator that removes heat in the amount of Q_L from the low – temperature reservoir and rejects heat in the amount of $Q_L + Q_H$ to the high-temperature reservoir. During this process, the high temperature reservoir receives a net amount of heat Q_L (the difference between $Q_L + Q_H$ and Q_H). Thus, the combination of these two devices can be viewed as a refrigerator, as shown in Fig. (4 – 10b), that transfers heat in an amount of Q_L from a cooler body to a warmer one without requiring any input from outside. This is clearly a violation of the Clausius statement. Therefore, a violation of the Kelvin – Planck statement results in the violation of the Clausius statement.

It can also be shown in a similar manner that a violation of the Clausius statement leads to the violation of the Kelvin – Planck statement. Therefore, the Clausius and the Kelvin – Planck statements are two equivalent expressions of the second law of thermodynamics.

PERPETUAL-MOTION MACHINES

We have repeatedly stated that a process cannot take place unless it satisfies both the first and second laws of thermodynamics. Any device that violates either law is called a perpetual-motion machine, and despite numerous attempts, no perpetual-motion machine is known to have worked. But this has not stopped inventors from trying to create new ones.

A device that violates the first law of thermodynamics is called a perpetual-motion machine of the first kind (PMM1), and a device that violates the second law of thermodynamics is called a perpetual-motion machine of the second kind (PMM2).

Consider the steam power plant shown in Fig. 4 – 3. It is proposed to heat the steam by resistance heaters placed inside the boiler, instead of by the exergy supplied from fossil or nuclear fuels. Part of the electricity generated by the plant is to be used to power the resistors as well as the pump. The rest of the electric exergy is to be supplied to the electric network as the net work output. The inventor claims that once the system is started, this power plant will produce electricity indefinitely without requiring any exergy input from the outside.

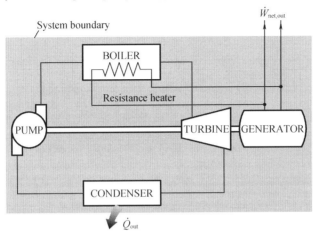

Fig. 4-3 A perpetual-motion machine that violates the first law of thermodynamics (PMM1)

Well, here is an invention that could solve the world's exergy problem—if it works, of course. A careful examination of this invention reveals that the system enclosed by the shaded area is continuously supplying exergy to the outside at a rate of $\dot{Q}_{out} + \dot{W}_{net,out}$ without receiving any exergy. That is, this system is creating exergy at a rate of $\dot{Q}_{out} + \dot{W}_{net,out}$, which is clearly a violation of the first law. Therefore, this wonderful device is nothing more than a PMM1 and does not warrant any further consideration.

Convinced that exergy cannot be created, the inventor suggests the following modification that will greatly improve the thermal efficiency of that power plant without violating the first law. Aware that more than one-half of the heat transferred to the steam in the furnace is discarded in the condenser to the environment, the inventor suggests getting rid of this wasteful component and sending the steam to the pump as soon as it leaves the turbine, as shown in Fig. 4-4. This way, all the heat transferred to the steam in the boiler will be converted to work, and thus the power plant will have a theoretical efficiency of 100 percent. The inventor realizes that some heat losses and friction between the moving components are unavoidable and that these effects will hurt the efficiency somewhat, but still expects the efficiency to be no less than 80 percent (as opposed to 40 percent in most actual power plants) for a carefully designed system.

Well, the possibility of doubling the efficiency would certainly be very tempting to plant managers and, if not properly trained, they would probably give this idea a chance, since intuitively they see nothing wrong with it. A student of thermodynamics, however, will immediately label this device as a PMM2, since it works on a cycle and does a net amount of work while exchanging heat with a single reservoir (the furnace) only. It satisfies the first law but violates the second law, and therefore it will not work.

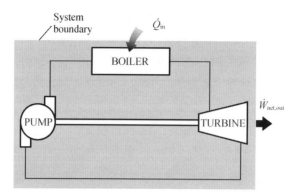

Fig. 4-4 A perpetual-motion machine that violates the second law of thermodynamics (PMM2)

Countless perpetual-motion machines have been proposed throughout history, and many more are being proposed. Some proposers have even gone so far as to patent their inventions, only to find out that what they actually have in their hands is a worthless piece of paper.

Some perpetual-motion machine inventors were very successful in fund raising. For example, a Philadelphia carpenter named J. W. Kelly collected millions of dollars between 1874 and 1898 from investors in his *hydropneumatic-pulsating-vacu-engine*, which supposedly could push a railroad train 3 000 miles on 1 L of water. Of course, it never did. Tired of applications for perpetual-motion machines, the U. S. Patent Office decreed in 1918 that it would no longer consider any perpetual motion machine applications. However, several such patent applications were still filed, and some made it through the patent office undetected. Some applicants whose patent applications were denied sought legal action. For example, in 1982 the U. S. Patent Office dismissed as just another perpetual motion machine a huge device that involves several hundred kilograms of rotating magnets and kilometers of copper wire that is supposed to be generating more electricity than it is consuming from a battery pack. However, the inventor challenged the decision, and in 1985 the National Bureau of Standards finally tested the machine just to certify that it is battery-operated. However, it did not convince the inventor that his machine will not work.

The proposers of perpetual-motion machines generally have innovative minds, but they usually lack formal engineering training, which is very unfortunate. No one is immune from being deceived by an innovative perpetual motion machine. As the saying goes, however, if something sounds too good to be true, it probably is.

4-2 Reversible and Irreversible Processes

The second law of thermodynamics states that no heat engine can have an efficiency of 100 percent. Therefore, What is the highest efficiency that a heat engine can possibly have? Before we can answer this question, we need to define an idealized process first, which is called the *reversible process*.

The processes that were discussed at the beginning of this chapter occurred in a certain direction. Once having taken place, these processes cannot reverse themselves spontaneously and restore the system to its initial state. For this reason, they are classified as *irreversible processes*. Once a cup of hot coffee cools, it will not heat up by retrieving the heat it lost from the surroundings. If it could, the surroundings, as well as the system (coffee), would be restored to their original condition, and this would be a reversible process.

A process is called *irreversible* if the system and all parts of its surroundings cannot be exactly restored to their respective initial states after the process has occurred. A process is *reversible* if both the system and surroundings can be returned to their initial states (Fig. 4-5).

Fig. 4-5 Two familiar reversible processes.
(a) Frictionless pendulum; (b) Quasi-equilibrium expansion and compression of a gas

It should be pointed out that a system can be restored to its initial state following a process, regardless of whether the process is reversible or irreversible. But for reversible processes, this restoration is made without leaving any net change on the surroundings, whereas for irreversible processes, the surroundings usually do some work on the system and therefore does not return to their original state.

Reversible processes actually do not occur in nature. They are merely *idealizations* of actual processes. Reversible processes can be approximated by actual devices, but they can never be achieved. That is, all the processes occurring in nature are irreversible. You may be wondering, then, *why* we are bothering with such fictitious processes. There are two reasons. First, they are easy to analyze, since a system passes through a series of equilibrium states during a reversible process; second, they serve as idealized models to which actual processes can be compared.

Engineers are interested in reversible processes because work-producing devices such as car engines and gas or steam turbines *deliver the most work*, and work-consuming devices such as compressors, fans, and pumps *consume the least work* when reversible processes are used instead of irreversible ones (Fig. 4-6).

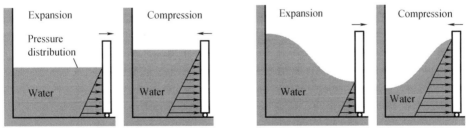

Fig. 4-6 Reversible processes deliver the most and consume the least work.
(a) Slow (reversible) process; (b) Fast (irreversible) process

Reversible processes can be viewed as *theoretical limits* for the corresponding irreversible ones. Some processes are more irreversible than others. We may never be able to have a reversible process, but we can certainly approach it. The more closely we approximate a reversible process, the more work delivered by a work-producing device or the less work required by a work-consuming device.

The concept of reversible processes leads to the definition of the second law efficiency for actual processes, which is the degree of approximation to the corresponding reversible processes. This enables us to compare the performance of different devices that are designed to do the same task on the basis of their efficiencies.

IRREVERSIBILITIES

The factors that cause a process to be irreversible are called irreversibilities. They include friction, unrestrained expansion, mixing of two fluids, heat transfer across a finite temperature difference, electric resistance, inelastic deformation of solids, and chemical reactions. The presence of any of these effects renders a process irreversible. A reversible process involves none of these. Some of the frequently encountered irreversibilities are discussed briefly below.

Friction is a familiar form of irreversibility associated with bodies in motion. When two bodies in contact are forced to move relative to each other (a piston in a cylinder, for example, as shown in Fig. 4-7), a friction force that opposes the motion develops at the interface of these two bodies, and some work is needed to overcome this friction force. The exergy supplied as work is eventually converted to heat during the process and is transferred to the bodies in contact, as evidenced by a temperature rise at the interface. When the direction of the motion is reversed, the bodies are restored to their original position, but the interface does not cool, and heat is not converted back to work. Instead, more of the work is converted to heat while overcoming the friction forces that also oppose the reverse motion. Since the system (the moving bodies) and the sur-roundings cannot be returned to their original states, this process is irreversible. As a consequence, any process that involves friction is irreversible. The larger the friction forces involved, the more irreversible the process is.

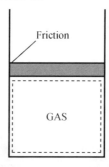

Fig. 4-7 Friction renders a process irreversible

Friction does not always involve two solid bodies in contact. It is also encountered between a fluid and solid and even between the layers of a fluid moving at different velocities. A considerable fraction of the power produced by a car engine is used to overcome the friction (the

Chapter 4 The Second Law of Thermodynamics

drag force) between the air and the external surfaces of the car, and it eventually becomes part of the internal exergy of the air. It is not possible to reverse this process and recover that lost power, even though doing so would not violate the conservation of exergy principle.

Another example of irreversibility is the unrestrained expansion of a gas separated from a vacuum by a membrane, as shown in Fig. 4 – 8. When the membrane is ruptured, the gas fills the entire tank. The only way to restore the system to its original state is to compress it to its initial volume, while transferring heat from the gas until it reaches its initial temperature. From the conservation of exergy considerations, it can easily be shown that the amount of heat transferred from the gas equals the amount of work done on the gas by the surroundings. The restoration of the surroundings involves conversion of this heat completely to work, which would violate the second law. Thus, unrestrained expansion of a gas is an irreversible process.

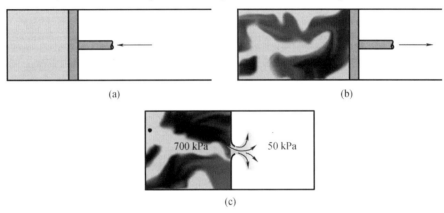

Fig. 4 – 8 Irreversible compression and expansion processes
(a) Fast compression; (b) Fast expansion; (c) Unrestrained expansion

A third form of irreversibility familiar to us all is heat transfer through a finite temperature difference. Consider a can of cold soda left in a warm room (Fig. 4 – 9). Heat is transferred from the warmer room air to the cooler soda. The only way this process can be reversed and the soda restored to its original temperature is to provide refrigeration, which requires some work input. At the end of the reverse process, the soda will be restored to its initial state, but the surroundings will not be. The internal exergy of the surroundings will increase by an amount equal in magnitude to the work supplied to the refrigerator. The restoration of the surroundings to the initial state can be done only by converting this excess internal exergy completely to work, which is impossible to do without violating the second law. Since only the system, not both the system and the surroundings, can be restored to its initial condition, heat transfer through a finite temperature difference is an irreversible process.

Heat transfer can occur only when there is a temperature difference between a system and its surroundings. Therefore, it is physically impossible to have a reversible heat transfer process. But a heat transfer process becomes less and less irreversible as the temperature difference between the two bodies approaches zero. Then heat transfer through a differential temperature difference dT can be considered to be reversible. As dT approaches zero, the process can be reversed in direction (at least theoretically) without requiring any refrigeration. Notice that reversible heat transfer is a conceptual process and cannot be duplicated in the real world.

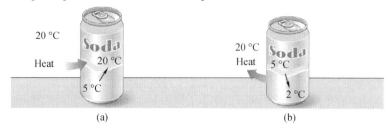

Fig. 4 – 9 (a) **Heat transfer through a temperature difference is irreversible, and** (b) **the reverse process is impossible**

(a) An irreversible heat transfer process; (b) An impossible heat transfer process

The larger the temperature difference between two bodies, the larger the heat transfer rate will be. Any significant heat transfer through a small temperature difference requires a very large surface area and a very longtime. Therefore, even though approaching reversible heat transfer is desirable from a thermodynamic point of view, it is impractical and not economically feasible.

INTERNALLY AND EXTERNALLY REVERSIBLE PROCESSES

A typical process involves interactions between a system and its surroundings, and a reversible process involves no irreversibilities associated with either of them.

A process is called internally reversible if no irreversibilities occur within the boundaries of the system during the process. During an internally reversible process, a system proceeds through a series of equilibrium states, and when the process is reversed, the system passes through exactly the same equilibrium states while returning to its initial state. That is, the paths of the forward and reverse processes coincide for an internally reversible process. The quasi-equilibrium process is an example of an internallyreversible process.

A process is called externally reversible if no irreversibilities occur outside the system boundaries during the process. Heat transfer between a reservoir and a system is an externally reversible process if the outer surface of the system is at the temperature of the reservoir.

A process is called totally reversible, or simply reversible, if it involves no irreversibilities within the system or its surroundings (Fig. 4 – 10). A totally reversible process involves no heat transfer through a finite temperature difference, no nonquasi-equilibrium changes, and no friction or other dissipative effects.

As an example, consider the transfer of heat to two identical systems that are undergoing a constant-pressure (thus constant-temperature) phase change process, as shown in Fig. 4 – 11. Both processes are internally reversible, since both take place isothermally and both pass through exactly the same equilibrium states. The first process shown is externally reversible also, since heat transfer for this process takes place through an infinitesimal temperature difference dT. The second process, however, is externally irreversible, since it involves heat transfer through a finite temperature difference ΔT.

Chapter 4 The Second Law of Thermodynamics

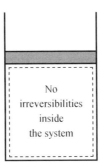

Fig. 4 – 10 A reversible process involves no internal and external irreversibilities

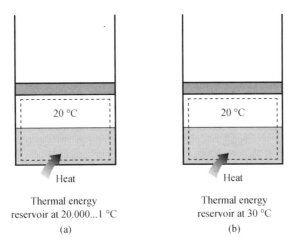

Fig. 4 – 11 Totally and internally reversible heat transfer processes
(a) Totally reversible; (b) Internally reversible

4 – 3 Heat Engines and Refrigerators

HEAT ENGINES

As pointed out earlier, work can easily be converted to other forms of exergy, but converting other forms of exergy to work is not that easy. The mechanical work done by the shaft shown in Fig. 4 – 12, for example, is first converted to the internal exergy of the water. This exergy may then leave the water as heat. We know from experience that any attempt to reverse this process will fail. That is, transferring heat to the water does not cause the shaft to rotate. From these observations, we conclude that work can be converted to heat directly and completely, but converting heat to work requires the use of some special devices. These devices are called heat engines.

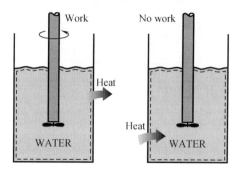

Fig. 4 – 12 Work can always be converted to heat directly and completely, but the reverse is not true

Heat engines differ considerably from one another, but all can be characterized by the following (Fig. 4 – 13):

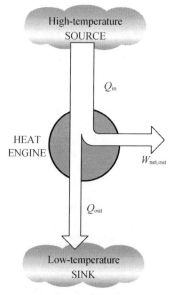

Fig. 4 – 13 Part of the heat received by a heat engine is converted to work, while the rest is rejected to a sink

   a. They receive heat from a high-temperature source (solar exergy, oil furnace, nuclear reactor, etc.).

   b. They convert part of this heat to work (usually in the form of a rotating shaft).

   c. They reject the remaining waste heat to a low-temperature sink (the atmosphere, rivers, etc.).

   d. They operate on a cycle.

Heat engines and other cyclic devices usually involve a fluid to and from which heat is transferred while undergoing a cycle. This fluid is called the working fluid. The term *heat engine* is often used in a broader sense to include work producing devices that do not operate in a thermodynamic cycle. Engines that involve internal combustion such as gas turbines and car engines fall into this category. These devices operate in a mechanical cycle but not in a thermodynamic cycle since the working fluid (the combustion gases) does not under go a complete cycle. Instead of being cooled to the initial temperature, the exhaust gases are purged and replaced by fresh air-and-fuel mixture at the end of the cycle.

The work-producing device that best fits into the definition of a heat engine is the *steam power plant*, which is an external-combustion engine. That is, combustion takes place outside the engine, and the thermal exergy released during this process is transferred to the steam as heat. The schematic of a basic steam power plant is shown in Fig. 4 – 14. This is a rather simplified diagram, and the discussion of actual steam power plants is given in later chapters. The various quantities shown on this figure are as follows:

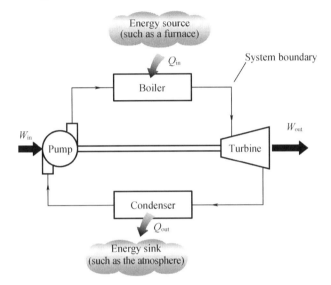

Fig. 4 – 14 Schematic of a steam power plant

Q_{in} = amount of heat supplied to steam in boiler from a high-temperature source (furnace);

Q_{out} = amount of heat rejected from steam in condenser to a low temperature sink (the atmosphere, a river, etc.);

W_{out} = amount of work delivered by steam as it expands in turbine;

W_{in} = amount of work required to compress water to boiler pressure.

Notice that the directions of the heat and work interactions are indicated by the subscripts *in* and *out*. Therefore, all four of the described quantities are always *positive*.

The net work output of this power plant is simply the difference between the total work output of the plant and the total work input (Fig. 4 – 15):

$$W_{net,out} = W_{out} - W_{in} \quad (\text{kJ}) \tag{4-1}$$

The net work can also be determined from the heat transfer data alone. The four components of the steam power plant involve mass flow in and out, and therefore they should be treated as open systems. These components, together with the connecting pipes, however, always contain the same fluid (not counting the steam that may leak out, of course). No mass enters or leaves this combination system, which is indicated by the shaded area on Fig. 4 – 14; thus, it can be analyzed as a closed system. Recall that for a closed system undergoing a cycle, the change in internal exergy U is zero, and therefore the net work output of the system is also equal to the net heat transfer to the system:

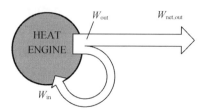

Fig. 4 – 15 **A portion of the work output of a heat engine is consumed internally to maintain continuous operation**

$$W_{net,out} = Q_{in} - Q_{out} \quad (kJ) \quad (4-2)$$

In Eq. (4 – 2), Q_{out} represents the magnitude of the exergy wasted in order to complete the cycle. But Q_{out} is never zero; thus, the net work output of a heat engine is always less than the amount of heat input. That is, only part of the heat transferred to the heat engine is converted to work. The fraction of the heat input that is converted to net work output is a measure of the performance of a heat engine and is called the thermal efficiency η_{th} (Fig. 4 – 16).

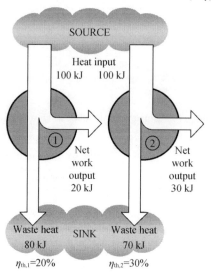

Fig. 4 – 16 **Some heat engines perform better than others**
(**convert more of the heat they receive to work**)

For heat engines, the desired output is the net work output, and the required input is the amount of heat supplied to the working fluid. Then the thermal efficiency of a heat engine can be expressed as

$$Thermal\ efficiency = \frac{Net\ work\ output}{Total\ heat\ input} \quad (4-3)$$

or

Chapter 4 The Second Law of Thermodynamics

$$\eta_{th} = \frac{W_{net,out}}{Q_{in}} \qquad (4-4)$$

It can also be expressed as

$$\eta_{th} = 1 - \frac{Q_{out}}{Q_{in}} \qquad (4-5)$$

since $W_{net,out} = Q_{in} - Q_{out}$.

Cyclic devices of practical interest such as heat engines, refrigerators, and heat pumps operate between a high-temperature medium (or reservoir) at temperature T_H and a low-temperature medium (or reservoir) at temperature T_L. To bring uniformity to the treatment of heat engines, refrigerators, and heat pumps, we define these two quantities:

Q_H = magnitude of heat transfer between the cyclic device and the high temperature medium at temperature T_H;

Q_L = magnitude of heat transfer between the cyclic device and the low temperature medium at temperature T_L.

Notice that both Q_L and Q_H are defined as *magnitudes* and therefore are positive quantities. The direction of Q_H and Q_L is easily determined by inspection. Then the net work output and thermal efficiency relations for any heat engine (shown in Fig. 4 – 17) can also be expressed as

$$W_{net,out} = Q_H - Q_L$$

and

$$\eta_{th} = \frac{W_{net,out}}{Q_H}$$

or

$$\eta_{th} = 1 - \frac{Q_L}{Q_H} \qquad (4-6)$$

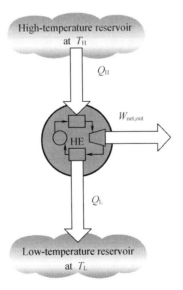

Fig. 4 – 17 Schematic of a heat engine

Thermodynamics

The thermal efficiency of a heat engine is always less than unity since both Q_L and Q_H are defined as positive quantities.

Thermal efficiency is a measure of how efficiently a heat engine converts the heat that it receives to work. Heat engines are built for the purpose of converting heat to work, and engineers are constantly trying to improve the efficiencies of these devices since increased efficiency means less fuel consumption and thus lower fuel bills and less pollution.

The thermal efficiencies of work-producing devices are relatively low. Ordinary spark – ignition automobile engines have a thermal efficiency of about 25 percent. That is, an automobile engine converts about 25 percent of the chemical exergy of the gasoline to mechanical work. This number is as high as 40 percent for diesel engines and large gas – turbine plants and as high as 60 percent for large combined gas – steam power plants. Thus, even with the most efficient heat engines available today, almost one-half of the exergy supplied ends up in the rivers, lakes, or the atmosphere as waste or useless exergy (Fig. 4 – 18).

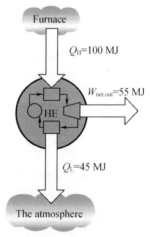

Fig. 4 – 18 Even the most efficient heat engines reject almost one-half of the exergy they receive as waste heat

EXAMPLE 4 – 1 Net Power Production of a Heat Engine

Heat is transferred to a heat engine from a furnace at a rate of 80 MW. If the rate of waste heat rejection to a nearby river is 50 MW, determine the net power output and the thermal efficiency for this heat engine.

Solution The rates of heat transfer to and from a heat engine are given. The net power output and the thermal efficiency are to be determined.

Assumptions Heat losses through the pipes and other components are negligible.

Analysis A schematic of the heat engine is given in Fig. 4 – 19. The furnace serves as the high-temperature reservoir for this heat engine and the river as the low-temperature reservoir. The given quantities can be expressed as

$$\dot{Q}_H = 80 \text{ MW and } \dot{Q}_L = 50 \text{ MW}$$

The net power output of this heat engine is

Chapter 4　The Second Law of Thermodynamics

$$\dot{W}_{net,out} = \dot{Q}_H - \dot{Q}_L = (80 - 50)\,\text{MW} = 30\,\text{MW}$$

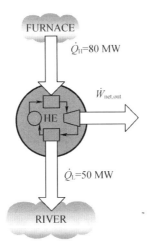

Fig. 4 – 19　Schematic for Example 4 – 1

Then the thermal efficiency is easily determined to be

$$\eta_{th} = \frac{\dot{W}_{net,out}}{\dot{Q}_H} = \frac{30\,\text{MW}}{80\,\text{MW}} = 0.375\,(\text{or } 37.5\%)$$

Discussion: Note that the heat engine converts 37.5 percent of the heat it receives to work.

REFRIGERATORS

We all know from experience that heat is transferred in the direction of decreasing temperature, that is, from high-temperature mediums to low-temperature ones. This heat transfer process occurs in nature without requiring any devices. The reverse process, however, cannot occur by itself. The transfer of heat from a low-temperature medium to a high-temperature one requires special devices called refrigerators.

Refrigerators, like heat engines, are cyclic devices. The working fluid used in the refrigeration cycle is called a refrigerant. The most frequently used refrigeration cycle is the *vapor – compression refrigeration cycle*, which involves four main components: a compressor, a condenser, an expansion valve, and an evaporator, as shown in Fig. 4 – 20.

The refrigerant enters the compressor as a vapor and is compressed to the condenser pressure. It leaves the compressor at a relatively high temperature and cools down and condenses as it flows through the coils of the condenser by rejecting heat to the surrounding medium. It then enters a capillary tube where its pressure and temperature drop drastically due to the throttling effect. The low – temperature refrigerant then enters the evaporator, where it evaporates by absorbing heat from the refrigerated space. The cycle is completed as the refrigerant leaves the evaporator and reenters the compressor.

Thermodynamics

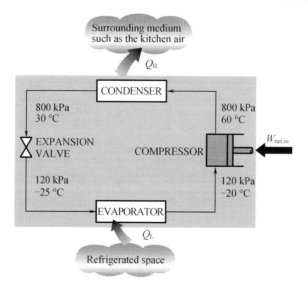

Fig. 4 – 20 Basic components of a refrigeration system and typical operating conditions

In a household refrigerator, the freezer compartment where heat is absorbed by the refrigerant serves as the evaporator, and the coils usually behind the refrigerator where heat is dissipated to the kitchen air serve as the condenser.

A refrigerator is shown schematically in Fig. 4 – 21. Here Q_L is the magnitude of the heat removed from the refrigerated space at temperature T_L, Q_H is the magnitude of the heat rejected to the warm environment at temperature T_H, and $W_{net,in}$ is the net work input to the refrigerator. As discussed before, Q_L and Q_H represent magnitudes and thus are positive quantities.

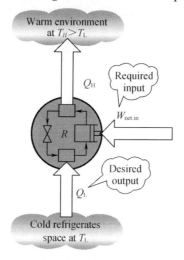

Fig. 4 – 21 The objective of a refrigerator is to remove Q_L from the cooled space

The *efficiency* of a refrigerator is expressed in terms of the coefficient of performance (COP), denoted by COP_R. The objective of a refrigerator is to remove heat (Q_L) from the refrigerated space. To accomplish this objective, it requires a work input of $W_{net,in}$. Then the COP of a refrigerator can be expressed as

Chapter 4 The Second Law of Thermodynamics

$$COP_R = \frac{Desired\ output}{Required\ input} = \frac{Q_L}{W_{net,in}} \qquad (4-7)$$

This relation can also be expressed in rate form by replacing Q_L by $\dot{Q}_L$ and $W_{net,in}$ by $\dot{W}_{net,in}$. The conservation of exergy principle for a cyclic device requires that

$$W_{net,in} = Q_H - Q_L\ (kJ) \qquad (4-8)$$

Then the *COP* relation becomes

$$COP_R = \frac{Q_L}{Q_H - Q_L} = \frac{1}{Q_H/Q_L - 1} \qquad (4-9)$$

Notice that the value of COP_R can be *greater than unity*. That is, the amount of heat removed from the refrigerated space can be greater than the amount of work input. This is in contrast to the thermal efficiency, which can never be greater than 1. In fact, one reason for expressing the efficiency of a refrigerator by another term—the coefficient of performance—is the desire to avoid the oddity of having efficiencies greater than unity.

EXAMPLE 4 – 2 Heat Rejection by a Refrigerator

The food compartment of a refrigerator, shown in Fig. 4 – 22, is maintained at 4 ℃ by removing heat from it at a rate of 360 kJ/min. If the required power input to the refrigerator is 2 kW, determine (a) the coefficient of performance of the refrigerator and (b) the rate of heat rejection to the room that houses the refrigerator.

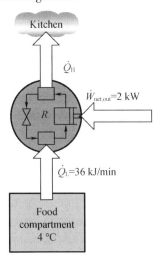

Fig. 4 – 22 Schematic for Example 4 – 2

Solution The power consumption of a refrigerator is given. The COP and the rate of heat rejection are to be determined.

Assumptions Steady operating conditions exist.

Analysis (a) The coefficient of performance of the refrigerator is

$$COP_R = \frac{\dot{Q}_L}{\dot{W}_{net,in}} = \frac{360\ kJ/min}{2\ kW} = \frac{360\ kJ/min}{2 \times (60\ kJ/min)} = 3$$

That is, 3 kJ of heat is removed from the refrigerated space for each kJ of work supplied.

(b) The rate at which heat is rejected to the room that houses the refrigerators determined from the conservation of exergy relation for cyclic devices,

$$\dot{Q}_H = \dot{Q}_L + \dot{W}_{net,in} = 360 \text{ kJ/min} + (2 \text{ kW})\frac{(60 \text{ kJ/min})}{(1 \text{ kW})} = 480 \text{ kJ/min}$$

Discussion Notice that both the exergy removed from the refrigerated space as heat and the exergy supplied to the refrigerator as electrical work eventually show up in the room air and become part of the internal exergy of the air. This demonstrates that exergy can change from one form to another, can move from one place to another, but is never destroyed during a process.

4 – 4 The Carnot Cycle

We mentioned earlier that heat engines are cyclic devices and that the working fluid of a heat engine returns to its initial state at the end of each cycle. Work is done by the working fluid during one part of the cycle and on the working fluid during another part. The difference between these two is the net work delivered by the heat engine. The efficiency of a heat engine cycle greatly depends on how the individual processes that make up the cycle are executed. The net work, thus the cycle efficiency, can be maximized by using processes that require the least amount of work and deliver the most, that is, by using *reversible processes*. Therefore, it is no surprise that the most efficient cycles are reversible cycles, that is, cycles that consist entirely of reversible processes.

Reversible cycles cannot be achieved in practice because the irreversibilities associated with each process cannot be eliminated. However, reversible cycles provide upper limits on the performance of real cycles. Heat engines and refrigerators that work on reversible cycles serve as models to which actual heat engines and refrigerators can be compared. Reversible cycles also serve as starting points in the development of actual cycles and are modified as needed to meet certain requirements.

In a *Carnot cycle*, the system executing the cycle undergoes a series of four internallyreversible processes: two adiabatic processes alternated with two isothermal processes.

Fig. 4 – 23 shows the $p - v$ diagram of a Carnot power cycle in which the system is a gas in a piston – cylinder assembly. Fig. 4 – 24 provides details of how the cycle is executed. The piston

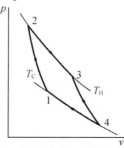

Fig. 4 – 23 $p - v$ diagram for a Carnot ga spower cycle

Chapter 4　The Second Law of Thermodynamics

and cylinder walls are nonconducting. The heat transfers are in the directions of the arrows. Also note that there are two reservoirs at temperatures T_H and T_C, respectively, and an insulating stand. Initially, the piston cylinder assembly is on the insulating stand and the system is at state 1, where the temperature is T_C.

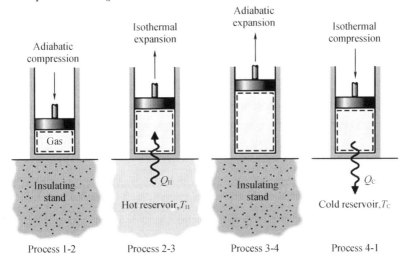

Fig. 4-24　Carnot power cycle executed by a gas in a piston-cylinder assembly

The four processes of the cycle are:

Process 1-2: The gas is compressed *adiabatically* to state 2, where the temperature is T_H.

Process 2-3: The assembly is placed in contact with the reservoir at T_H. The gas expands *isothermally* while receiving exergy Q_H from the hot reservoir by heat transfer.

Process 3-4: The assembly is again placed on the insulating stand and the gas is allowed to continue to expand *adiabatically* until the temperature drops to T_C.

Process 4-1: The assembly is placed in contact with the reservoir at T_C. The gas is compressed *isothermally* to its initial state while it discharges exergy Q_C to the cold reservoir by heat transfer.

For the heat transfer during Process 2-3 to be reversible, the difference between the gas temperature and the temperature of the hot reservoir must be vanishingly small. Since the reservoir temperature remains constant, this implies that the temperature of the gas also remains constant during Process 2-3. The same can be concluded for Process 4-1.

For each of the four internally reversible processes of the Carnot cycle, the work can be represented as an area on Fig. 4-23. The area under the adiabatic process line 1-2 represents the work done per unit of mass to compress the gas in this process. The areas under process lines 2-3 and 3-4 represent the work done per unit of mass by the gas as it expands in the processes. The area under process line 4-1 is the work done per unit of mass to compress the gas in this process. The enclosed area on the $p-v$ diagram, shown shaded, is the net work developed by the cycle per unit of mass.

The Carnot cycle can also be executed in a steady-flow system. It is discussed in later chapters in conjunction with other power cycles. Being a reversible cycle, the Carnot cycle is the most efficient cycle operating between two specified temperature limits. Even though the Carnot cycle cannot be achieved in reality, the efficiency of actual cycles can be improved by attempting to approximate the Carnot cycle more closely.

THE REVERSED CARNOT CYCLE

The Carnot heat engine cycle just described is a totally reversible cycle. Therefore, all the processes that comprise it can be *reversed*, in which case it becomes the Carnot refrigeration cycle. This time, the cycle remains exactly the same, except that the directions of any heat and work interactions are reversed: Heat in the amount of Q_L is absorbed from the low-temperature reservoir, heat in the amount of Q_H is rejected to a high-temperature reservoir, and a work input of $W_{net,in}$ is required to accomplish all this.

The $P-v$ diagram of the reversed Carnot cycle is the same as the one given for the Carnot cycle, except that the directions of the processes are reversed, as shown in Fig. 4 – 25.

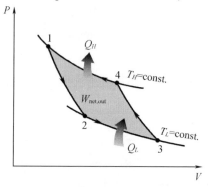

Fig. 4 – 25 $P-v$ **diagram of the reversed Carnot cycle**

4 – 5   The Carnot Principles

The second law of thermodynamics puts limits on the operation of cyclic devices as expressed by the Kelvin – Planck and Clausius statements. A heat engine cannot operate by exchanging heat with a single reservoir, and a refrigerator cannot operate without a net exergy input from an external source.

We can draw valuable conclusions from these statements. Two conclusions pertain to the thermal efficiency of reversible and irreversible (i. e., actual) heat engines, and they are known as the Carnot principles (Fig. 4 – 26), expressed as follows:

a. The efficiency of an irreversible heat engine is always less than the efficiency of are versible one operating between the same two reservoirs.

b. The efficiencies of all reversible heat engines operating between the same two reservoirs are the same.

Chapter 4 The Second Law of Thermodynamics

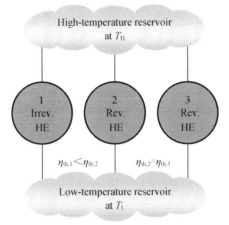

Fig. 4 – 26 The Carnot principles

These two statements can be proved by demonstrating that the violation of either statement results in the violation of the second law of thermodynamics. To prove the first statement, consider two heat engines operating between the same reservoirs, as shown in Fig. 4 – 27. One engine is reversible and the other is irreversible. Now each engine is supplied with the same amount of heat Q_H. The amount of work produced by the reversible heat engine is W_{rev}, and the amount produced by the irreversible one is W_{irrev}.

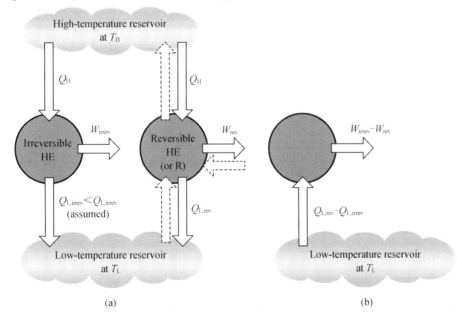

Fig. 4 – 27 Proof of the first Carnot principle.
(a) A reversible and an irreversible heat engine operating between the same two reservoirs (the reversible heat engine is then reversed to run as a refrigerator); (b) The equivalent combined system

In violation of the first Carnot principle, we assume that the irreversible heat engine is more efficient than the reversible one (that is, $\eta_{th,ir} > \eta_{th,r}$) and thus delivers more work than the reversible one. Now let the reversible heat engine be reversed and operate as a refrigerator. This refrigerator will receive a work input of W_{rev} and reject heat to the high-temperature reservoir. Since the refrigerator is rejecting heat in the amount of Q_H to the high temperature reservoir and the irreversible heat engine is receiving the same amount of heat from this reservoir, the net heat exchange for this reservoir is zero. Thus, it could be eliminated by having the refrigerator discharge Q_H directly into the irreversible heat engine. Now considering the refrigerator and the irreversible engine together, we have an engine that produces a net work in the amount of $W_{irrev} - W_{rev}$ while exchanging heat with a single reservoir, a violation of the Kelvin-Planck statement of the second law. Therefore, our initial assumption that $\eta_{th,ir} > \eta_{th,r}$ is incorrect. Then we conclude that no heat engine can be more efficient than a reversible heat engine operating between the same reservoirs.

The second Carnot principle can also be proved in a similar manner. This time, let us replace the irreversible engine by another reversible engine that is more efficient and thus delivers more work than the first reversible engine. By following through the same reasoning, we end up having an engine that produces a net amount of work while exchanging heat with a single reservoir, which is a violation of the second law. Therefore, we conclude that no reversible heat engine can be more efficient than a reversible one operating between the same two reservoirs, regardless of how the cycle is completed or the kind of working fluid used.

The second Carnot principle discussed above states that all reversible heat engines have the same thermal efficiency when operating between the same two reservoirs (Fig. 4-28). That is, the efficiency of a reversible engine is independent of the working fluid employed and its properties, the way the cycle is executed, or the type of reversible engine used. Since exergy reservoirs are characterized by their temperatures, the thermal efficiency of reversible heat engines is a function of the reservoir temperatures only. That is,

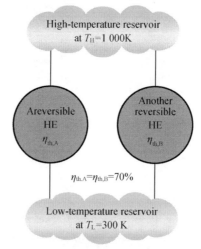

Fig. 4-28 All reversible heat engines operating between the same two reservoirs have the same efficiency (the second Carnot principle)

Chapter 4 The Second Law of Thermodynamics

$$\eta_{th,r} = g(T_H - T_L)$$

or

$$\frac{Q_H}{Q_L} = f(T_H, T_L) \tag{4-7}$$

since $\eta_{th} = 1 - Q_L/Q_H$. In these relations T and T are the temperatures of the high-and low-temperature reservoirs, respectively.

The functional form of $f(T, T)$ can be developed with the help of the three reversible heat engines shown in Fig. 4 – 29. Engines A and C are supplied with the same amount of heat Q_1 from the high-temperature reservoir at T_1. Engine C rejects Q_3 to the low-temperature reservoir at T_3. Engine B receives the heat Q_2 rejected by engine A at temperature T_2 and rejects heat in the amount of Q_3 to a reservoir at T_3.

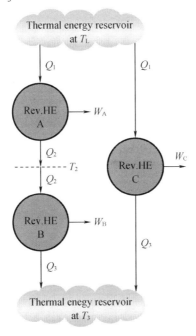

Fig. 4 – 29 The arrangement of heat engines used to develop the thermodynamic temperature scale

The amounts of heat rejected by engines B and C must be the same since engines A and B can be combined into one reversible engine operating between the same reservoirs as engine C and thus the combined engine will have the same efficiency as engine C. Since the heat input to engine C is the same as the heat input to the combined engines A and B, both systems must reject the same amount of heat.

Applying Eq. (4 –7) to all three engines separately, we obtain

$$\frac{Q_1}{Q_2} = f(T_1, T_2), \quad \frac{Q_2}{Q_3} = f(T_2, T_3) \text{ and } \frac{Q_1}{Q_3} = f(T_1, T_3)$$

Now consider the identity

$$\frac{Q_1}{Q_3} = \frac{Q_1 Q_2}{Q_2 Q_3}$$

which corresponds to
$$f(T_1,T_3) = f(T_1,T_2) \cdot f(T_2,T_3)$$
A careful examination of this equation reveals that the left-hand side is a function of T_1 and T_3, and therefore the right-hand side must also be a function of T_1 and T_3 only, and not T_2. That is, the value of the product on the right-hand side of this equation is independent of the value of T_2. This condition will be satisfied only if the function f has the following form:
$$f(T_1,T_2) = \frac{\varphi(T_1)}{\varphi(T_2)} \text{ and } f(T_2,T_3) = \frac{\varphi(T_2)}{\varphi(T_3)}$$
so that $\varphi(T_2)$ will cancel from the product of $f(T_1, T_2)$ and $f(T_2, T_3)$, yielding
$$\frac{Q_1}{Q_3} = f(T_1,T_3) = \frac{\varphi(T_1)}{\varphi(T_3)} \qquad (4-8)$$
This relation is much more specific than Eq. (4 – 7) for the functional form of Q_1/Q_3 in terms of T_1 and T_3.

For a reversible heat engine operating between two reservoirs at temperatures T_H and T_L, Eq. (4 – 8) can be written as
$$\frac{Q_H}{Q_L} = \frac{\varphi(T_H)}{\varphi(T_L)} \qquad (4-9)$$
This is the only requirement that the second law places on the ratio of heat transfers to and from the reversible heat engines. Several functions $\varphi(T)$ satisfy this equation, and the choice is completely arbitrary. Lord Kelvin first proposed taking $\varphi(T) = T$ to define a thermodynamic temperature scale as (Fig. 4 – 30)
$$\left(\frac{Q_H}{Q_L}\right)_{rev} = \frac{T_H}{T_L} \qquad (4-10)$$

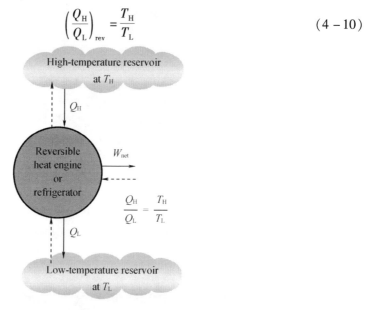

Fig. 4 – 30 For reversible cycles, the heat transfer ratio Q_H/Q_L can be replaced by the absolute temperature ratio T_H/T_L

This temperature scale is called the Kelvin scale, and the temperatures on this scale are called absolute temperatures. On the Kelvin scale, the temperature ratios depend on the ratios of

Chapter 4 The Second Law of Thermodynamics

heat transfer between a reversible heat engine and the reservoirs and are independent of the physical properties of any substance. On this scale, temperatures vary between zero and infinity.

The thermodynamic temperature scale is not completely defined by Eq. (4 – 10) since it gives us only a ratio of absolute temperatures. We also need to know the magnitude of a Kelvin. At the International Conference on Weights and Measures held in 1954, the triple point of water (the state at which all three phases of water exist in equilibrium) was assigned the value 273.16 K (Fig. 4 – 31). The temperatures on these two scales differ by a constant 273.15:

$$T(℃) = T(K) - 273.15 \qquad (4-11)$$

Even though the thermodynamic temperature scale is defined with the help of the reversible heat engines, it is not possible, nor is it practical, to actually operate such an engine to determine numerical values on the absolute temperature scale. Absolute temperatures can be measured accurately by other means, such as the constant-volume ideal gas thermometer together with extrapolation techniques as discussed before. The validity of Eq. (4 – 10) can be demonstrated from physical considerations for a reversible cycle using an ideal gas as the working fluid.

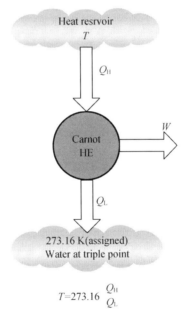

Fig. 4 – 31 A conceptual experimental set up to determine thermodynamic temperatures on the Kelvin scale by measuring heat transfers Q_H and Q_L.

THE CARNOT HEAT ENGINE

The hypothetical heat engine that operates on the reversible Carnot cycle is called the Carnot heat engine. The thermal efficiency of any heat engine, reversible or irreversible, is given by Eq. (4 – 6) as

$$\eta_{th} = 1 - \frac{Q_L}{Q_H}$$

where Q_H is heat transferred to the heat engine from a high-temperature reservoir at T_H, and Q_L is heat rejected to a low – temperature reservoir at T_L. For reversible heat engines, the heat transfer ratio in the above relation can be replaced by the ratio of the absolute temperatures of the two reservoirs, as given by Eq. (4 – 10). Then the efficiency of a Carnot engine, or any reversible heat engine, becomes

$$\eta_{th,r} = 1 - \frac{T_L}{T_H} \qquad (4-12)$$

This relation is often referred to as the Carnot efficiency, since the Carnot heat engine is the best known reversible engine. *This is the highest efficiency a heat engine operating between the two thermal exergy reservoirs at temperatures T_L and T_H can have* (Fig. 4 – 32). All irreversible (i.e., actual) heat engines operating between these temperature limits (T_L and T_H) have lower efficiencies. An actual heat engine cannot reach this maximum theoretical efficiency value because it is impossible to completely eliminate all the irreversibilities associated with the actual cycle.

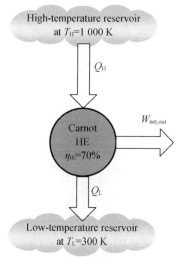

Fig. 4 – 32 The Carnot heat engine is the most efficient of all heat engines operating between the same high-and low-temperature reservoirs

Note that T_L and T_H in Eq. (4 – 12) are *absolute temperatures*. Using ℃ or ℉ for temperatures in this relation gives results grossly in error. The thermal efficiencies of actual and reversible heat engines operating between the same temperature limits compare as follows (Fig. 4 – 33):

$$\eta_{th} \begin{cases} < \eta_{th,rev} & \text{irreversible heat engine} \\ = \eta_{th,rev} & \text{reversible heat engine} \\ > \eta_{th,rev} & \text{impossible heat engine} \end{cases} \qquad (4-13)$$

Most work-producing devices (heat engines) in operation today have efficiencies under 40 percent, which appear low relative to 100 percent. However, when the performance of actual heat engines is assessed, the efficiencies should not be compared to 100 percent; instead, they should be compared to the efficiency of a reversible heat engine operating between the same temperature limits—because this is the true theoretical upper limit for the efficiency, not 100 percent.

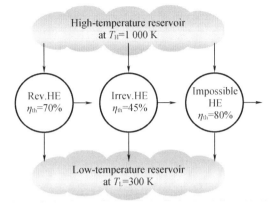

Fig. 4 – 33 No heat engine can have a higher efficiency than a reversible heat engine operating between the same high and low-temperature reservoirs

The maximum efficiency of a steam power plant operating between $T_H = 1\ 000$ K and $T_L = 400$ K is 60 percent, as determined from Eq. (4 – 12). Compared with this value, an actual efficiency of 40 percent does not seem so bad, even though there is still plenty of room for improvement.

It is obvious from Eq. (4 – 12) that the efficiency of a Carnot heat engine increases as T_H is increased, or as T_L is decreased. This is to be expected since as T_L decreases, so does the amount of heat rejected, and as T_L approaches zero, the Carnot efficiency approaches unity. This is also true for actual heat engines. *The thermal efficiency of actual heat engines can be maximized by supplying heat to the engine at the highest possible temperature and rejecting heat from the engine at the lowest possible temperature.*

THE CARNOT REFRIGERATOR

A refrigerator or a heat pump that operates on the reversed Carnot cycle is called a Carnot refrigerator. The coefficient of performance of any refrigerator or heat pump, reversible or irreversible, is given by Eq. (4 – 9) as

$$COP_R = \frac{1}{Q_H/Q_L - 1}$$

where Q_L is the amount of heat absorbed from the low-temperature medium and Q_H is the amount of heat rejected to the high-temperature medium. The COP_s of all reversible refrigerators or heat pumps can be determined by replacing the heat transfer ratios in the above relations by the ratios of the absolute temperatures of the high-and low-temperature reservoirs, as expressed by Eq. (4 – 10). Then the COP relations for reversible refrigerators and heat pumps become

$$COP_R = \frac{1}{T_H/T_L - 1} \qquad (4-14)$$

These are the highest coefficients of performance that a refrigerator or a heat pump operating between the temperature limits of T_L and T_H can have. All actual refrigerators or heat pumps operating between these temperature limits (T_L and T_H) have lower coefficients of performance (Fig. 4-34).

The coefficients of performance of actual and reversible refrigerators operating between the same temperature limits can be compared as follows:

$$COP_R \begin{cases} < COP_{R,r} \text{ irreversible refrigerator} \\ = COP_{R,r} \text{ reversible refrigerator} \\ > COP_{R,r} \text{ impossible refrigerator} \end{cases} \quad (4-15)$$

The COP of a reversible refrigerator or heat pump is the maximum theoretical value for the specified temperature limits. Actual refrigerators may approach these values as their designs are improved, but they can never reach them. As a final note, the COP_s of the refrigerators decrease as T_L decreases. That is, it requires more work to absorb heat from lower-temperature media. As the temperature of the refrigerated space approaches zero, the amount of work required to produce a finite amount of refrigeration approaches infinity and COP_R approaches zero.

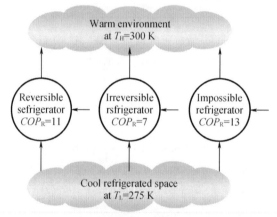

Fig. 4-34 No refrigerator can have a higher COP than a reversible refrigerator operating between the same temperature limits

4-6 Expressions of The Second Law of Thermodynamic

CLAUSIUS INEQUALITY

The second law of thermodynamics often leads to expressions that involve inequalities. An irreversible (i.e., actual) heat engine, for example, is less efficient than a reversible one operating between the same two thermal exergy reservoirs. Likewise, an irreversible refrigerator or a heat pump has a lower coefficient of performance (COP) than a reversible one operating between the same temperature limits. The Clausius inequality provides the basis for introducing two ideas instrumental for analyses of both closed systems and control volumes from a second law perspective: the property entropy and the entropy balance.

The Clausius inequality states that for any thermodynamic cycle

$$\oint \frac{\delta Q}{T} \leqslant 0$$

over the entire cycle. The equality and inequality have the same interpretation as in the Kelvin – Planck statement: the equality applies when there are no internal irreversibilities as the system executes the cycle, and the inequality applies when internal irreversibilities are present. The Clausius inequality can be demonstrated using the Kelvin – Planck statement of the second law.

To demonstrate the validity of the Clausius inequality, consider a system connected to a thermal exergy reservoir at a constant thermodynamic (i.e., absolute) temperature of T_R through a *reversible* cyclic device (Fig. 4 – 35). The cyclic device receives heat δQ_R from the reservoir and supplies heat δQ to the system whose temperature at that part of the boundary is T (a variable) while producing work δW_{rev}. The system produces work δW_{sys} as a result of this heat transfer. Applying the exergy balance to the combined system identified by dashed lines yields

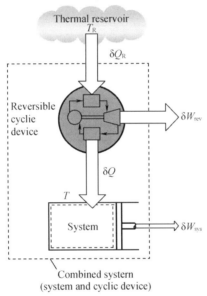

Fig. 4 – 35   The system considered in the development of the Clausius inequality

$$\delta W_C = \delta Q_R - dE_C$$

where δW_C is the total work of the combined system ($\delta W_{rev} + \delta W_{sys}$) and dE_C is the change in the total exergy of the combined system. Considering that the cyclic device is a *reversible* one, we have

$$\frac{\delta Q_R}{T_R} = \frac{\delta Q}{T}$$

where the sign of δQ is determined with respect to the system (positive if *to* the system and negative if *from* the system) and the sign of δQ_R is determined with respect to the reversible cyclic device. Eliminating δQ_R from the two relations above yields,

$$\delta W_C = T_R \frac{\delta Q}{T} - dE_C$$

We now let the system undergo a cycle while the cyclic device undergoes an integral number of cycles. Then the preceding relation becomes

$$W_C = T_R \oint \left(\frac{\delta Q}{T}\right)_b$$

since the cyclic integral of exergy (the net change in the exergy, which is a property, during a cycle) is zero. Here W_C is the cyclic integral of δW_C, and it represents the net work for the combined cycle.

It appears that the combined system is exchanging heat with a single thermal exergy reservoir while involving (producing or consuming) work W_C during a cycle. On the basis of the Kelvin – Planck statement of the second law, which states that *no system can produce a net amount of work while operating in a cycle and exchanging heat with a single thermal exergy reservoir*, we reason that W_C cannot be a work output, and thus it cannot be a positive quantity. Considering that T_R is the thermodynamic temperature and thus a positive quantity, we must have

$$\oint \left(\frac{\delta Q}{T}\right)_b \leq 0 \qquad (4-16)$$

which is the *Clausius inequality*. This inequality is valid for all thermodynamic cycles, reversible or irreversible, including the refrigeration cycles.

If no irreversibilities occur within the system as well as the reversible cyclic device, then the cycle undergone by the combined system is internally reversible. As such, it can be reversed. In the reversed cycle case, all the quantities have the same magnitude but the opposite sign. Therefore, the work W_C, which could not be a positive quantity in the regular case, cannot be a negative quantity in the reversed case. Then it follows that $W_{C,\text{int rev}} = 0$ since it cannot be a positive or negative quantity, and therefore

$$\oint \left(\frac{\delta Q}{T}\right)_{\text{int r}} \leq 0 \qquad (4-17)$$

for internally reversible cycles. Thus, we conclude that *the equality in the Clausius inequality holds for totally or just internally reversible cycles and the inequality for the irreversible ones*.

THE INCREASE OF ENTROPY PRINCIPLE

Consider a cycle that is made up of two processes: process 1 – 2, which is arbitrary (reversible or irreversible), and process 2 – 1, which is internally reversible, as shown in Figure 4 – 36.

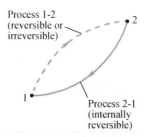

Fig. 4 – 36 A cycle composed of a reversible and an irreversible process

Chapter 4 The Second Law of Thermodynamics

From the Clausius inequality,

$$\oint \frac{\delta Q}{T} \leq 0$$

or

$$\int_1^2 \frac{\delta Q}{T} + \int_1^2 \left(\frac{\delta Q}{T}\right)_{\text{int rev}} \leq 0$$

The second integral in the previous relation is recognized as the entropy change $S_1 = S_2$. Therefore,

$$\int_1^2 \frac{\delta Q}{T} + S_1 - S_2 \leq 0$$

which can be rearranged as

$$S_2 - S_1 \geq \int_1^2 \frac{\delta Q}{T} \qquad (4-18)$$

It can also be expressed in differential form as

$$dS \geq \frac{\delta Q}{T} \qquad (4-19)$$

where the equality holds for an internally reversible process and the inequality for an irreversible process. We may conclude from these equations that the entropy change of a closed system during an irreversible process is greater than the integral of $\delta Q/T$ evaluated for that process. In the limiting case of a reversible process, these two quantities become equal. We again emphasize that T in these relations is the *thermodynamic temperature* at the *boundary* where the differential heat δQ is transferred between the system and the surroundings.

The quantity $\Delta S = S_2 - S_1$ represents the *entropy change* of the system. For a reversible process, it becomes equal to $\int_1^2 \delta Q/T$, which represents the *entropy transfer* with heat.

The inequality sign in the preceding relations is a constant reminder that the entropy change of a closed system during an irreversible process is always greater than the entropy transfer. That is, some entropy is *generated* or *created* during an irreversible process, and this generation is due entirely to the presence of irreversibilities. The entropy generated during a process is called entropy generation and is denoted by $S]_{\text{gen}}$. Noting that the difference between the entropy change of a closed system and the entropy transfer is equal to entropy generation, Eq. (4-18) can be rewritten as an equality as

$$\Delta S_{\text{sys}} = S_2 - S_1 = \int_1^2 \frac{\delta Q}{T} + S_{\text{gen}} \qquad (4-20)$$

Note that the entropy generation S_{gen} is always a *positive* quantity or zero. Its value depends on the process, and thus it is *not* a property of the system. Also, in the absence of any entropy transfer, the entropy change of a system is equal to the entropy generation.

Eq. (4-18) has far-reaching implications in thermodynamics. For an isolated system (or simply an adiabatic closed system), the heat transfer is zero, and Eq. (4-18) reduces to

$$\Delta S_{\text{iso}} \geq 0 \qquad (4-21)$$

This equation can be expressed as *the entropy of an isolated system during a process always increases or*, *in the limiting case of a reversible process*, *remains constant*. In other words, it *never* decreases. Since entropy is produced in all actual processes, the only processes that can occur are those for which the entropy of the isolated system increases. This is known as the increase of entropy principle. The increase of entropy principle is sometimes considered an alternativestatement of the second law.

Entropy is an extensive property, and thus the total entropy of a system is equal to the sum of the entropies of the parts of the system. An isolated system may consist of any number of subsystems. A system and its surroundings, for example, constitute an isolated system since both can be enclosed by a sufficiently large arbitrary boundary across which there is no heat, work, or mass transfer. Therefore, a system and its surroundings can be viewed as the two subsystems of an isolated system, and the entropy change of this isolated system during a process is the sum of the entropy changes of the system and its surroundings, which is equal to the entropy generation since an isolated system involves no entropy transfer.

That is,

$$S_{gen} = \Delta S_{total} = \Delta S_{sys} + \Delta S_{surr} \geq 0 \qquad (4-22)$$

where the equality holds for reversible processes and the inequality for irreversible ones. Note that S_{surr} refers to the change in the entropy of the surroundings as a result of the occurrence of the process under consideration.

Since no actual process is truly reversible, we can conclude that some entropy is generated during a process, and therefore the entropy of the universe, which can be considered to be an isolated system, is continuously increasing. The more irreversible a process, the larger the entropy generated during that process. No entropy is generated during reversible processes ($S_{gen} = 0$).

Entropy increase of the universe is a major concern not only to engineers but also to philosophers, theologians, economists, and environmentalists since entropy is viewed as a measure of the disorder in the universe.

The increase of entropy principle does not imply that the entropy of a system cannot decrease. The entropy change of a system *can* be negative during a process (Fig. 4-37), but entropy generation cannot. The increase of entropy principle can be summarized as follows:

$$S]_{gen} \begin{cases} > 0 & \text{Irreversible process} \\ = 0 & \text{Reversible process} \\ < 0 & \text{Impossible process} \end{cases}$$

This relation serves as a criterion in determining whether a process is reversible, irreversible, or impossible.

Things in nature have a tendency to change until they attain a state of equilibrium. The increase of entropy principle dictates that the entropy of an isolated system increases until the entropy of the system reaches a *maximum* value. At that point, the system is said to have reached an equilibrium state since the increase of entropy principle prohibits the system from undergoing any change of state that results in a decrease in entropy.

Chapter 4   The Second Law of Thermodynamics

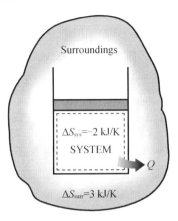

$S_{gen} = \Delta S_{total} = \Delta S_{sys} + \Delta S_{surr} = 1 \text{ kJ/K}$

Fig. 4–37   The entropy change of a system can be negative, but the entropy generation cannot

In light of the preceding discussions, we draw the following conclusions:

a. Processes can occur in a *certain* direction only, not in *any* direction. A process must proceed in the direction that complies with the increase of entropy principle, that is, $S_{gen} = 0$. A process that violates this principle is impossible. This principle often forces chemical reactions to come to a halt before reaching completion.

b. Entropy is a *nonconserved property*, and there is *no* such thing as the *conservation of entropy principle*. Entropy is conserved during the idealized reversible processes only and increases during *all* actual processes.

c. The performance of engineering systems is degraded by the presence of irreversibilities, and *entropy generation* is a measure of the magnitudes of the irreversibilities present during that process. The greater the extent of irreversibilities, the greater the entropy generation. As a result, entropy generation can be used as a quantitative measure of irreversibilities associated with a process. It is also used to establish criteria for the performance of engineering devices. This point is illustrated further in Example 4–3.

EXAMPLE 4–3 Entropy Generation during Heat Transfer Processes

A heat source at 800 K loses 2000 kJ of heat to a sink at (a) 500 K and (b) 750 K. Determine which heat transfer process is more irreversible.

Solution: Heat is transferred from a heat source to two heat sinks at different temperatures. The heat transfer process that is more irreversible is to be determined.

Analysis: A sketch of the reservoirs is shown in Fig. 4–38. Both cases involve heat transfer through a finite temperature difference, and therefore both are irreversible. The magnitude of the irreversibility associated with each process can be determined by calculating the total entropy change for each case. The total entropy change for a heat transfer process involving two reservoirs (a source and a sink) is the sum of the entropy changes of each reservoir since the two reservoirs form an adiabatic system.

Thermodynamics

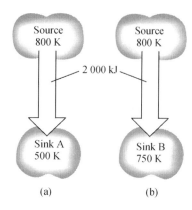

Fig. 4-38 Schematic for Example 4-3

Or do they? The problem statement gives the impression that the two reservoirs are in direct contact during the heat transfer process. But this cannot be the case since the temperature at a point can have only one value, and thus it cannot be 800 K on one side of the point of contact and 500 K on the other side. In other words, the temperature function cannot have a jump discontinuity. Therefore, it is reasonable to assume that the two reservoirs are separated by a partition through which the temperature drops from 800 K on one side to 500 K (or 750 K) on the other. Therefore, the entropy change of the partition should also be considered when evaluating the total entropy change for this process. However, considering that entropy is a property and the values of properties depend on the state of a system, we can argue that the entropy change of the partition is zero since the partition appears to have undergone a *steady* process and thus experienced no change in its properties at any point. We base this argument on the fact that the temperature on both sides of the partition and thus throughout remains constant during this process. Therefore, we are justified to assume that $\Delta S_{partition} = 0$ since the entropy (as well as the exergy) content of the partition remains constant during this process.

The entropy change for each reservoir can be determined from $\Delta S = Q/T_0$ since each reservoir undergoes an internally reversible, isothermal process.

(a) For the heat transfer process to a sink at 500 K:

$$\Delta S_{source} = \frac{Q_{source}}{T_{source}} = \frac{-2\,000 \text{ kJ}}{800 \text{ K}} = -2.5 \text{ kJ/K}$$

$$\Delta S_{sink} = \frac{Q_{sink}}{T_{sink}} = \frac{2\,000 \text{ kJ}}{500 \text{ K}} = +4.0 \text{ kJ/K}$$

and

$$S_{gen} = \Delta S_{total} = \Delta S_{source} + \Delta S_{sink} = (-2.5 + 4.0) \text{ kJ/K} = 1.5 \text{ kJ/K}$$

Therefore, 1.5 kJ/K of entropy is generated during this process. Noting that both reservoirs have undergone internally reversible processes, the entire entropy generation took place in the partition.

(b) Repeating the calculations in part (a) for a sink temperature of 750 K, we obtain

$$\Delta S_{source} = -2.5 \text{ kJ/K}$$
$$\Delta S_{sink} = +2.7 \text{ kJ/K}$$

and
$$S_{gen} = \Delta S_{total} = (-2.5 + 2.7)\,\text{kJ/K} = 0.2\ \text{kJ/K}$$

The total entropy change for the process in part (b) is smaller, and therefore it is less irreversible. This is expected since the process in (b) involves a smaller temperature difference and thus a smaller irreversibility.

Discussion: The irreversibilities associated with both processes could be eliminated by operating a Carnot heat engine between the source and the sink. For this case it can be shown that $\Delta S_{total} = 0$.

4–7 Entropy Balance

The property *entropy* is a measure of molecular disorder or randomness of a system, and the second law of thermodynamics states that entropy can be created but it cannot be destroyed. Therefore, the entropy change of a system during a process is greater than the entropy transfer by an amount equal to the entropy generated during the process within the system, and the *increase of entropy principle* for any system is expressed as

$$\begin{pmatrix}\text{Total}\\ \text{entropy}\\ \text{entering}\end{pmatrix} - \begin{pmatrix}\text{Total}\\ \text{entropy}\\ \text{leaving}\end{pmatrix} + \begin{pmatrix}\text{Total}\\ \text{entropy}\\ \text{generated}\end{pmatrix} = \begin{pmatrix}\text{Change in the}\\ \text{total entropy}\\ \text{of the system}\end{pmatrix}$$

or

$$S_{in} - S_{out} + S_{gen} = \Delta S_{sys} \qquad (4-23)$$

which is a verbal statement of Eq. (4–20). This relation is often referred to as the entropy balance and is applicable to any system undergoing any process. The entropy balance relation above can be stated as: *the entropy change of a system during a process is equal to the net entropy transfer through the system boundary and the entropy generated within the system*. Next we discuss the various terms in that relation.

ENTROPY CHANGE OF A SYSTEM, ΔS_{sys}

Despite the reputation of entropy as being vague and abstract and the intimidation associated with it, entropy balance is actually easier to deal with than exergy balance since, unlike exergy, entropy does not exist in various forms. Therefore, the determination of entropy change of a system during a process involves evaluating entropy of the system at the beginning and at the end of the process and taking their difference. That is,

Entropy change = Entropy at final state − Entropy at initial state

or

$$\Delta S_{sys} = S_{final} - S_{initial} = S_2 - S_1 \qquad (4-24)$$

Note that entropy is a property, and the value of a property does not change unless the state of the system changes. Therefore, the entropy change of a system is zero if the state of the system does not change during the process. For example, the entropy change of steady-flow devices such as nozzles, compressors, turbines, pumps, and heat exchangers is zero during steady operation.

When the properties of the system are not uniform, the entropy of the system can be determined by integration from

$$S_{sys} = \int s\delta m = \int_V s\rho dV \qquad (4-25)$$

where V is the volume of the system and ρ is density.

MECHANISMS OF ENTROPY TRANSFER, S_{in} and S_{out}

Entropy can be transferred to or from a system by two mechanisms: *heat transfer* and *mass flow*. Entropy transfer is recognized at the system boundary as it crosses the boundary, and it represents the entropy gained or lost by a system during a process. The only form of entropy interaction associated with a fixed mass or closed system is *heat transfer*, and thus the entropy transfer for an adiabatic closed system is zero.

1. Heat Transfer

Heat is, in essence, a form of disorganized exergy, and some disorganization (entropy) will flow with heat. Heat transfer to a system increases the entropy of that system and thus the level of molecular disorder or randomness, and heat transfer from a system decreases it. In fact, heat rejection is the only way the entropy of a fixed mass can be decreased. The ratio of the heat transfer Q at a location to the absolute temperature T at that location is called the *entropy flow* or *entropy transfer* and is expressed as (Fig. 4-39)

Entropy transfer by heat transfer:

$$S_{heat} = \frac{Q}{T} \quad (T = \text{constant}) \qquad (4-26)$$

The quantity Q/T represents the entropy transfer accompanied by heat transfer, and the direction of entropy transfer is the same as the direction of heat transfer since thermodynamic temperature T is always a positive quantity.

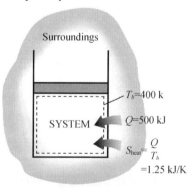

Fig. 4-39 Heat transfer is always accompanied by entropy transfer in the amount of Q/T, where T is the boundary temperature

When the temperature T is not constant, the entropy transfer during a process 1-2 can be determined by integration (or by summation if appropriate) as

Chapter 4 The Second Law of Thermodynamics

$$S_{\text{heat}} = \int_1^2 \frac{\delta Q}{T} \cong \sum \frac{Q_k}{T_k} \qquad (4-27)$$

where Q_k is the heat transfer through the boundary at temperature T_k at location k.

When two systems are in contact, the entropy transfer from the warmer system is equal to the entropy transfer into the cooler one at the point of contact. That is, no entropy can be created or destroyed at the boundary since the boundary has no thickness and occupies no volume.

Note that work is entropy-free, and no entropy is transferred by work. Exergy is transferred by both heat and work, whereas entropy is transferred only by heat. That is,

Entropy transfer by work:

$$S_{\text{work}} = 0 \qquad (4-28)$$

The first law of thermodynamics makes no distinction between heat transfer and work; it considers them as *equals*. The distinction between heat transfer and work is brought out by the second law: *an exergy interaction that is accompanied by entropy transfer is heat transfer, and an exergy interaction that is not accompanied by entropy transfer is work*. That is, no entropy is exchanged during a work interaction between a system and its surroundings. Thus, only *exergy* is exchanged during work interaction whereas both *exergy* and *entropy* are exchanged during heat transfer.

2. Mass Flow

Mass contains entropy as well as exergy, and the entropy and exergy contents of a system are proportional to the mass. Both entropy and exergy are carried into or out of a system by streams of matter, and the rates of entropy and exergy transport into or out of a system are proportional to the mass flow rate. Closed systems do not involve any mass flow and thus any entropy transfer by mass. When a mass in the amount of m enters or leaves a system, entropy in the amount of ms, where s is the specific entropy (entropy per unit mass entering or leaving), accompanies it. That is,

Entropy transfer by mass flow:

$$S_{\text{mass}} = ms \qquad (4-29)$$

Therefore, the entropy of a system increases by ms when mass in the amount of m enters and decreases by the same amount when the same amount of mass at the same state leaves the system. When the properties of the mass change during the process, the entropy transfer by mass flow can be determined by integration from

$$\dot{S}_{\text{mass}} = \int_{A_c} s\rho V_n dA_c \quad \text{and} \quad S_{\text{mass}} = \int s\delta m = \int_{A_c} \dot{S}_{\text{mass}} dt \qquad (4-30)$$

where A_c is the cross-sectional area of the flow and V_n is the local velocity normal to dA_c.

ENTROPY GENERATION, S_{gen}

Irreversibilities such as friction, mixing, chemical reactions, heat transfer through a finite temperature difference, unrestrained expansion, nonquasi-equilibrium compression, or expansion always cause the entropy of a system to increase, and entropy generation is a measure of the entropy created by such effects during a process.

For a *reversible process*, the entropy generation is zero and thus the *entropy change* of a system is equal to the *entropy transfer*. Therefore, the entropy balance relation in the reversible case becomes analogous to the exergy balance relation, which states that *exergy change* of a system during a process is equal to the *exergy transfer* during that process. However, note that the exergy change of a system equals the exergy transfer for *any* process, but the entropy change of a system equals the entropy transfer only for a *reversible* process.

The entropy transfer by heat Q/T is zero for adiabatic systems, and the entropy transfer by mass ms is zero for systems that involve no mass flow across their boundary (i. e., closed systems). Entropy balance for *any system* undergoing *any process* can be expressed more explicitly as

$$\underbrace{S_{in} - S_{out}}_{\text{Net entropy transfer by heat and mass}} + \underbrace{S_{gen}}_{\text{Entropy generation}} = \underbrace{\Delta S_{sys}}_{\text{Change in entropy}} \quad (\text{kJ/K}) \qquad (4-31)$$

or, in the rate form, as

$$\underbrace{\dot{S}_{in} - \dot{S}_{out}}_{\substack{\text{Rate of net entropy} \\ \text{transfer by heat} \\ \text{and mass}}} + \underbrace{\dot{S}_{gen}}_{\substack{\text{Rate of entropy} \\ \text{generation}}} = \underbrace{dS_{sys}/dt}_{\substack{\text{Rate of change} \\ \text{in entropy}}} (\text{kW/K}) \qquad (4-32)$$

where the rates of entropy transfer by heat transferred at a rate of $\dot{Q}$ and mass flowing at a rate of $\dot{m}$ are $\dot{S}_{heat} = \dot{Q}/T$ and $\dot{S}_{mass} = \dot{m}s$. The entropy balance can also be expressed on a unit-mass basis as

$$S_{in} - S_{out} + S_{gen} = \Delta S_{sys} \quad (\text{kJ/kg} \cdot \text{K}) \qquad (4-33)$$

where all the quantities are expressed per unit mass of the system. Note that for a *reversible process*, the entropy generation term S_{gen} drops out from all of the relations above.

The term S_{gen} represents the entropy generation *within the system boundary* only (Fig. 4-40), and not the entropy generation that may occur outside the system boundary during the process as a result of external irreversibilities. Therefore, a process for which $S_{gen} = 0$ is *internally reversible*, but not necessarily *totally* reversible. The *total* entropy generated during a process can be determined by applying the entropy balance to an *extended system* that includes the system itself and its immediate surroundings where external irreversibilities might be occurring. Also, the entropy change in this case is equal to the sum of the entropy change of the system and the entropy change of the immediate surroundings. Note that under steady conditions, the state and thus the entropy of the immediate surroundings (let us call it the "buffer zone") at any point does not change during the process, and the entropy change of the buffer zone is zero. The entropy change of the buffer zone, if any, is usually small relative to the entropy change of the system, and thus it is usually disregarded.

When evaluating the entropy transfer between an extended system and the surroundings, the boundary temperature of the extended system is simply taken to be the *environment temperature*.

Chapter 4 The Second Law of Thermodynamics

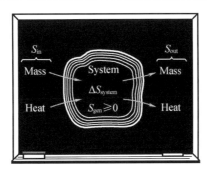

Fig. 4 – 40 Mechanisms of entropy transfer for a general system

CLOSED SYSTEMS

A closed system involves *no mass flow* across its boundaries, and its entropy change is simply the difference between the initial and final entropies of the system. The *entropy change* of a closed system is due to the *entropy transfer* accompanying heat transfer and the *entropy generation* within the system boundaries. Taking the positive direction of heat transfer to be *to* the system, the general entropy balance relation Eq. (4 – 31) can be expressed for a closed system as

Closed system:

$$\sum \frac{Q_k}{T_k} + S_{gen} = \Delta S_{sys} = S_2 - S_1 \quad (\text{kJ/K}) \qquad (4-34)$$

The entropy balance relation above can be stated as: the entropy change of a closed system during a process is equal to the sum of the net entropy transferred through the system boundary by heat transfer and the entropy generated within the system boundaries.

For an *adiabatic process* ($Q=0$), the entropy transfer term in the above relation drops out and the entropy change of the closed system be comes equal to the entropy generation within the system boundaries. That is,

Adiabatic closed system:

$$S_{gen} = \Delta S_{\text{adiabatic system}} \qquad (4-35)$$

Noting that any closed system and its surroundings can be treated as an adiabatic system and the total entropy change of a system is equal to the sum of the entropy changes of its parts, the entropy balance for a closed system and its surroundings can be written as

System + Surroundings:

$$S_{gen} = \sum \Delta S = \Delta S_{sys} + \Delta S_{sur} \qquad (4-36)$$

where $S_{sys} = m(s_2 - s_1)$ and the entropy change of the surroundings can be determined from $\Delta S_{surr} = Q_{surr}/T_{surr}$ if its temperature is constant. At initial stages of studying entropy and entropy transfer, it is more instructive to start with the general form of the entropy balance Eq. (4 – 31) and to simplify it for the problem under consideration. The specific relations above are convenient to use after a certain degree of intuitive understanding of the material is achieved.

CONTROL VOLUMES

The entropy balance relations for control volumes differ from those for closed systems in that they involve one more mechanism of entropy exchange: *mass flow across the boundaries*. As mentioned earlier, mass possesses entropy as well as exergy, and the amounts of these two extensive properties are proportional to the amount of mass.

Taking the positive direction of heat transfer to be *to* the system, the general entropy balance relations Eqs. (4 – 31) and (4 – 32) can be expressed for control volumes as

$$\sum \frac{Q_k}{T_k} + \sum m_i s_i - \sum m_e s_e + S_{gen} = (S_2 - S_1)_{CV} \quad (kJ/K) \qquad (4-37)$$

or, in the rate form, as

$$\sum \frac{\dot{Q}_k}{T_k} + \sum \dot{m}_i s_i - \sum \dot{m}_e s_e + \dot{S}_{gen} = dS_{CV}/dt \quad (kW/K) \qquad (4-38)$$

This entropy balance relation can be stated as:

The rate of entropy change within the control volume during a process is equal to the sum of the rate of entropy transfer through the control volume boundary by heat transfer, the net rate of entropy transfer into the control volume by mass flow, and the rate of entropy generation within the boundaries of the control volume as a result of irreversibilities.

Most control volumes encountered in practice such as turbines, compressors, nozzles, diffusers, heat exchangers, pipes, and ducts operate steadily, and thus they experience no change in their entropy. Therefore, the entropy balance relation for a general steady-flow process can be obtained from Eq. (4 – 38) by setting $dS_{CV}/dt = 0$ and rearranging to give

Steady – flow:

$$\dot{S}_{gen} = \sum \dot{m}_e s_e - \sum \dot{m}_i s_i - \sum \frac{\dot{Q}_k}{T_k} \qquad (4-39)$$

For *single-stream* (one inlet and one exit) steady-flow devices, the entropy balance relation simplifies to

Steady-flow, single – stream:

$$\dot{S}_{gen} = \dot{m}_e (s_e - s_i) - \sum \frac{\dot{Q}_k}{T_k} \qquad (4-40)$$

For the case of an *adiabatic* single-stream device, the entropy balance relation further simplifies to steady-flow, single – stream, adiabatic:

$$\dot{S}_{gen} = \dot{m}_e (s_e - s_i) \qquad (4-41)$$

which indicates that the specific entropy of the fluid must increase as it flows through an adiabatic device since $\dot{S}_{gen} \geq 0$ (Fig. 4 – 41). If the flow through the device is *reversible* and *adiabatic*, then the entropy remains constant, $s_e = s_i$, regardless of the changes in other properties.

Chapter 4 The Second Law of Thermodynamics

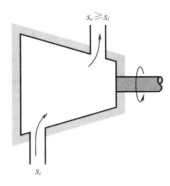

Fig. 4 – 41 The entropy of a substance always increases (or remains constant in the case of a reversible process) as it flow through a single-stream, adiabatic, steady-flow device

EXAMPLE 4 – 4 Entropy Generation in a Wall

Consider steady heat transfer through a 5 m × 7 m brick wall of a house of thickness 30 cm. On a day when the temperature of the outdoors is 0 ℃, the house is maintained at 27 ℃. The temperatures of the inner and outer surfaces of the brick wall are measured to be 20 ℃ and 5 ℃, respectively, and the rate of heat transfer through the wall is 1 035 W. Determine the rate of entropy generation in the wall, and the rate of total entropy generation associated with this heat transfer process.

Solution Steady heat transfer through a wall is considered. For specified heat transfer rate, wall temperatures, and environment temperatures, the entropy generation rate within the wall and the total entropy generation rate are to be determined.

Assumptions ①The process is steady, and thus the rate of heat transfer through the wall is constant. ②Heat transfer through the wall is one dimensional.

Analysis We first take the *wall* as the system (Fig. 4 – 42). This is a *closed system* since no mass crosses the system boundary during the process. We note that the entropy change of the wall is zero during this process since the state and thus the entropy of the wall do not change anywhere in the wall. Heat and entropy are entering from one side of the wall and leaving from the other side. The rate form of the entropy balance for the wall simplifies to

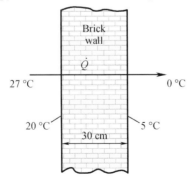

Fig. 4 – 42 Schematic for Example 4 – 4

$$\underbrace{\dot{S}_{in} - \dot{S}_{out}}_{\text{Rate of net entropy transfer by heat and mass}} + \underbrace{\dot{S}_{gen}}_{\text{Rate of entropy generation}} = \underbrace{dS_{sys} \rightarrow 0 (\text{steady})/dt}_{\text{Rate of change in entropy}}$$

$$\left(\frac{\dot{Q}}{T}\right)_{in} - \left(\frac{\dot{Q}}{T}\right)_{out} + \dot{S}_{gen} = 0$$

$$\frac{1\ 035\ \text{W}}{293\ \text{K}} - \frac{1\ 035\ \text{W}}{278\ \text{K}} + \dot{S}_{gen} = 0$$

Therefore, the rate of entropy generation in the wall is

$$\dot{S}_{gen,wall} = 0.191\ \text{W/K}$$

Note that entropy transfer by heat at any location is Q/T at that location, and the direction of entropy transfer is the same as the direction of heat transfer.

To determine the rate of total entropy generation during this heat transfer process, we extend the system to include the regions on both sides of the wall that experience a temperature change. Then one side of the system boundary becomes room temperature while the other side becomes the temperature of the outdoors. The entropy balance for this *extended system* (system + immediate surroundings) is the same as that given above, except the two boundary temperatures are now 300 K and 273 K instead of 293 K and 278 K, respectively. Then the rate of total entropy generation becomes

$$\frac{1\ 035\ \text{W}}{300\ \text{K}} - \frac{1\ 035\ \text{W}}{273\ \text{K}} + \dot{S}_{gen} = 0 \rightarrow \dot{S}_{gen,wall} = 0.341\ \text{W/K}$$

Discussion Note that the entropy change of this extended system is also zero since the state of air does not change at any point during the process. The difference between the two entropy generations is 0.150 W/K, and it represents the entropy generated in the air layers on both sides of the wall. The entropy generation in this case is entirely due to irreversible heat transfer through a finite temperature difference.

4-8 Exergy

When a new exergy source, such as a geothermal well, is discovered, the first thing the explorers do is estimate the amount of exergy contained in the source. This information alone, however, is of little value in deciding whether to build a power plant on that site. What we really need to know is the *work potential* of the source—that is, the amount of exergy we can extract as useful work. The rest of the exergy is eventually discarded as waste exergy and is not worthy of our consideration. Thus, it would be very desirable to have a property to enable us to determine the useful work potential of a given amount of exergy at some specified state. This property is *exergy*, which is also called the *availability* or *available exergy*.

The work potential of the exergy contained in a system at a specified state is simply the maximum useful work that can be obtained from the system. You will recall that the work done during a process depends on the initial state, the final state, and the process path. That is,

Chapter 4 The Second Law of Thermodynamics

$Work = f$ (initial state, process path, final state)

In an exergy analysis, the *initial state* is specified, and thus it is not a variable. The work output is maximized when the process between two specified states is executed in a *reversible manner*. Therefore, all the irreversibilities are disregarded in determining the work potential. Finally, the system must be in the *dead state* at the end of the process to maximize the work output.

Let us consider next the concept of the *dead state*, which is also important in completing our understanding of the property exergy. If the state of a fixed quantity of matter, a closed system, departs from that of the environment, an opportunity exists for developing work. However, as the system changes state toward that of the environment, the opportunity diminishes, ceasing to exist when the two are in equilibrium with one another. This state of the system is called the *dead state*. At the dead state, the fixed quantity of matter under consideration is imagined to be sealed in an envelope impervious to mass flow, at rest relative to the environment, and internally in equilibrium at the temperature T_0 and pressure p_0 of the environment. At the dead state, both the system and environment possess exergy, but the value of exergy is zero because there is no possibility of a spontaneous change within the system or the environment, nor can there bean interaction between them.

Distinction should be made between the *surroundings*, *immediate surroundings*, and the *environment*. By definition, surroundings are everything outside the system boundaries. The immediate surroundings refer to the portion of the surroundings that is affected by the process, and environment refers to the region beyond the immediate surroundings whose properties are not affected by the process at any point. Therefore, any irreversibilities during a process occur within the system and its immediate surroundings, and the environment is free of any irreversibilities. When analyzing the cooling of a hot baked potato in a room at 25 ℃, for example, the warm air that surrounds the potato is the immediate surroundings, and the remaining part of the room air at 25 ℃ is the environment. Note that the temperature of the immediate surroundings changes from the temperature of the potato at the boundary to the environment temperature of 25 ℃ (Fig. 4 –43).

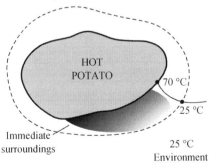

Fig. 4 –43 The immediate surroundings of a hot potato are simply the temperature gradient zone of the air next to the potato

The notion that a system must go to the dead state at the end of the process to maximize the work output can be explained as follows: If the system temperature at the final state is greater than (or less than) the temperature of the environment it is in, we can always produce additional work by running a heat engine between these two temperature levels. If the final pressure is greater than (or less than) the pressure of the environment, we can still obtain work by letting the system expand to the pressure of the environment. If the final velocity of the system is not zero, we can catch that extra kinetic exergy by a turbine and convert it to rotating shaft work, and so on. No work can be produced from a system that is initially at the dead state. The atmosphere around us contains a tremendous amount of exergy. However, the atmosphere is in the dead state, and the exergy it contains has no work potential.

Therefore, It can be concluded that a *system delivers the maximum possible work as it undergoes a reversible process from the specified initial state to the state of its environment, that is, the dead state.* This represents the *useful work potential* of the system at the specified state and is called exergy. It is important to realize that exergy does not represent the amount of work that a work producing device will actually deliver upon installation. Rather, it represents the *upper limit on the amount of work a device can deliver without violating any thermodynamic laws.* There will always be a difference, large or small, between exergy and the actual work delivered by a device. This difference represents the room engineers have for improvement.

Note that the exergy of a system at a specified state depends on the conditions of the environment (the dead state) as well as the properties of the system. Therefore, exergy is a property of the *system – environment combination* and not of the system alone. Altering the environment is another way of increasing exergy, but it is definitely not an easy alternative. Today, an equivalent term, *exergy*, introduced in Europe in the 1950s, has found global acceptance partly because it is shorter, it rhymes with exergy and entropy, and it can be adapted without requiring translation. In this text the preferred term is *exergy*.

Kinetic exergy is a form of *mechanical energy*, and thus it can be converted to work entirely. Therefore, the *work potential* or *exergy* of the kinetic exergy of a system is equal to the kinetic exergy itself regardless of the temperature and pressure of the environment. That is,

Exergy of kinetic exergy:

$$x_{e_k} = e_k = \frac{V^2}{2} \quad (\text{kJ/kg}) \qquad (4-42)$$

where V is the velocity of the system relative to the environment.

Potential exergy is also a form of *mechanical energy*, and thus it can be converted to work entirely. Therefore, the *exergy* of the potential exergy of a system is equal to the potential exergy itself regardless of the temperature and pressure of the environment (Fig. 4 – 44). That is,

Exergy of potential exergy:

$$x_{pe} = pe = gz \quad (\text{kJ/kg}) \qquad (4-43)$$

where g is the gravitational acceleration and z is the elevation of the system relative to a reference level in the environment. Thus, the energies of kinetic and potential energies are equal to themselves, and they are entirely available for work.

Chapter 4 The Second Law of Thermodynamics

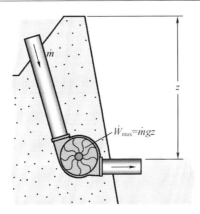

Fig. 4 – 44 The *work potential* or *exergy* of potential exergy is equal to the potential exergy itself

REVERSIBLE WORK AND IRREVERSIBILITY

The property exergy serves as a valuable tool in determining the quality of exergy and comparing the work potentials of different exergy sources or systems. The evaluation of exergy alone, however, is not sufficient for studying engineering devices operating between two fixed states. This is because when evaluating exergy, the final state is always assumed to be the *dead state*, which is hardly ever the case for actual engineering systems. The is entropic efficiencies discussed are also of limited use because the exit state of the model (isentropic) process is not the same as the actual exit state and it is limited to adiabatic processes.

In this section, we describe two quantities that are related to the actual initial and final states of processes and serve as valuable tools in the thermodynamic analysis of components or systems. These two quantities are the *reversible work* and *irreversibility* (or *exergy destruction*). But first we examine the surroundings work, which is the work done by or against the surroundings during a process.

The work done by work-producing devices is not always entirely in a usable form. For example, when a gas in a piston – cylinder device expands, part of the work done by the gas is used to push the atmospheric air out of the way of the piston (Fig. 4 – 45). This work, which cannot be recovered and utilized for any useful purpose, is equal to the atmospheric pressure P_0 times the volume change of the system,

$$W_{\text{surr}} = P_0 (V_2 - V_1) \tag{4-44}$$

The difference between the actual work W and the surroundings work W_{surr} is called the useful work W_u:

$$W_u = W - W_{\text{surr}} = W - P_0 (V_2 - V_1) \tag{4-45}$$

When a system is expanding and doing work, part of the work done is used to overcome the atmospheric pressure, and thus W_{surr} represents a loss. When a system is compressed, however, the atmospheric pressure helps the compression process, and thus W_{surr} represents a gain.

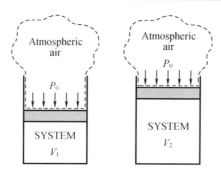

Fig. 4 – 45 As a closed system expands, some work needs to be done to push the atmospheric air out of the way (W_{surr})

Note that the work done by or against the atmospheric pressure has significance only for systems whose volume changes during the process (i. e., systems that involve moving boundary work). It has no significance for cyclic devices and systems whose boundaries remain fixed during a process such as rigid tanks and steady-flow devices (turbines, compressors, nozzles, heat exchangers, etc.), as shown in Fig. 4 – 46.

Reversible work W_{rev} is defined as *the maximum amount of useful work that can be produced (or the minimum work that needs to be supplied) as a system undergoes a process between the specified initial and final states*. This is the useful work output (or input) obtained (or expended) when the process between the initial and final states is executed in a totally reversible manner. When the final state is the dead state, the reversible work equals exergy. For processes that require work, reversible work represents the minimum amount of work necessary to carry out that process. For convenience in presentation, the term *work* is used to denote both work and power throughout this chapter.

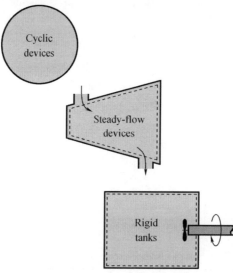

Fig. 4 – 46 For constant – volume systems, the total actual and useful works are identical ($W_u = W$)

Any difference between the reversible work W_{rev} and the useful work W_u is due to the

Chapter 4 The Second Law of Thermodynamics

irreversibilities present during the process, and this difference is called irreversibility I. It is expressed as

$$I = W_{rev,out} - W_{u,out} \text{ or } I = W_{u,in} - W_{rev,in} \qquad (4-46)$$

The irreversibility is equivalent to the *exergy destroyed*. For a totally reversible process, the actual and reversible work terms are identical, and thus the irreversibility is zero. This is expected since totally reversible processes generate no entropy. Irreversibility is a *positive quantity* for all actual (irreversible) processes since $W_{rev} \geq W_u$ for work producing devices and $W_{rev} \leq W_u$ for work − consuming devices.

Irreversibility can be viewed as the *wasted work potential* or the *lost opportunity* to do work. It represents the exergy that could have been converted to work but was not. The smaller the irreversibility associated with a process, the greater the work that is produced (or the smaller the work that is consumed). The performance of a system can be improved by minimizing the irreversibility associated with it.

SECOND − LAW EFFICIENCY, η_{II}

We have defined the *thermal efficiency* and the *coefficient of performance* for devices as a measure of their performance. They are defined on the basis of the first law only, and they are sometimes referred to as the *first law efficiencies*. The first law efficiency, however, makes no reference to the best possible performance, and thus it may be misleading.

Consider two heat engines, both having a thermal efficiency of 30 percent, as shown in Fig. 4 − 47. One of the engines (engine A) is supplied with heat from a source at 600 K, and the other one (engine B) from a source at 1 000 K. Both engines reject heat to a medium at 300 K. At first glance, both engines seem to convert to work the same fraction of heat that they receive; thus they are performing equally well. When we take a second look at these engines in light of the second law of thermodynamics, however, we see a totally different picture. These engines, at best, can perform as reversible engines, in which case their efficiencies would be

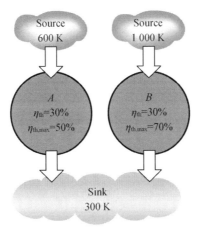

Fig. 4 − 47 Two heat engines that have the same thermal efficiency, but different maximum thermal efficiencies

$$\eta_{\text{rev},A} = \left(1 - \frac{T_L}{T_H}\right)_A = 1 - \frac{300 \text{ K}}{600 \text{ K}} = 50\%$$

$$\eta_{\text{rev},B} = \left(1 - \frac{T_L}{T_H}\right)_B = 1 - \frac{300 \text{ K}}{1\,000 \text{ K}} = 70\%$$

Now it is becoming apparent that engine B has a greater work potential available to it (70 percent of the heat supplied as compared to 50 percent for engine A), and thus should do a lot better than engine A. Therefore, we can say that engine B is performing poorly relative to engine A even though both have the same thermal efficiency.

It is obvious from this example that the first law efficiency alone is not a realistic measure of performance of engineering devices. To overcome this deficiency, we define a second law efficiency η_{II} as the ratio of the actual thermal efficiency to the maximum possible (reversible) thermal efficiency under the same conditions:

$$\eta_{\text{II}} = \frac{\eta_{\text{th}}}{\eta_{\text{th},r}} \quad \text{(heat engines)} \tag{4-47}$$

Based on this definition, the second law efficiencies of the two heat engines discussed above are

$$\eta_{\text{II},A} = \frac{0.3}{0.5} = 0.6 \text{ and } \eta_{\text{II}} = \frac{0.3}{0.7} = 0.43$$

That is, engine A is converting 60 percent of the available work potential to useful work. This ratio is only 43 percent for engine B.

The second law efficiency can also be expressed as the ratio of the useful work output and the maximum possible (reversible) work output:

$$\eta_{\text{II}} = \frac{W_u}{W_{\text{rev}}} \quad \text{(work - producing devices)} \tag{4-48}$$

This definition is more general since it can be applied to processes (in turbines, piston - cylinder devices, etc.) as well as to cycles. Note that the second law efficiency cannot exceed 100 percent (Fig. 4-48).

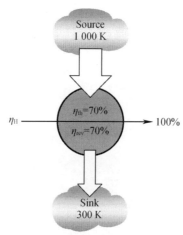

Fig. 4-48 Second law efficiency of all reversible devices is 100 percent

Chapter 4 The Second Law of Thermodynamics

We can also define a second law efficiency for work-consuming noncyclic (such as compressors) and cyclic (such as refrigerators) devices as the ratio of the minimum (reversible) work input to the useful work input:

$$\eta_{\text{II}} = \frac{W_{\text{rev}}}{W_u} \quad \text{(work-consuming devices)} \tag{4-49}$$

For cyclic devices such as refrigerators and heat pumps, it can also be expressed in terms of the coefficients of performance as

$$\eta_{\text{II}} = \frac{COP}{COP_{\text{rev}}} \quad \text{(refrigerators and heat pumps)} \tag{4-50}$$

In the above relations, the reversible work W_{rev} should be determined by using the same initial and final states as in the actual process.

The definitions above for the second law efficiency do not apply to devices that are not intended to produce or consume work. As a consequence, we need a more general definition. However, there is some disagreement on a general definition of the second law efficiency, and thus a person may encounter different definitions for the same device. The second law efficiency is intended to serve as a measure of approximation to reversible operation, and thus its value should range from zero in the worst case (complete destruction of exergy) to one in the best case (no destruction of exergy). With this in mind, we define the second law efficiency of a system during a process as

$$\eta_{\text{II}} = \frac{\textit{Exergy recoverd}}{\textit{Exergy supplied}} = 1 - \frac{\textit{Exergy destroyed}}{\textit{Exergy supplied}} \tag{4-51}$$

Therefore, when determining the second law efficiency, the first thing we need to do is determine how much exergy or work potential is consumed during a process. In a reversible operation, we should be able to recover entirely the exergy supplied during the process, and the irreversibility in this case should be zero. The second law efficiency is zero when we recover none of the exergy supplied to the system. Note that the exergy can be supplied or recovered at various amounts in various forms such as heat, work, kinetic exergy, potential exergy, internal exergy, and enthalpy. Sometimes there are differing (though valid) opinions on what constitutes supplied exergy, and this causes differing definitions for second law efficiency.

For *heat engine*, the exergy supplied is the decrease in the exergy of the heat transferred to the engine, which is the difference between the exergy of the heat supplied and the exergy of the heat rejected. (The exergy of the heat rejected at the temperature of the surroundings is zero.) The net work output is the recovered exergy.

For a *refrigerator* or *heat pump*, the exergy supplied is the work input since the work supplied to a cyclic device is entirely available. The recovered exergy is the exergy of the heat transferred to the high-temperature medium (which is the reversible work) for a heat pump, and the exergy of the heat transferred from the low-temperature medium for a refrigerator.

For a heat exchanger with two unmixed fluid streams, normally the exergy supplied is the

Thermodynamics

decrease in the exergy of the higher-temperature fluid stream, and the exergy recovered is the increase in the exergy of the lower temperature fluid stream.

EXERGY CHANGE OF A SYSTEM

Unlike exergy, the value of exergy depends on the state of the environment as well as the state of the system. Therefore, exergy is a combination property. The exergy of a system that is in equilibrium with its environment is zero. The state of the environment is referred to as the "dead state" since the system is practically "dead", which means doing no work, from a thermodynamic point of view when it reaches that state.

In this section we limit the discussion to thermo – mechanical exergy, and thus disregard any mixing and chemical reactions. Therefore, a system at this "restricted dead state" is at the temperature and pressure of the environment and it has no kinetic or potential energies relative to the environment. However, it may have a different chemical composition than the environment. Exergy associated with different chemical compositions and chemical reactions is discussed in later chapters.

Below we develop relations for the exergies and exergy changes for a fixed mass and a flow stream.

Exergy of a fixed mass: nonflow (or closed system) exergy

In general, internal exergy consists of *sensible*, *latent*, *chemical*, and *nuclear* energies. However, in the absence of any chemical or nuclear reactions, the chemical and nuclear energies can be disregarded. the internal exergy can be considered to consist of only sensible and latent energies that can be transferred to or from a system as *heat* whenever there is a temperature difference across the system boundary. The second law of thermodynamics states that heat cannot be converted to work entirely, and thus the work potential of internal exergy must be less than the internal exergy itself. But how much less?

In order to answer that question, we need to consider a stationary closed system at a specified state which undergoes a *reversible* process to the state of the environment (that is, the final temperature and pressure of the system should be T_0 and P_0, respectively). The useful work delivered during this process is the exergy of the system at its initial state (Fig. 4 – 49).

Consider a piston – cylinder device that contains a fluid of mass m at temperature T and pressure P. The system (the mass inside the cylinder) has a volume V, internal exergy U, and entropy S. The system is now allowed to undergo a differential change of state during which the volume changes by a differential amount dV and heat is transferred in the differential amount of dQ. Taking the direction of heat and work transfers to be *from* the system (heat and work outputs), the exergy balance for the system during this differential process can be expressed as

Chapter 4 The Second Law of Thermodynamics

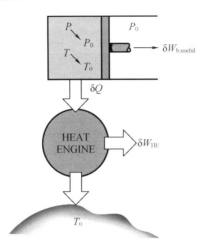

Fig. 4 – 49 The *exergy* of a specified mass at a specified state is the useful work that can be produced as the mass undergoes a reversible process to the state of the environment

$$\underbrace{\delta E_{in} - \delta E_{out}}_{\substack{\text{Net exergy transfer} \\ \text{by heat, work, and mass}}} = \underbrace{dE_{sys}}_{\substack{\text{Change in internal, kinetic,} \\ \text{potential, etc., energies}}} \quad (4-52)$$

$$-\delta Q - \delta W = dU$$

since the only form of exergy the system contains is *internal exergy*, and the only forms of exergy transfer a fixed mass can involve are heat and work. Also, the only form of work a simple compressible system can involve during a reversible process is the boundary work, which is given to be $\delta W = PdV$ when the direction of work is taken to be from the system (otherwise it would be $-P\,dV$). The pressure P in the $P\,dV$ expression is the absolute pressure, which is measured from absolute zero. Any useful work delivered by a piston – cylinder device is due to the pressure above the atmospheric level.

Therefore,

$$\delta W = PdV = (P - P_0)dV + P_0 dV \delta W_{b,\text{useful}} + P_0 dV \quad (4-53)$$

A reversible process cannot involve any heat transfer through a finite temperature difference, and thus any heat transfer between the system at temperature T and its surroundings at T_0 must occur through a reversible heat engine. Noting that $dS = \delta Q/T$ for a reversible process, and the thermal efficiency of a reversible heat engine operating between the temperatures of T and T_0 is $\eta_{th} = 1 - T_0/T$, the differential work produced by the engine as a result of this heat transfer is

$$\delta W_{HE} = \left(1 - \frac{T_0}{T}\right)\delta Q = \delta Q - \frac{T_0}{T}\delta Q = \delta Q - (-T_0 dS) \rightarrow \delta Q = \delta W_{HE} - T_0 dS \quad (4-54)$$

Substituting the δW and δQ expressions in Eqs. (52 – 54) into the exergy balance relation gives, after rearranging,

$$\delta W_{\text{total useful}} = \delta W_{HE} + \delta W_{b,\text{useful}} = -dU - P_0 dV + T_0 dS$$

Integrating from the given state (no subscript) to the dead state (0 subscript) we obtain

$$W_{\text{total useful}} = (U - U_0) + P_0(V - V_0) - T_0(S - S_0) \quad (4-55)$$

where $W_{\text{total useful}}$ is the total useful work delivered as the system undergoes a reversible process from the given state to the dead state, which is *exergy* by definition.

A closed system, in general, may possess kinetic and potential energies, and the total exergy of a closed system is equal to the sum of its internal, kinetic, and potential energies. Noting that kinetic and potential energies themselves are forms of exergy, the exergy of a closed system of mass m is

$$X = (U - U_0) + P_0(V - V_0) - T_0(S - S_0) + m\frac{V^2}{2} + mgz \qquad (4-56)$$

On a unit mass basis, the closed system (or non-flow) exergy φ is expressed as

$$\varphi = (u - u_0) + P_0(v - v_0) - T_0(s - s_0) + \frac{V^2}{2} + gz = (e - e_0) + P_0(v - v_0) - T_0(s - s_0)$$

$$(4-57)$$

where u_0, v_0, and s_0 are the properties of the *system* evaluated at the dead state. Note that the exergy of a system is zero at the dead state since $e = e_0$, $v = v_0$, and $s = s_0$ at that state.

The exergy change of a closed system during a process is simply the difference between the final and initial exergies of the system,

$$\begin{aligned}\Delta X &= X_2 - X_1 \\ &= m(\varphi_2 - \varphi_1) \\ &= (E_2 - E_1) + P_0(V_2 - V_1) - T_0(S_2 - S_1) \\ &= (U_2 - U_1) + P_0(V_2 - V_1) - T_0(S_2 - S_1) + m\frac{V_2^2 - V_1^2}{2} + mgz\end{aligned} \qquad (4-58)$$

or, on a unit mass basis,

$$\begin{aligned}\Delta \varphi &= \varphi_2 - \varphi_1 \\ &= (u_2 - u_1) + P_0(v_2 - v_1) - T_0(s_2 - s_1) + \frac{V_2^2 - V_1^2}{2} + gz \\ &= (e_2 - e_1) + P_0(v_2 - v_1) - T_0(s_2 - s_1)\end{aligned} \qquad (4-59)$$

For *stationary* closed systems, the kinetic and potential exergy terms drop out. When the properties of a system are not uniform, the exergy of the system can be determined by integration from

$$X]_{sys} = \int \varphi \delta m = \int_V \varphi \rho dV \qquad (4-60)$$

where V is the volume of the system and ρ is density.

Note that exergy is a property, and the value of a property does not change unless the *state* changes. Therefore, the *exergy change* of a system is zero if the state of the system or the environment does not change during the process. For example, the exergy change of steady flow devices such as nozzles, compressors, turbines, pumps, and heat exchangers in a given environment is zero during steady operation.

The exergy of a closed system is either *positive* or *zero*. It is never negative. Even a medium at *low temperature* ($T < T_0$) and/or *low pressure* ($P < P_0$) contains exergy since a cold medium can serve as the heat sink to a heat engine that absorbs heat from the environment at T_0, and an evacuated space makes it possible for the atmospheric pressure to move a piston and do useful work (Fig. 4-50).

Chapter 4 The Second Law of Thermodynamics

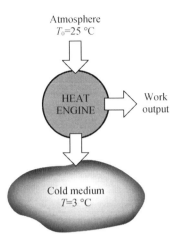

Fig. 4 – 50 The *exergy* of a cold medium is also a *positive* quantity since work can be produced by transferring heat to it

Exergy of a flow stream: flow(or stream) exergy

It was shown that a flowing fluid has an additional form of exergy, called the *flow exergy*, which is the exergy needed to maintain flow in a pipe or duct, and was expressed as $w_{flow} = Pv$ where v is the specific volume of the fluid, which is equivalent to the *volume change* of a unit mass of the fluid as it is displaced during flow. The flow work is essentially the boundary work done by a fluid on the fluid downstream, and thus the exergy associated with flow work is equivalent to the exergy associated with the boundary work, which is the boundary work in excess of the work done against the atmospheric air at P_0 to displace it by a volume v (Fig. 4 – 51). Noting that the flow work is Pv and the work done against the atmosphere is $P_0 v$, the *exergy* associated with flow exergy can be expressed as

$$X_{flow} = Pv - P_0 v = (P - P_0)v \qquad (4-61)$$

Therefore, the exergy associated with flow exergy is obtained by replacing the pressure P in the flow work relation by the pressure in excess of the atmospheric pressure, $P - P_0$. Then the exergy of a flow stream is determined by simply adding the flow exergy relation above to the exergy relation in Eq. (4 – 57) for a nonflowing fluid,

$$x_{flowing\ fluid} = x_{nonflowing\ fluid} + x_{flow}$$

$$= (u - u_0) + P_0(v - v_0) - T_0(s - s_0) + \frac{V^2}{2} + gz + (P - P_0)v$$

$$= (u + Pv) - (u_0 + P_0 v_0) - T_0(s - s_0) + \frac{V^2}{2} + gz$$

$$= (h - h_0) - T_0(s - s_0) + \frac{V^2}{2} + gz \qquad (4-62)$$

Thermodynamics

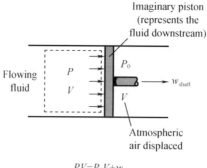

$PV = P_0 V + W_{shaft}$

Fig. 4 – 51 The *exergy* associated with *flow exergy* is the useful work that would be delivered by an imaginary piston in the flow section

The final expression is called flow (or stream) exergy, and is denoted by ψ.
Flow exergy:

$$\psi = (h - h_0) - T_0(s - s_0) + \frac{V^2}{2} + gz \qquad (4-63)$$

Then the *exergy change* of a fluid stream as it undergoes a process from state 1 to state 2 becomes

$$\Delta\psi = \psi_2 - \psi_1 = (h_2 - h_1) - T_0(s_2 - s_1) + \frac{V_2^2 - V_1^2}{2} + g(z_2 - z_1) \qquad (4-64)$$

For fluid streams with negligible kinetic and potential energies, the kinetic and potential exergy terms drop out.

Note that the *exergy change* of a closed system or a fluid stream represents the *maximum* amount of useful work that can be done (or the *minimum* amount of useful work that needs to be supplied if it is negative) as the system changes from state 1 to state 2 in a specified environment, and represents the *reversible work* W_{rev}. It is independent of the type of process executed, the kind of system used, and the nature of exergy interactions with the surroundings. Also note that the exergy of a closed system cannot be negative, but the exergy of a flow stream can at pressures below the environment pressure P_0.

EXERGY TRANSFER BY HEAT, WORK, AND MASS

Exergy, like exergy, can be transferred to or from a system in three forms: *heat*, *work*, and *mass flow*. Exergy transfer is recognized at the system boundary as exergy crosses it, and it represents the exergy gained or lost by a system during a process. The only two forms of exergy interactions associated with a fixed mass or closed system are *heat transfer* and *work*.

Exergy by heat transfer, Q

The work potential of the exergy transferred from a heat source at temperature T is the maximum work that can be obtained from that exergy in an environment at temperature T_0 and is equivalent to the work produced by a Carnot heat engine operating between the source and the environment. Therefore, the Carnot efficiency $\eta_c = 1 - T_0/T$ represents the fraction of exergy of a heat source at temperature T that can be converted to work. For example, only 70 percent of the exergy transferred from a heat source at $T = 1\,000$ K can be converted to work in an environment at $T_0 = 300$ K.

Heat is a form of disorganized exergy, and thus only a portion of it can be converted to work, which is a form of organized exergy (the second law). We can always produce work from heat at a temperature above the environment temperature by transferring it to a heat engine that rejects the waste heat to the environment. Therefore, heat transfer is always accompanied by exergy transfer. Heat transfer Q at a location at thermodynamic temperature T is always accompanied by *exergy transfer* X_{heat} in the amount of

Exergy transfer by heat:

$$X_{heat} = \left(1 - \frac{T_0}{T}\right)Q \quad (kJ) \qquad (4-65)$$

This relation gives the exergy transfer accompanying heat transfer Q whether T is greater than or less than T_0. When $T > T_0$, heat transfer to a system increases the exergy of that system and heat transfer from a system decreases it. But the opposite is true when $T < T_0$. In this case, the heat transfer Q is the heat rejected to the cold medium (the waste heat), and it should not be confused with the heat supplied by the environment at T_0. The exergy transferred with heat is zero when $T = T_0$ at the point of transfer.

Perhaps you are wondering what happens when $T < T_0$. That is, what if we have a medium that is at a lower temperature than the environment? In this case it is conceivable that we can run a heat engine between the environment and the "cold" medium, and thus a cold medium offers us an opportunity to produce work. However, In this time, the environment serves as the heat source and the cold medium as the heat sink. In this case, the relation above gives the negative of the exergy transfer associated with the heat Q transferred to the cold medium. For example, for $T = 100$ K and a heat transfer of $Q = 1$ kJ to the medium, Eq. (4-65) gives $X_{heat}(1 - 300/100)(1\text{ kJ}) = -2$ kJ, which means that the exergy of the cold medium decreases by 2 kJ. It also means that this exergy can be recovered, and the cold medium – environment combination has the potential to produce 2 units of work for each unit of heat rejected to the cold medium at 100 K. That is, a Carnot heat engine operating between $T_0 = 300$ K and $T = 100$ K produces 2 units of work while rejecting 1 unit of heat for each 3 units of heat it receives from the environment.

When $T > T_0$, the exergy and heat transfer are in the same direction. That is, both the exergy and exergy content of the medium to which heat is transferred increase. When $T < T_0$ (cold medium), however, the exergy and heat transfer are in opposite directions. That is, the exergy of the cold medium increases as a result of heat transfer, but its exergy decreases. The exergy of the cold medium eventually becomes zero when its temperature reaches T_0. Eq. (4-65) can also be viewed as the *exergy associated with thermal exergy* Q at temperature T.

When the temperature T at the location where heat transfer is taking place is not constant, the exergy transfer accompanying heat transfer is determined by integration to be

$$X_{heat} = \int \left(1 - \frac{T_0}{T}\right)\delta Q \qquad (4-66)$$

Note that heat transfer through a finite temperature difference is irreversible, and some entropy is generated as a result. The entropy generation is always accompanied by exergy destruction, as illustrated in Fig. 4 – 52. Note that the *heat transfer* Q at a location at temperature T is always accompanied by *entropy transfer* in the amount of Q/T and *exergy transfer* in the amount of $(1 - T_0/T)Q$.

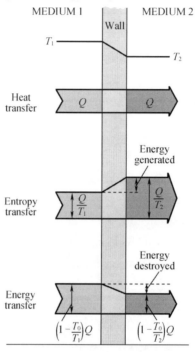

Fig. 4 – 52 The transfer and destruction of exergy during a heat transfer process through a finite temperature difference

Exergy transfer by work, W

Exergy is the useful work potential, and the exergy transfer by work can simply be expressed as Exergy transfer by work:

$$X_{\text{heat}} = \begin{cases} W - W_{\text{surr}} & (\text{for boundary work}) \\ W & (\text{for other forms of work}) \end{cases} \quad (4-67)$$

where $W_{\text{surr}} = P_0(V_2 - V_1)$, P_0 is atmospheric pressure, and V_1 and V_2 are the initial and final volumes of the system. Therefore, the exergy transfer with work such as shaft work and electrical work is equal to the work W itself. In the case of a system that involves boundary work, such as a piston – cylinder device, the work done to push the atmospheric air out of the way during expansion cannot be transferred, and thus it must be subtracted. Also, during a compression process, part of the work is done by the atmospheric air, and thus we need to supply less useful work from an external source.

To clarify this point further, consider a vertical cylinder fitted with a weightless and frictionless piston. The cylinder is filled with a gas that is maintained at the atmospheric pressure P_0 at all times. Heat is now transferred to the system and the gas in the cylinder expands. As a result, the piston rises and boundary work is done. However, this work cannot be used for any useful purpose since it is just enough to push the atmospheric air aside. When the gas is cooled, the piston moves down, compressing the gas. Again, no work is needed from an external source to accomplish this compression process. Thus we conclude that the work done by or against the atmosphere is not available for any useful purpose, and should be excluded from available work.

Exergy transfer by mass, m

Mass contains *exergy* as well as exergy and entropy, and the exergy, exergy, and entropy contents of a system are proportional to mass. Also, the rates of exergy, entropy, and exergy transport into or out of a system are proportional to the mass flow rate. Mass flow is a mechanism to transport exergy, entropy, and exergy into or out of a system. When mass in the amount of m enters or leaves a system, exergy in the amount of $m\psi$, where $\psi = (h - h_0) - T_0(s - s_0) + V^2/2 + gz$, accompanies it. That is, Exergy transfer by mass:

$$X_{mass} = m\psi \qquad (4-68)$$

Therefore, the exergy of a system increases by mc when mass in the amount of m enters, and decreases by the same amount when the same amount of mass at the same state leaves the system.

Exergy flow associated with a fluid stream when the fluid properties are variable can be determined by integration from

$$\dot{X}_{mass} = \int_{A_c} \psi \rho V_n dA_c \quad \text{and} \quad X_{mass} = \int \psi \delta m = \int_{\Delta t} \dot{X}_{mass} dt \qquad (4-69)$$

where A_c is the cross-sectional area of the flow and V_n is the local velocity normal to dA_c.

Note that exergy transfer by heat $X]_{heat}$ is zero for adiabatic systems, and the exergy transfer by mass $X]_{mass}$ is zero for systems that involve no mass flow across their boundaries (i.e., closed systems). The total exergy transfer is zero for isolated systems since they involve no heat, work, or mass transfer.

THE DECREASE OF EXERGY PRINCIPLE AND EXERGY DESTRUCTION

We presented the *conservation of exergy principle* and indicated that exergy cannot be created or destroyed during a process and established the *increase of entropy principle*, which can be regarded as one of the statements of the second law, and indicated that entropy can be created but cannot be destroyed. That is, entropy generation $S]_{gen}$ must be positive (actual processes) or zero (reversible processes), but it cannot be negative. Now we are about to establish an alternative statement of the second law of thermodynamics, called the *decrease of exergy principle*, which is the counterpart of the increase of entropy principle.

Consider an *isolated system* shown in Fig. (4-53). By definition, no heat, work, or mass can cross the boundaries of an isolated system, and thus there is no exergy and entropy transfer. Then the *exergy* and *entropy* balances for an isolated system can be expressed as

Exergy balance:
$$(E_{in} \rightarrow 0) - (E_{out} \rightarrow 0) = \Delta E_{sys} \rightarrow 0 = E_2 - E_1 \qquad (4-70)$$

Entropy balance:
$$(S_{in} \rightarrow 0) - (S_{out} \rightarrow 0) + S_{gen} = \Delta S_{sys} \rightarrow S_{gen} = S_2 - S_1 \qquad (4-71)$$

Multiplying the second relation by T_0 and subtracting it from the first one gives

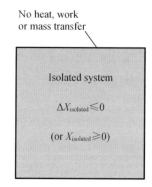

Fig. 4−53 The isolated system considered in the development of the decrease of exergy principle

$$-T_0 S_{gen} = E_2 - E_1 - T_0(S_2 - S_1) \tag{4-72}$$

From Eq. 4−58 we have

$$X_2 - X_1 = (E_2 - E_1) + P_0(V_2 - V_1)^{\to 0} - T_0(S_2 - S_1) = (E_2 - E_1) - T_0(S_2 - S_1) \tag{4-73}$$

since $V_2 = V_1$ for an isolated system (it cannot involve any moving boundary and thus any boundary work). Combining Eqs. (4−72) and (4−73) gives

$$-T_0 S_{gen} = X_2 - X_1 \leq 0 \tag{4-74}$$

since T_0 is the thermodynamic temperature of the environment and thus a positive quantity, $S_{gen} \geq 0$, and thus $T_0 S_{gen} \geq 0$. Then we conclude that

$$\Delta X_{iso} = (X_2 - X_1)_{iso} \leq 0 \tag{4-75}$$

This equation can be expressed as *the exergy of an isolated system during a process always decreases or, in the limiting case of a reversible process, remains constant*. In other words, it *never* increases and *exergy is destroyed* during an actual process. This is known as the decrease of exergy principle. For an isolated system, the decrease in exergy equals exergy destroyed.

Exergy Destruction

Irreversibilities such as friction, mixing, chemical reactions, heat transfer through a finite temperature difference, unrestrained expansion, nonquasi-equilibrium compression or expansion always *generate entropy*, and anything that generates entropy always *destroys exergy*. The exergy destroyed is proportional to the entropy generated, as can be seen from Eq. (4−74), and is expressed as

$$X_{destroyed} = T_0 S_{gen} \geq 0 \tag{4-76}$$

Note that exergy destroyed is a *positive quantity* for any actual process and becomes *zero* for a reversible process. Exergy destroyed represents the lost work potential and is also called the *irreversibility* or *lost work*.

Eqs. (4−75) and (4−76) for the decrease of exergy and the exergy destruction are applicable to *any kind of system* undergoing *any kind of process* since any system and its surroundings can be enclosed by a sufficiently large arbitrary boundary across which there is no heat, work, and mass transfer, and thus any system and its surroundings constitute an *isolated system*.

No actual process is truly reversible, and thus some exergy is destroyed during a process. Therefore, the exergy of the universe, which can be considered to be an isolated system, is continuously decreasing. The more irreversible a process is, the larger the exergy destruction during that process. No exergy is destroyed during a reversible process ($X_{destroyed, rev} = 0$).

The decrease of exergy principle does not imply that the exergy of a system cannot increase. The exergy change of a system *can* be positive or negative during a process, but exergy destroyed cannot be negative. The decrease of exergy principle can be summarized as follows:

$$X_{destroyed} \begin{cases} > 0 & \text{Irreversible process} \\ = 0 & \text{Reversible process} \\ < 0 & \text{Impossible process} \end{cases} \quad (4-77)$$

This relation serves as an alternative criterion to determine whether a process is reversible, irreversible, or impossible.

Summary

The *second law of thermodynamics* states that processes occur in a certain direction, not in any direction. A process does not occur unless it satisfies both the first and the second laws of thermodynamics. Bodies that can absorb or reject finite amounts of heat isothermally are called *thermal exergy reservoirs* or *heat reservoirs*. We introduce two devices in this chapter, heat engine and refrigerator. Work can be converted to heat directly, but heat can be converted to work only by some devices called *heat engines*. Refrigerators are devices that absorb heat from low-temperature media and reject it to higher-temperature ones.

The *Kelvin – Planck statement* of the second law of thermodynamics states that no heat engine can produce a net amount of work while exchanging heat with a single reservoir only. The *Clausius statement* of the second law states that no device can transfer heat from a cooler body to a warmer one without leaving an effect on the surroundings. Any device that violates the first or the second law of thermodynamics is called a *perpetual-motion machine*.

A process is said to be *reversible* if both the system and the surroundings can be restored to their original conditions. Any other process is *irreversible*. The effects such as friction, nonquasi-equilibrium expansion or compression, and heat transfer through a finite temperature difference render a process irreversible and are called *irreversibilities*. The *Carnot cycle* is a reversible cycle that is composed of four reversible processes, two isothermal and two adiabatic. The *Carnot principles* state that the thermal efficiencies of all reversible heat engines operating between the same two reservoirs are the same, and that no heat engine is more efficient than a reversible one operating between the same two reservoirs.

The second law of thermodynamics leads to the definition of a new property called *entropy*, which is a quantitative measure of microscopic disorder for a system. Most steady flow devices operate under adiabatic conditions, and the ideal process for these devices is the is entropic process. The parameter that describes how efficiently a device approximates a corresponding isentropic device is called *isentropic* or *adiabatic efficiency*. We have also introduced two important principles, *The increase of entropy principle* and *the decrease of exergy principle*.

Problems

4 – 1   A Carnot heat engine, shown in Fig. P4 – 1, receives 500 kJ of heat per cycle from

a high-temperature source at 652 ℃ and rejects heat to a low-temperature sink at 30 ℃. Determine (a) the thermal efficiency of this Carnot engine and (b) the amount of heat rejected to the sink per cycle.

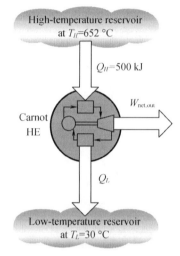

Fig. P4 – 1

4 – 2 An inventor claims to have developed a refrigerator that maintains the refrigerated space at 35 ℉ while operating in a room where the temperature is 75 ℉ and that has a COP of 13.5. Is this claim reasonable?

4 – 3 A 50 kg block of iron casting at 500 K is thrown into a large lake that is at a temperature of 285 K. The iron block eventually reaches thermal equilibrium with the lake water. Assuming an average specific heat of 0.45 kJ/(kg · K) for the iron, determine (a) the entropy change of the iron block, (b) the entropy change of the lake water, and (c) the entropy generated during this process.

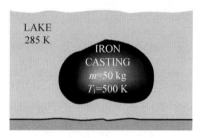

Fig. P4 – 3

4 – 4 A frictionless piston – cylinder device contains a saturated liquid – vapor mixture of water at 100 ℃. During a constant – pressure process, 600 kJ of heat is transferred to the surrounding air at 25 ℃. As a result, part of the water vapor contained in the cylinder condenses. Determine (a) the entropy change of the water and (b) the total entropy generation during this heat transfer process.

Chapter 4　The Second Law of Thermodynamics

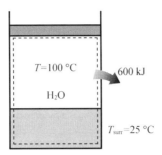

Fig. P4-4

4-5　A heat engine receives heat from a source at 1 200 K at a rate of 500 kJ/s and rejects the waste heat to a medium at 300 K (Fig. P4-5). The power output of the heat engine is 180 kW. Determine the reversible power and the irreversibility rate for this process.

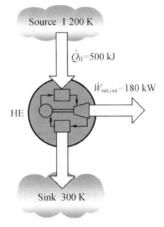

Fig. P4-5

4-6 Refrigerant-134a is to be compressed from 0.14 MPa and 10 ℃ to 0.8 MPa and 50 ℃ steadily by a compressor. Taking the environment conditions to be 20 ℃ and 95 kPa, determine the energy change of the refrigerant during this process and the minimum work input that needs to be supplied to the compressor per unit mass of the refrigerant.

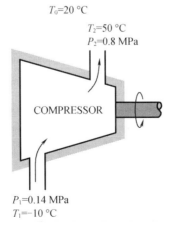

Fig. P4-6

Chapter 5 Properties of Real Gas and Vapor

We start this chapter with the introduction of the concept and properties of real gas and pure substance, and a discussion of the physics of phase change processes. *ideal gas* and the *ideal gas equation of state* are discussed in this chapter. The *compressibility factor*, which accounts for the deviation of real gases from ideal gas behavior, is introduced, and some of the best-known equations of state such as the van der Waals, Beattie – Bridgeman, and Benedict – Webb – Rubin equations are presented.

5 – 1 The Real Gas Equation of State

COMPRESSIBILITY FACTOR

Any equation that relates the pressure, temperature, and specific volume of a substance is called an equation of state. Property relations that involve other properties of a substance at equilibrium states are also referred to as equations of state. There are several equations of state, some simple and others very complex. *Gas* and *vapor* are often used as synonymous words. The vapor phase of a substance is customarily called a *gas* when it is above the critical temperature. *Vapor* usually implies a gas that is not far from a state of condensation.

The ideal gas equation is very simple and thus very convenient to use. However, gases deviate from ideal gas behavior significantly at states near the saturation region and the critical point. This deviation from ideal gas behavior at a given temperature and pressure can accurately be accounted for by the introduction of a correction factor called the compressibility factor Z defined as

$$Z = \frac{Pv}{RT} \qquad (5-1)$$

or

$$Pv = ZRT \qquad (5-2)$$

It can also be expressed as

$$Z = \frac{v_{actual}}{v_{ideal}} \qquad (5-3)$$

where $v_{ideal} = RT/P$. Obviously, $Z = 1$ for ideal gases. For real gases Z can be greater than or less than unity. The farther away Z is from unity, the more the gas deviates from ideal gas behavior.

Gases behave differently at a given temperature and pressure, but they behave very much the same at temperatures and pressures normalized with respect to their critical temperatures and pressures. The normalization is done as

$$P_R = \frac{P}{P_{cr}} \text{ and } T_R = \frac{T}{T_{cr}} \qquad (5-4)$$

Chapter 5 Properties of Real Gas and Vapor

Here P_R is called the reduced pressure and T_R the reduced temperature. The Z factor for all gases is approximately the same at the same reduced pressure and temperature. This is called the principle of corresponding states. In Fig. 5-1, the experimentally determined Z values are plotted against P_R and T_R for several gases. The gases seem to obey the principle of corresponding states reasonably well. By curve-fitting all the data, we obtain the generalized compressibility chart that can be used for all gases (Fig. 5-1).

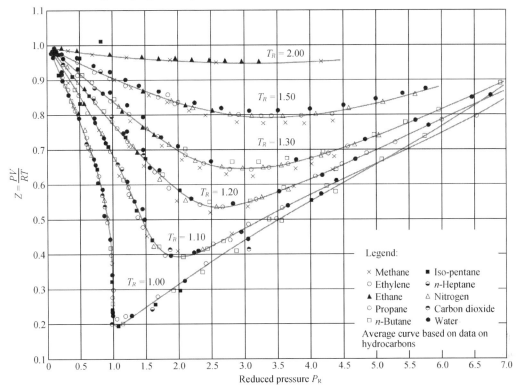

Fig. 5-1 Comparison of Z factors for various gases

The following observations can be made from the generalized compressibility chart:

a. At very low pressures ($P_R \ll 1$), gases behave as an ideal gas regardless of temperature (Fig. 5-3).

b. At high temperatures ($T_R > 2$), ideal gas behavior can be assumed with good accuracy regardless of pressure (except when $P_R \gg 1$).

c. The deviation of a gas from ideal gas behavior is greatest in the vicinity of the critical point (Fig. 5-2).

When P and v, or T and v, are given instead of P and T, the generalized compressibility chart can still be used to determine the third property, but it would involve tedious trial and error. Therefore, it is necessary to define one more reduced property called the pseudo-reduced specific volume v_R as

$$v_R = \frac{v_{\text{actual}}}{RT_{\text{cr}}/P_{\text{cr}}} \tag{5-5}$$

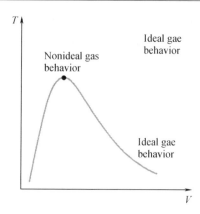

Fig. 5 –2 gases deviate from the ideal gas behavior the most in the neighborhood of the critical point

Note that v_R is defined differently from P_R and T_R. It is related to T_{cr} and P_{cr} instead of v_{cr}. Lines of constant v_R are also added to the compressibility charts, and this enables one to determine T or P without having to resort to time-consuming iterations.

VAN DER WAALS EQUATION OF STATE

An improvement over the ideal gas equation of state based on elementary molecular arguments was suggested in 1873 by van der Waals, who noted that gas molecules actually occupy more than the negligibly small volume presumed by the ideal gas model and also exert long – range attractive forces on one another. Thus, not all of the volume of a container would be available to the gas molecules, and the force they exert on the container wall would be reduced because of the attractive forces that exist between molecules. Based on these elementary molecular arguments, the *van der Waals equation of state* is

$$\left(P + \frac{a}{v^2}\right)(v - b) = RT \tag{5-6}$$

Van der Waals intended to improve the ideal gas equation of state by including two of the effects not considered in the ideal gas model: the *intermolecular attraction forces* and the *volume occupied by the molecules themselves*. The term a/v^2 accounts for the intermolecular forces, and b accounts for the volume occupied by the gas molecules. In a room at atmospheric pressure and temperature, the volume actually occupied by molecules is only about one-thousandth of the volume of the room. As the pressure increases, the volume occupied by the molecules becomes an increasingly significant part of the total volume. Van der Waals proposed to correct this by replacing v in the ideal gas relation with the quantity $v - b$, where b represents the volume occupied by the gas molecules per unit mass.

The van der Waals equation gives pressure as a function of temperature and specific volume and thus is explicit in pressure. Since the equation can be solved for temperature as afunction of pressure and specific volume, it is also explicit in temperature. However, the equation is cubic in specific volume, so it cannot generally be solved for specific volume in terms of temperature and pressure. The van der Waals equation is not explicit in specificvolume.

The determination of the two constants appearing in this equation is based on the observation

Chapter 5 Properties of Real Gas and Vapor

that the critical isotherm on a $P - v$ diagram has a horizontal inflection point at the critical point (Fig. 5 – 3). Thus, the first and the second derivatives of P with respect to v at the critical point must be zero.

That is,

$$\left(\frac{\partial P}{\partial v}\right)_{T=T_{cr}=const} = 0 \text{ and } \left(\frac{\partial^2 P}{\partial v^2}\right)_{T=T_{cr}=const} = 0$$

By performing the differentiations and eliminating v_{cr}, the constants a and b are determined to be

$$a = \frac{27R^2 T_{cr}^2}{64P_{cr}} \text{ and } a = \frac{RT_{cr}}{8P_{cr}} \quad (5-7)$$

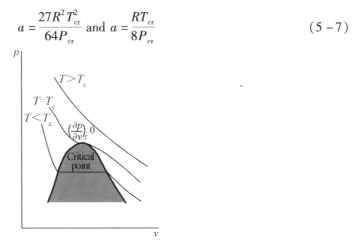

Fig. 5 – 3 Critical isotherm of a pure substance as an inflection point at the critical state

The constants a and b can be determined for any substance from the critical point data alone (Table A. 2).

The accuracy of the van der Waals equation of state is often inadequate, However, it can be improved by using values of a and b that are based on the actual behavior of the gas over a wider range instead of a single point. Despite its limitations, the van der Waals equation of state has a historical value in that it was one of the first attempts to model the behavior of real gases. The van der Waals equation of state can also be expressed on a unit mole basis by replacing the $\bar{v}$ in Eq. 5 – 6 by $\bar{v}$ and the R in Eqs. (5 – 6) and (5 – 7) by R_0.

EXAMPLE 5 – 1 Evaluating Gas Pressure

Predict the pressure of nitrogen gas at $T = 175$ K and $v = 0.003\ 75$ m³/kg on the basis of the van der Waals equation of state. Compare the values obtained to the experimentally determined value of 10 000 kPa.

Solution: The pressure of nitrogen gas is to be determined using four different equations of state.

Properties: The gas constant of nitrogen gas is 0.296 8 kPa · m³/kg · K (Table A – 2).

Analysis: The van der Waals constants for nitrogen are determined from Eq. (5 – 6) to be
$$a = 0.175 \text{ m}^6 \cdot \text{kPa/kg}^2$$
$$b = 0.001\ 38 \text{ m}^3/\text{kg}$$

From Eq. (5 – 5),

$$P = \frac{RT}{v-b} - \frac{a}{v^2} = 9\ 471 \text{ kPa}$$

which is in error by 5.3 percent.

5 – 2 Property Diagrams and Saturation State

A substance that has a fixed chemical composition throughout is called a pure substance. Water, nitrogen, helium, and carbon dioxide, for example, are all pure substances. A pure substance does not have to be of a single chemical element or compound, however. A mixture of various chemical elements or compounds also qualifies as a pure substance as long as the mixture is homogeneous. Air, for example, is a mixture of several gases, but it is often considered to be a pure substance because it has a uniform chemical composition.

A mixture of two or more phases of a pure substance is still a pure substance as long as the chemical composition of all phases is the same. A mixture of ice and liquid water, for example, is a pure substance because both phases have the same chemical composition. A mixture of liquid air and gaseous air, however, is not a pure substance since the composition of liquid air is different from the composition of gaseous air, and thus the mixture is no longer chemically homogeneous. This is due to different components in air condensing at different temperatures at a specified pressure.

PROPERTY DIAGRAMS

The variations of properties during phase-change processes are best studied and understood with the help of property diagrams. Next, we develop and discuss the $T-v$, $P-v$, and $P-T$ diagrams for pure substances.

1 The $T-v$ diagram

The phase-change process of water at 1 atm pressure was described in detail in the last section and plotted on a $T-v$ diagram. Now we repeat this process at different pressures to develop the $T-v$ diagram.

Let us add weights on top of the piston until the pressure inside the cylinder reaches 1 MPa. At this pressure, water has a somewhat smaller specific volume than it does at 1 atm pressure. As heat is transferred to the water at this new pressure, the process follows a path that looks very much like the process path at 1 atm pressure, as shown in Fig. 5 – 4, but there are some noticeable differences. First, water starts boiling at a much higher temperature (179.9 ℃) at this pressure. Second, the specific volume of the saturated liquid is larger and the specific volume of the saturated vapor is smaller than the corresponding values at 1 atm pressure. That is, the horizontal line that connects the saturated liquid and saturated vapor states is much shorter.

Chapter 5 Properties of Real Gas and Vapor

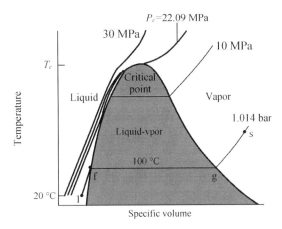

Fig. 5 – 4 $T-v$ **diagram of constant – pressure phase – change processes of a pure substance at various pressures (numerical values are for water).**

As the pressure is increased further, this saturation line continues to shrink, as shown in Fig. 5 – 4, and it becomes a point when the pressure reaches 22.06 MPa for the case of water. This point is called the critical point, and it is defined as *the point at which the saturated liquid and saturated vapor states are identical*. The temperature, pressure, and specific volume of a substance at the critical point are called, respectively, the *critical temperature* T_{cr}, *critical pressure* P_{cr}, and *critical specific volume* v_{cr}. The critical point properties of water are $P_{cr} = 22.06$ MPa, $T_{cr} = 373.95$ ℃, and $v_{cr} = 0.003\ 106$ m³/kg.

At pressures above the critical pressure, there is not a distinct phase change process (Fig. 5 – 5). Instead, the specific volume of the substance continually increases, and at all times there is only one phase present. Eventually, it resembles a vapor, but we can never tell when the change has occurred. Above the critical state, there is no line that separates the compressed liquid region and the super heated vapor region. However, it is customary to refer to the substance as super heated vapor at temperatures above the critical temperature and as compressed liquid at temperatures below the critical temperature.

The saturated liquid states in Fig. 5 – 4 can be connected by a line called the saturated liquid line, and saturated vapor states in the same figure can be connected by another line, called the saturated vapor line. These two lines meet at the critical point, forming a dome as shown in Fig. 5 – 6. All the compressed liquid states are located in the region to the left of the saturated liquid line, called the compressed liquid region. All the super heated vapor states are located to the right of the saturated vapor line, called the superheated vapor region. In these two regions, the substance exists in a single phase, a liquid or a vapor. All the states that involve both phases in equilibrium are located under the dome, called the saturated liquid – vapor mixture region, or the wet region.

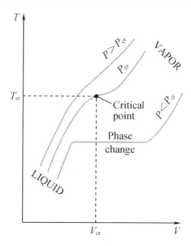

Fig. 5-5 At super critical pressures ($P > P_{cr}$), there is no distinct phase-change (boiling) process

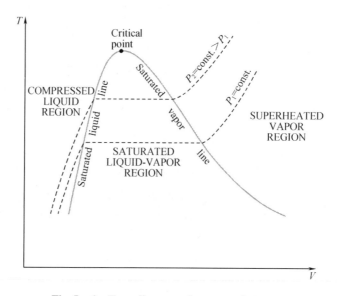

Fig. 5-6 $T-v$ diagram of a pure substance

2 The $P-v$ diagram

The general shape of the $P-v$ diagram of a pure substance is very much like the $T-v$ diagram, but the $T=$ const lines on this diagram have a downward trend, as shown in Fig. 5-7.

Consider again a piston – cylinder device that contains liquid water at 1 MPa and 150 ℃. Water at this state exists as a compressed liquid. Now the weights on top of the piston are removed one by one so that the pressure inside the cylinder decreases gradually (Fig. 5-8). The water is allowed to exchange heat with the surroundings so its temperature remains constant. As the pressure decreases, the volume of the water increases slightly. When the pressure reaches the saturation – pressure value at the specified temperature (0.476 2 MPa), the water starts to boil. During this vaporization process, both the temperature and the pressure remain constant, but the specific volume increases. Once the last drop of liquid is vaporized, further reduction in pressure results in a further increase in specific volume. Notice that during the phase – change process, we did not remove any weights. Doing so would cause the pressure and therefore the temperature to drop, since $T_{sat}=f(P_{sat})$, and the process would no longer be isothermal.

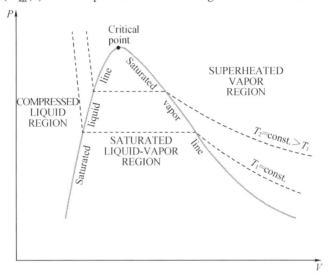

Fig. 5 – 7 $P-v$ diagram of a pure substance

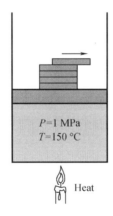

Fig. 5 – 8   The pressure in a piston – cylinderdevice can be reduced by reducing the weight of the piston

When the process is repeated for other temperatures, similar paths are obtained for the phase – change processes. Connecting the saturated liquid and the saturated vapor states by a curve, we obtain the $P-v$ diagram of a pure substance, as shown in Fig. 5 – 7.

The two equilibrium diagrams developed so far represent the equilibrium states involving the liquid and the vapor phases only. But these diagrams can easily be extended to include the solid phase as well as the solid-liquid and the solid – vapor saturation regions. The basic principles discussed in conjunction with the liquid – vapor phase – change process apply equally to the solid – liquid and solid – vapor phase – change processes. Most substances contract during a solidification process. Others, like water, expand as they freeze. The $P-v$ diagrams for both groups of substances are given in Figs. 5 – 9 and 5 – 10. These two diagrams differ only in the solid – liquid saturation region. The $T - v$ diagrams look very much like the $P - v$ diagrams, especially for substances that contract on freezing.

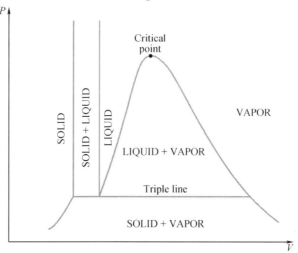

Fig. 5 – 9 $P - v$ **diagram of a substance that contracts on freezing**

The fact that water expands upon freezing has vital consequences in nature. If water contracted on freezing as most other substances do, the ice formed would be heavier than the liquid water, and it would settle to the bottom of rivers, lakes, and oceans instead of floating at the top. The sun's rays would never reach these ice layers, and the bottoms of many rivers, lakes, and oceans would be covered with ice at times, seriously disrupting marine life.

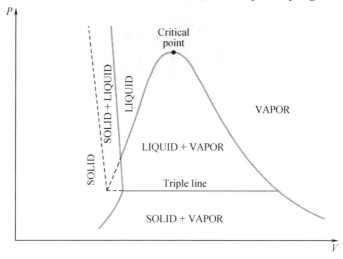

Fig. 5 – 10 $P - v$ **diagram of a substance that expands on freezing (such as water)**

We are all familiar with two phases being in equilibrium, but under some conditions all three phases of a pure substance coexist in equilibrium (Fig. 5 – 11). On $P - v$ or $T - v$ diagrams, these triple-phase states form a line called the triple line. The states on the triple line of a substance have the same pressure and temperature but different specific volumes. The triple line appears as a point on the $P - T$ diagrams and, therefore, is often called the triple point. The triple-point temperatures and pressures of various substances are given in Table 5 – 1. For water, the triple-point temperature and pressure are 0.01 ℃ and 0.611 7 kPa, respectively. That is, all three phases of water coexist in equilibrium only if the temperature and pressure have precisely these values. No substance can exist in the liquid phase in stable equilibrium at pressures below the triple-point pressure. The same can be said for temperature for substances that contract on freezing. However, substances at high pressures can exist in the liquid phase at temperatures below the triple-point temperature.

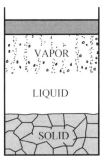

Fig. 5 – 11 At triple-point pressure and temperature, a substance exists in three phases in equilibrium

TABLE 5 – 1   Triple-point temperatures and pressures of various substances

Substance	Formula	T_{tp}/K	P_{tp}/kPa
Acetylene	C_2H_2	192.4	120
Ammonia	NH_3	195.40	6.076
Argon	A	83.81	68.9
Carbon(graphite)	C	3900	10.100
Carbon dioxide	CO_2	216.55	517
Carbon monoxide	CO	68.10	15.37
Deuterium	D_2	18.63	17.1
Ethane	C_2H_6	89.89	8×10^{-4}
Ethylene	C_2H_4	104.0	0.12
Helium 4(λ point)	He	2.19	5.1
Hydrogen	H_2	13.84	7.04
Hydrogen chloride	HCl	158.96	13.9
Mercury	Hg	234.2	1.65×10^{-7}
Methane	CH_4	90.68	11.7
Neon	Ne	24.57	43.2
Nitric oxide	NO	109.50	21.92
Nitrogen	N_2	63.18	12.6
Nitrous oxide	N_2O	182.34	87.85
Oxygen	O_2	54.36	0.152

TABLE 5 – 1 (Continued)

Substance	Formula	T_{tp}/K	P_{tp}/kPa
Palladium	Pd	1825	3.5×10^{-3}
Platinum	Pt	2045	2.0×10^{-4}
Sulfur dioxide	SO_2	197.69	1.67
Titanium	Ti	1941	5.3×10^{-3}
Uranium hexafluoride	UF_6	337.17	151.7
Water	H_2O	273.16	0.61
Xenon	Xe	161.3	81.5
Zinc	Zn	692.65	0.065

Source: Data from National Bureau of Standards(U. S.) Circ., 500(1952).

There are two ways a substance can pass from the solid to vapor phase: either it melts first into a liquid and subsequently evaporates, or it evaporates directly without melting first. The latter occurs at pressures below the triple point value, since a pure substance cannot exist in the liquid phase at those pressures (Fig. 5 – 12). Passing from the solid phase directly into the vapor phase is called sublimation. For substances that have a triple-point pressure above the atmospheric pressure such as solid CO_2 (dry ice), sublimation is the only way to change from the solid to vapor phase at atmospheric conditions.

Fig. 5 – 12 At low pressures (below the triple-point value), solids evaporate without melting first (*sublimation*)

3 The *P* - *T* diagram

Fig. 5 – 13 shows the *P* - *T* diagram of a pure substance. This diagram is often called the phase diagram since all three phases are separated from each other by three lines. The sublimation line separates the solid and vapor regions, the vaporization line separates the liquid and vapor regions, and the melting (or fusion) line separates the solid and liquid regions. These three lines meet at the triple point, where all three phases coexist in equilibrium. The vaporization line ends at the critical point because no distinction can be made between liquid and vapor phases above the critical point. Substances that expand and contract on freezing differ only in the melting line on the *P* - *T* diagram.

Chapter 5 Properties of Real Gas and Vapor

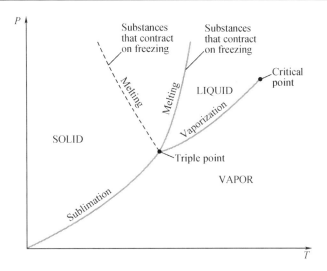

Fig. 5 – 13 $P-T$ diagram of pure substances

The state of a simple compressible substance is fixed by any two independent, intensive properties. Once the two appropriate properties are fixed, all the other properties become dependent properties. Remembering that any equation with two independent variables in the form $z = z(x, y)$ represents a surface in space, we can represent the $P-v-T$ behavior of a substance as a surface in space, as shown in Figs. 5 – 14 and 5 – 15. Here T and v may be viewed as the independent variables (the base) and P as the dependent variable (the height).

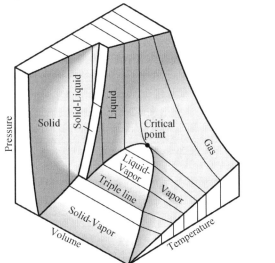 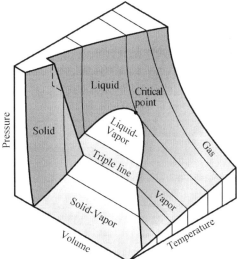

Fig. 5 – 14 $P-v-T$ surface of a substance that *contracts* on freezing

Fig. 5 – 15 $P-v-T$ surface of a substance that *expands* on freezing (like water)

All the points on the surface represent equilibrium states. All states along the path of a quasi-equilibrium process lie on the $P-v-T$ surface since such a process must pass through equilibrium states. The single-phase regions appear as curved surfaces on the $P-v-T$ surface, and the two-phase regions as surfaces perpendicular to the $P-T$ plane. This is expected since the projections of two-phase regions on the $P-T$ plane are lines.

All the two-dimensional diagrams we have discussed so far are merely projections of this three-dimensional surface onto the appropriate planes. A $P-v$ diagram is just a projection of the $P-v-T$ surface on the $P-v$ plane, and a $T-v$ diagram is nothing more than the bird's eye view of this surface. The $P-v-T$ surfaces present a great deal of information at once, but in a thermodynamic analysis it is more convenient to work with two-dimensional diagrams, such as the $P-v$ and $T-v$ diagrams.

4 The $h-s$ diagram

The other diagram commonly used in engineering is the enthalpy-entropy diagram, which is quite valuable in the analysis of steady-flow devices such as turbines, compressors, and nozzles. The coordinates of an $h-s$ diagram represent two properties of major interest: enthalpy, which is a primary property in the first-law analysis of the steady-flow devices, and entropy, which is the property that accounts for irreversibilities during adiabatic processes. In analyzing the steady flow of steam through an adiabatic turbine, for example, the vertical distance between the inlet and the exit states h is a measure of the work output of the turbine, and the horizontal distance s is a measure of the irreversibilities associated with the process.

The $h-s$ diagram is also called a Mollier diagram after the German scientist R. Mollier (1863–1935). An $h-s$ diagram is given in the appendix for steam in Fig. B–2.

Note the location of the critical point and the appearance of lines of constant temperature and constant pressure. Lines of constant quality are shown in the two phase liquid-vapor region (some figures give lines of constant percent moisture). The figure is intended for evaluating properties at superheated vapor states and for two – phase liquid-vapor mixtures. Liquid data are seldom shown. In the superheated vapor region, constant temperature lines become nearly horizontal as pressure is reduced.

SATURATION STATE

Consider a piston – cylinder device containing liquid water at 20 ℃ and 1 atm pressure (state 1, Fig. 5–16). Under these conditions, water exists in the liquid phase, and it is called a compressed liquid, or a subcooled liquid, meaning that it is *not about to vaporize*. Heat is now transferred to the water until its temperature rises to 40 ℃. As the temperature rises, the liquid water expands slightly, and so its specific volume increases. To accommodate this expansion, the piston moves up slightly. The pressure in the cylinder remains constant at 1 atm during this process since it depends on the outside barometric pressure and the weight of the piston, both of which are constant. Water is still a compressed liquid at this state since it has not started to vaporize.

As more heat is transferred, the temperature keeps rising until it reaches 100 ℃ (state 2, Fig. 5–17). At this point water is still a liquid, but any heat addition will cause some of the liquid to vaporize. That is, a phase-change process from liquid to vapor is about to take place. A liquid that is *about to vaporize* is called a saturated liquid. Therefore, state 2 is a saturated liquid state.

Chapter 5 Properties of Real Gas and Vapor

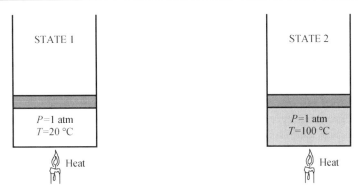

Fig. 5 – 16 At 1 atm and 20 ℃, water exists in the liquid phase (*compressed liquid*)

Fig. 5 – 17 At 1 atm pressure and 100 ℃, water exists as a liquid that is ready to vaporize (*saturated liquid*)

Once boiling starts, the temperature stops rising until the liquid is completely vaporized. That is, the temperature will remain constant during the entire phase-change process if the pressure is held constant. This can easily be verified by placing a thermometer into boiling pure water on top of a stove. At sea level ($P = 1$ atm), the thermometer will always read 100 ℃ if the pan is uncovered or covered with a light lid. During a boiling process, the only change we will observe is a large increase in the volume and a steady decline in the liquid level as a result of more liquid turning to vapor.

Midway about the vaporization line (state 3, Fig. 5 – 18), the cylinder contains equal amounts of liquid and vapor. As we continue transferring heat, the vaporization process continues until the last drop of liquid is vaporized (state 4, Fig. 5 – 19). At this point, the entire cylinder is filled with vapor that is on the borderline of the liquid phase. Any heat loss from this vapor will cause some of the vapor to condense (phase change from vapor to liquid). A vapor that is *about to condense* is called a saturated vapor. Therefore, state 4 is a saturated vapor state. A substance at states between 2 and 4 is referred to as a saturated liquid – vapor mixture since the *liquid and vapor phases coexist* in equilibrium at these states.

Once the phase-change process is completed, we are back to a single phase region again, and further transfer of heat results in an increase in both the temperature and the specific volume (Fig. 5 – 20). At state 5, the temperature of the vapor is, let us say, 300 ℃; and if we transfer some heat from the vapor, the temperature may drop somewhat but no condensation will take place as long as the temperature remains above 100 ℃ (for $P = 1$ atm). A vapor that is *not about to condense* (i. e., not a saturated vapor) is called a super heated vapor. Therefore, water at state 5 is a superheated vapor. This constant-pressure phase-change process is illustrated on a $T - v$ diagram in Fig. 5 – 21.

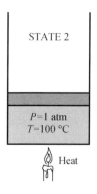

Fig. 5-18 As more heat is transferred, part of the saturated liquid vaporizes (*saturated liquid – vapor mixture*).

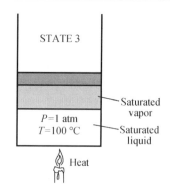

Fig. 5-19 At 1 atm pressure, the temperature remains constant at 100 ℃ until the last drop of liquid is vaporized (*saturated vapor*)

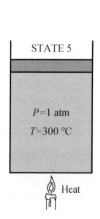

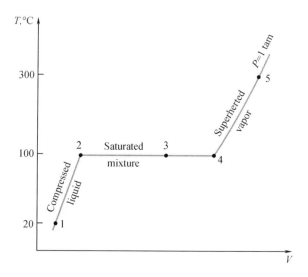

Fig. 5-20 As more heat is transferred, the temperature of the vapor starts to rise (*superheated vapor*)

Fig. 5-21 $T - v$ diagram for the heating process of water at constant pressure

If the entire process described here is reversed by cooling the water while maintaining the pressure at the same value, the water will go back to state 1, retracing the same path, and in so doing, the amount of heat released will exactly match the amount of heat added during the heating process.

In our daily life, water implies liquid water and steam implies water vapor. In thermodynamics, however, both water and steam usually mean only one thing: H_2O.

It probably came as no surprise to you that water started to boil at 100 ℃. Strictly speaking, the statement "water boils at 100 ℃" is incorrect. The correct statement is "water boils at 100 ℃ at 1 atm pressure." The only reason water started boiling at 100 ℃ was because we held the pressure constant at 1 atm (101.325 kPa). If the pressure inside the cylinder were raised to

500 kPa by adding weights on top of the piston, water would start boiling at 151.8 ℃. That is, *the temperature at which water starts boiling depends on the pressure*; therefore, *if the pressure is fixed, so is the boiling temperature.*

At a given pressure, the temperature at which a pure substance changes phase is called the saturation temperature T_{sat}. Likewise, at a given temperature, the pressure at which a pure substance changes phase is called the saturation pressure P_{sat}. At a pressure of 101.325 kPa, T_{sat} is 99.97 ℃. Conversely, at a temperature of 99.97 ℃, P_{sat} is 101.325 kPa. (At 100.00 ℃, P_{sat} is 101.42 kPa in the ITS −90 discussed in Chap. 1.)

Saturation tables that list the saturation pressure against the temperature (or the saturation temperature against the pressure) are available for practically all substances. A partial listing of such a table is given in Table 5−2 for water. This table indicates that the pressure of water changing phase (boiling or condensing) at 25 ℃ must be 3.17 kPa, and the pressure of water must be maintained at 3 976 kPa (about 40 atm) to have it boil at 250 ℃. Also, water can be frozen by dropping its pressure below 0.61 kPa.

TABLE 5−2 Saturation (boiling) pressure of water at various temperatures

Temperature, T/℃	Saturation pressure, P_{sat}/kPa
−10	0.26
−5	0.40
0	0.61
5	0.87
10	1.23
15	1.71
20	2.34
25	3.17
30	4.25
40	7.39
50	12.35
100	101.4
150	476.2
200	1 555
250	3 976
300	8 588

It takes a large amount of exergy to melt a solid or vaporize a liquid. The amount of exergy absorbed or released during a phase – change process is called the latent heat. More specifically, the amount of exergy absorbed during melting is called the latent heat of fusion and is equivalent to the amount of exergy released during freezing. Similarly, the amount of exergy absorbed during vaporization is called the latent heat of vaporization and is equivalent to the exergy released during condensation. The magnitudes of the latent heats depend on the temperature or pressure at which the phase change occurs. At 1 atm pressure, the latent heat of fusion of water is 333.7 kJ/kg and the latent heat of vaporization is 2 256.5 kJ/kg.

During a phase – change process, pressure and temperature are obviously dependent properties, and there is a definite relation between them, that is, $T_{sat} = f(P_{sat})$. A plot of T_{sat} versus P_{sat}, such as the one given for water in Fig. 5 – 22, is called a liquid – vapor saturation curve. A curve of this kind is characteristic of all pure substances.

It is clear from Fig. 5 – 22 that T_{sat} increases with P_{sat}. Thus, a substance at higher pressures boils at higher temperatures. In the kitchen, higher boiling temperatures mean shorter cooking times and exergy savings.

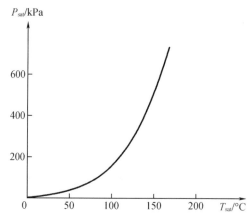

Fig. 5 – 22 The liquid – vapor saturation curve of a pure substance (numerical values are for water)

The atmospheric pressure, and the boiling temperature of water, decreases with elevation. As a consequence, it takes longer to cook at higher altitudes than it does at sea level. For example, the standard atmospheric pressure at an elevation of 2 000 m is 79.50 kPa, which corresponds to a boiling temperature of 93.3 ℃ as opposed to 100 ℃ at sea level (zero elevation). The variation of the boiling temperature of water with altitude at standard atmospheric conditions is given in Table 5 – 3. For each 1 000 m increase in elevation, the boiling temperature drops by a little over 3 ℃. Note that the atmospheric pressure at a location, and thus the boiling temperature, changes slightly with the weather conditions. But the corresponding change in the boiling temperature is no more than about 1 ℃.

TABLE 5-3 Variation of the standard atmospheric pressure and the boiling (saturation) temperature of water with altitude

Elevation/m	Atmospheric pressure/kPa	Boiling temperature/℃
0	101.33	100.0
1 000	89.55	96.5
2 000	79.50	93.3
5 000	54.05	83.3
10 000	26.50	66.3
20 000	5.53	34.7

5-3 Property Tables

For most substances, the relationships among thermodynamic properties are too complex to be expressed by simple equations. Therefore, properties are frequently presented in the form of tables. Some thermodynamic properties can be measured easily, but others cannot and are calculated by using the relations between them and measurable properties. The results of these measurements and calculations are presented in tables in a convenient format.

SATURATED LIQUID AND SATURATED VAPOR STATES

The properties of saturated liquid and saturated vapor for water are listed in Tables A-4 and A-5. Both tables give the same information. The only difference is that in Table A-4 properties are listed under temperature and in Table A-5 under pressure. Therefore, it is more convenient to use Table A-4 when *temperature* is given and Table A-5 when *pressure* is given. The use of Table A-4 is illustrated in Fig. 5-23.

The subscript f is used to denote properties of a saturated liquid, and the subscript g to denote the properties of saturated vapor. These symbols are commonly used in thermodynamics and originated from German. Another subscript commonly used is fg, which denotes the difference between the saturated vapor and saturated liquid values of the same property. For example,

v_f = specific volume of saturated liquid

v_g = specific volume of saturated vapor

v_{fg} = difference between v_g and v_f (that is, $v_{fg} = v_g - v_f$)

The quantity *hfg* is called the enthalpy of vaporization (or latent heat of vaporization). It represents the amount of exergy needed to vaporize a unit mass of saturated liquid at a given temperature or pressure. It decreases as the temperature or pressure increases and becomes zero at the critical point.

Thermodynamics

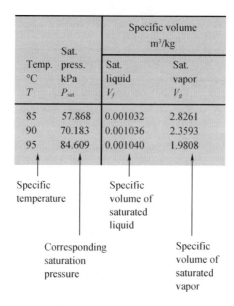

Fig. 5 – 23 A partial list of Table A – 4

EXAMPLE 5 – 2 Pressure of Saturated Liquid in a Tank

A rigid tank contains 50 kg of saturated liquid water at 90 ℃. Determine the pressure in the tank and the volume of the tank.

Solution A rigid tank contains saturated liquid water. The pressure and volume of the tank are to be determined.

Analysis The state of the saturated liquid water is shown on a $T-v$ diagram in Fig. 5 – 24. Since saturation conditions exist in the tank, the pressure must be the saturation pressure at 90 ℃:

$$P = P_{\text{sat @ 90℃}} = 70.183 \text{ kPa}$$

The specific volume of the saturated liquid at 90 ℃ is

$$v = v_{\text{sat @ 90℃}} = 0.001\ 036 \text{ m}^3/\text{kg}$$

Then the total volume of the tank becomes

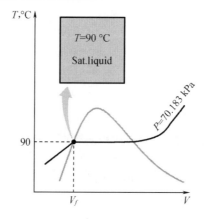

Fig. 5 – 24 Schematic and $T-v$ diagram for Example 5 – 2

Chapter 5 Properties of Real Gas and Vapor

$$V = mv = (150 \text{ kg})(10.001\ 036 \text{ m}^3/\text{kg}^2) = 0.051\ 8 \text{ m}^3$$

SATURATED LIQUID – VAPOR MIXTURE

During a vaporization process, a substance exists as part liquid and part vapor. That is, it is a mixture of saturated liquid and saturated vapor. To analyze this mixture properly, we need to know the proportions of the liquid and vapor phases in the mixture. This is done by defining a new property called the quality x as the ratio of the mass of vapor to the total mass of the mixture:

$$x = \frac{m_{\text{vapor}}}{m_{\text{total}}} \quad (5-8)$$

where

$$m_{\text{total}} = m_{\text{liquid}} + m_{\text{vapor}} = m_f + m_g$$

Quality has significance for *saturated mixtures* only. It has no meaning in the compressed liquid or superheated vapor regions. Its value is between 0 and 1. The quality of a system that consists of *saturated liquid* is 0 (or 0 percent), and the quality of a system consisting of *saturated vapor* is 1 (or 100 percent). In saturated mixtures, quality can serve as one of the two independent intensive properties needed to describe a state. Note that *the properties of the saturated liquid are the same whether it exists alone or in a mixture with saturated vapor*. During the vaporization process, only the amount of saturated liquid changes, not its properties. The same can be said about a saturated vapor.

A saturated mixture can be treated as a combination of two subsystems: the saturated liquid and the saturated vapor. However, the amount of mass for each phase is usually not known. Therefore, it is often more convenient to imagine that the two phases are mixed well, forming a homogeneous mixture. Then the properties of this "mixture" will simply be the average properties of the saturated liquid – vapor mixture under consideration. Here is how it is done.

Consider a tank that contains a saturated liquid – vapor mixture. The volume occupied by saturated liquid is V_f, and the volume occupied by saturated vapor is V_g. The total volume V is the sum of the two:

$$V = V_f + V_g$$
$$V = mv \rightarrow m_t v_{\text{avg}} = m_f v_f + m_g v_g$$
$$m_f = m_t - m_g \rightarrow m_t v_{\text{avg}} = (m_t - m_g) v_f + m_g v_g$$

Dividing by m_t yields

$$v_{\text{avg}} = (1-x) v_f + x v_g$$

since $x = m_g/m_t$. This relation can also be expressed as

$$v_{\text{avg}} = v_f + x v_{fg} \quad (5-9)$$

where $v_{fg} = v_g - v_f$. Solving for quality, we obtain

$$x = \frac{v_{\text{avg}} - v_f}{v_{fg}} \quad (5-10)$$

Based on this equation, quality can be related to the horizontal distances on a $P-v$ or $T-v$ diagram (Fig. 5-25). At a given temperature or pressure, the numerator of Eq. (5-10) is the distance between the actual state and the saturated liquid state, and the denominator is the length of the entire horizontal line that connects the saturated liquid and saturated vapor states. A state of 50 percent quality lies in the middle of this horizontal line.

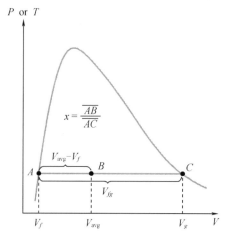

Fig. 5 – 25 Quality is related to the horizontal distances on $P-v$ and $T-v$ diagrams

The analysis given above can be repeated for internal exergy and enthalpy with the following results:

$$u_{avg} = u_f + xu_{fg} \quad (\text{kJ/kg}) \tag{5–11}$$

$$h_{avg} = h_f + xh_{fg} \quad (\text{kJ/kg}) \tag{5–12}$$

All the results are of the same format, and they can be summarized in a single equation as

$$y_{avg} = y_f + xy_{fg}$$

where y is v, u, or h. The subscript "avg" (for "average") is usually dropped for simplicity. The values of the average properties of the mixtures are always *between* the values of the saturated liquid and the saturated vapor properties (Fig. 5 – 26).

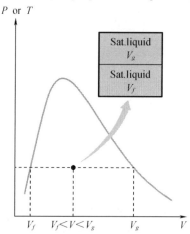

Fig. 5 – 26 The v value of a saturated liquid – vapor mixture lies between the v_f and v_g values at the specified T or P

That is,

Chapter 5 Properties of Real Gas and Vapor

$$y_f \leq y_{avg} \leq y_g$$

Finally, all the saturated - mixture states are located under the saturation curve, and to analyze saturated mixtures, all we need are saturated liquid and saturated vapor data (Tables A - 4 and A - 5 in the case of water).

EXAMPLE 5 - 3 Pressure and Volume of a Saturated Mixture

A rigid tank contains 10 kg of water at 90 ℃. If 8 kg of the water is in the liquid form and the rest is in the vapor form, determine (a) the pressure in the tank and (b) the volume of the tank.

Solution A rigid tank contains saturated mixture. The pressure and the volume of the tank are to be determined.

Analysis (a) The state of the saturated liquid - vapor mixture is shown in Fig. 5 - 27. Since the two phases coexist in equilibrium, we have a saturated mixture, and the pressure must be the saturation pressure at the given temperature:

$$P = P_{sat\,@\,90℃} = 70.183 \text{ kPa} \qquad (\text{Table A} - 4)$$

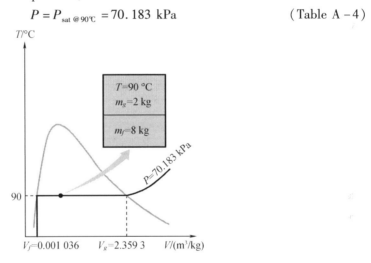

Fig. 5 - 27 Schematic and $T - v$ diagram for Example 5.3

(b) At 90 ℃, we have vf = 0.001 036 m³/kg and vg = 2.3593 m³/kg (Table A - 4). One way of finding the volume of the tank is to determine the volume occupied by each phase and then add them:

$$V = V_f + V_g = mv_f + mv_g$$
$$= (8 \text{ kg})(0.001\,036 \text{ m}^3/\text{kg}) + (2 \text{ kg})(2.359\,3 \text{ m}^3/\text{kg})$$
$$= 4.73 \text{ m}^3$$

Another way is to first determine the quality x, then the average specific volume v, and finally the total volume:

$$x = \frac{m_g}{m_t} = \frac{2 \text{ kg}}{10 \text{ kg}} = 0.2$$

$$v = v_f + xv_{fg}$$
$$= 0.001\,036 \text{ m}^3/\text{kg} + (2)[(2.359\,3 - 0.001\,036) \text{ m}^3/\text{kg}]$$
$$= 0.473 \text{ m}^3/\text{kg}$$

and
$$V = mv = (10 \text{ kg})(0.473 \text{ m}^3/\text{kg}) = 4.73 \text{ m}^3$$

Discussion The first method appears to be easier in this case since the masses of each phase are given. In most cases, however, the masses of each phase are not available, and the second method becomes more convenient.

Property tables are also available for saturated solid – vapor mixtures. Properties of saturated ice – water vapor mixtures, are listed in Table A – 8. Saturated solid – vapor mixtures can be handled just as saturated liquid – vapor mixtures.

SUPERHEATED VAPOR

In the region to the right of the saturated vapor line and at temperatures above the critical point temperature, a substance exists as super heated vapor. Since the superheated region is a single – phase region (vapor phase only), temperature and pressure are no longer dependent properties and they can conveniently be used as the two independent properties in the tables.

In these tables, the properties are listed against temperature for selected pressures starting with the saturated vapor data. The saturation temperature is given in parentheses following the pressure value.

Compared to saturated vapor, superheated vapor is characterized by

Lower pressures ($P < P_{sat}$ at a given T)

Higher temperatures ($T > T_{sat}$ at a given P)

Higher specific volumes ($v > v_g$ at a given P or T)

Higher internal energies ($u > u_g$ at a given P or T)

Higher enthalpies ($h > h_g$ at a given P or T)

EXAMPLE 5.4 Internal Exergy of Superheated Vapor

Determine the internal exergy of water at 20 psia and 400 °F.

Solution The internal exergy of water at a specified state is to be determined.

Analysis At 20 psia, the saturation temperature is 227.92 °F. Since $T > T_{sat}$, the water is in the superheated vapor region. Then the internal exergy at the given temperature and pressure is determined from the superheated vapor table (Table A – 12) to be
$$u = 1 \ 145.1 \text{ Btu/lbm}$$

COMPRESSED LIQUID

Compressed liquid tables are not as commonly available, and Table A – 7 is the only compressed liquid table in this text. The format of Table A – 7 is very much like the format of the superheated vapor tables. These states are referred to as compressed liquid states because the pressure at each state is higher than the saturation pressure corresponding to the temperature at the state. One reason for the lack of compressed liquid data is the relative independence of compressed liquid properties from pressure. Variation of properties of compressed liquid with pressure is very mild. Increasing the pressure 100 times often causes properties to change less than 1 percent.

Chapter 5 Properties of Real Gas and Vapor

In the absence of compressed liquid data, a general approximation is *to treat compressed liquid as saturated liquid at the given temperature*. This is because the compressed liquid properties depend on temperature much more strongly than they do on pressure. Thus,

$$y \cong y_{f@T} \tag{5-13}$$

for compressed liquids, where y is v, u, or h. Of these three properties, the property whose value is most sensitive to variations in the pressure is the enthalpy h. Although the above approximation results in negligible error in v and u, the error in h may reach undesirable levels. However, the error in h at low to moderate pressures and temperatures can be reduced significantly by evaluating it from

$$h \cong h_{f@T} + v_{f@T}(P - P_{sat@T}) \tag{5-14}$$

instead of taking it to be just h_f. Note, however, that the approximation in Eq. (5-14) does not yield any significant improvement at moderate to high temperatures and pressures, and it may even backfire and result in greater error due to over correction at very high temperatures and pressures (*see* Kostic, Ref. 1).

In general, a compressed liquid is characterized by

Higher pressures ($P > P_{sat}$ at a given T)

Lower temperatures ($T < T_{sat}$ at a given P)

Lower specific volumes ($v < v_f$ at a given P or T)

Lower internal energies ($u < u_f$ at a given P or T)

Lower enthalpies ($h < h_f$ at a given P or T)

But unlike superheated vapor, the compressed liquid properties are not much different from the corresponding saturated liquid values.

EXAMPLE 5-4 The Use of Steam Tables to Determine Properties

Determine the missing properties and the phase descriptions in the following table for water:

	$T/℃$	P/kPa	$u/(kJ/kg)$	x	Phase description
(a)		200		0.6	
(b)	125		1 600		
(c)		1 000	2 950		
(d)	75	500			
(e)		850		0.0	

Solution: Properties and phase descriptions of water are to be determined at various states.

Analysis: (a) The quality is given to be $x = 0.6$, which implies that 60 percent of the mass is in the vapor phase and the remaining 40 percent is in the liquid phase. Therefore, we have saturated liquid - vapor mixture at a pressure of 200 kPa. Then the temperature must be the saturation temperature at the given pressure:

$$T = T_{sat@200kPa} = 120.21 ℃ \quad \text{(Table A-5)}$$

At 200 kPa, we also read from Table A-5 that $u_f = 504.50$ kJ/kg and $u_{fg} = 2\,024.6$ kJ/kg.

Then the average internal exergy of the mixture is

$$u = u_f + xu_{fg}$$
$$= 504.5 \text{ kJ/kg} + (0.6)(2\ 024.6 \text{ kJ/kg})$$
$$= 1\ 719.26 \text{ kJ/kg}$$

(b) This time the temperature and the internal exergy are given, but we do not know which table to use to determine the missing properties because we have no clue as to whether we have saturated mixture, compressed liquid, or superheated vapor. To determine the region we are in, we first go to the saturation table (Table A-4) and determine the u_f and u_g values at the given temperature. At 125 ℃, we read $u_f = 524.83$ kJ/kg and $u_g = 2\ 534.3$ kJ/kg. Next we compare the given u value to these u_f and u_g values, keeping in mind that

if $\quad u < u_f \quad$ we have *compressed liquid*

if $\quad u_f \leqslant u \leqslant u_g \quad$ we have *saturated mixture*

if $\quad u > u_g \quad$ we have *superheated vapor*

In our case the given u value is 1 600, which falls between the u_f and u_g values at 125 ℃. Therefore, we have saturated liquid – vapor mixture. Then the pressure must be the saturation pressure at the given temperature:

$$P = P_{\text{sat @ 125 ℃}} = 232.23 \text{ kPa} \quad\quad\quad\quad (\text{Table A}-4)$$

The quality is determined from

$$x = \frac{u - u_f}{u_{fg}} = \frac{1\ 600 - 524.83}{2\ 009.5} = 0.535$$

The criteria above for determining whether we have compressed liquid, saturated mixture, or superheated vapor can also be used when enthalpy h or specific volume v is given instead of internal exergy u, or when pressure is given instead of temperature.

(c) This is similar to case (b), except pressure is given instead of temperature. Following the argument given above, we read the u_f and u_g values at the specified pressure. At 1 MPa, we have $u_f = 761.39$ kJ/kg and $u_g = 2\ 582.8$ kJ/kg. The specified u value is 2 950 kJ/kg, which is greater than the u_g value at 1 MPa. Therefore, we have superheated vapor, and the temperature at this state is determined from the superheated vapor table by interpolation to be

$$T = 395.2 \text{ ℃} \quad\quad\quad\quad (\text{Table A}-6)$$

We would leave the quality column blank in this case since quality has no meaning for a superheated vapor.

(d) In this case the temperature and pressure are given, but again we cannot tell which table to use to determine the missing properties because we do not know whether we have saturated mixture, compressed liquid, or superheated vapor. To determine the region we are in, we go to the saturation table (Table A-5) and determine the saturation temperature value at the given pressure. At 500 kPa, we have $T_{\text{sat}} = 151.83$ ℃. We then compare the given T value to this T_{sat} value, keeping in mind that

if $\quad T < T_{\text{sat @ given P}} \quad$ we have *compressed liquid*

if $\quad T = T_{\text{sat @ given P}} \quad$ we have *saturated mixture*

if $\quad T > T_{\text{sat @ given P}} \quad$ we have *superheated vapor*

Chapter 5 Properties of Real Gas and Vapor

In our case, the given T value is 75 ℃, which is less than the T_{sat} value at the specified pressure. Therefore, we have compressed liquid (Fig. 5 – 28), and normally we would determine the internal exergy value from the compressed liquid table. But in this case the given pressure is much lower than the lowest pressure value in the compressed liquid table (which is 5 MPa), and therefore we are justified to treat the compressed liquid as saturated liquid at the given temperature (*not* pressure):

$$u \cong u_{f\,@\,75\,℃} = 313.99 \text{ kJ/kg} \qquad (\text{Table A} - 4)$$

We would leave the quality column blank in this case since quality has no meaning in the compressed liquid region.

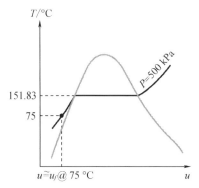

Fig. 5 – 28 At a given P and T, a pure substance will exist as a compressed liquid if $T < T_{sat}\,@\,P$.

(e) The quality is given to be $x = 0$, and thus we have saturated liquid at the specified pressure of 850 kPa. Then the temperature must be the saturation temperature at the given pressure, and the internal exergy must have the saturated liquid value:

$$T = T_{sat\,@\,850\,kPa} = 172.94 \text{ ℃}$$
$$u = u_{f\,@\,850\,kPa} = 731 \text{ kJ/kg} \qquad (\text{Table A} - 5)$$

ENTROPY CHANGE OF PURE SUBSTANCE

A pure substance is one that is uniform and invariable in chemical composition. A pure substance can exist in more than one phase, but its chemical composition must be the same in each phase.

Entropy is a property, and thus the value of entropy of a system is fixed once the state of the system is fixed. Specifying two intensive independent properties fixes the state of a simple compressible system, and thus the value of entropy, as well as the values of other properties at that state. Starting with its defining relation, the entropy change of a substance can be expressed in terms of other properties. But in general, these relations are too complicated and are not practical to use for hand calculations. Therefore, using a suitable reference state, the entropies of substances are evaluated from measurable property data following rather involved computations, and the results are tabulated in the same manner as the other properties such as v, u, and h.

The value of entropy at a specified state is determined just like any other property. In the compressed liquid and superheated vapor regions, it can be obtained directly from the tables at the specified state. In the saturated mixture region, it is determined from

$$s = s_f + x s_{fg} \quad (\text{kJ/kg} \cdot \text{K})$$

where x is the quality and sf and sfg values are listed in the saturation tables. In the absence of compressed liquid data, the entropy of the compressed liquid can be approximated by the entropy of the saturated liquid at the given temperature:

$$s_{@\ T,P} \cong s_{f@\ T} \quad (\text{kJ/kg} \cdot \text{K})$$

The entropy change of a specified mass m (a closed system) during a process is simply

$$\Delta S = m \Delta s = m(s_2 - s_1) \quad (\text{kJ/K}) \quad\quad (5-15)$$

which is the difference between the entropy values at the final and initial states.

When studying the second-law aspects of processes, entropy is commonly used as a coordinate on diagrams such as the $T - s$ and $h - s$ diagrams. The general characteristics of the $T - s$ diagram of pure substances are shown in Fig. 5 – 29 using data for water. Notice from this diagram that the constant volume lines are steeper than the constant – pressure lines and the constant pressure lines are parallel to the constant – temperature lines in the saturatedliquid – vapor mixture region. Also, the constant – pressure lines almost coincide with the saturated liquid line in the compressed liquid region.

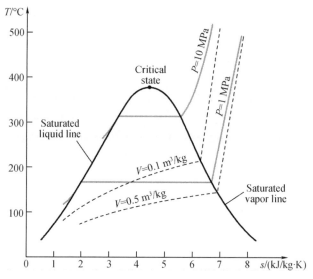

Fig. 5 – 29 Schematic of the $T - s$ diagram for water

5 – 4 Properties of Refrigerant

From about 1940 to the early 1990s, the most common class of refrigerants used in vapor compression.

refrigeration systems was the chlorine – containing CFCs (chlorofluorocarbons). Refrigerant 12 (CCl_2F_2) is one of these. Owing to concern about the effects of chlorine in refrigerants on the earth's protective ozone layer, international agreements have been implemented to phase out the use of CFCs. Classes of refrigerants containing various amounts of hydrogen in place of chlorine atoms have been developed that have less potential to deplete atmospheric ozone than do more fully chlorinated ones, such as Refrigerant 12. One such class, the HFCs, contain no chlorine. Refrigerant 134a (CF_3CH_2F) is the HFC considered by many to be an environmentally acceptable substitute for Refrigerant 12, and Refrigerant 134ahas replaced Refrigerant 12 in many applications.

Refrigerant 22 ($CHClF_2$) is in the class called HCFCs that contains some hydrogen in place of the chlorine atoms. Although Refrigerant 22 and other HCFCs are widely used today, discussions are under way that will likely result in phasing out their use at some time in the future. Ammonia (NH_3), which was widely used in the early development of vapor compression refrigeration, is again receiving some interest as an alternative to the CFCsbecause it contains no chlorine. Ammonia is also important in the absorption refrigerationsystems discussed in Section 10.5. Hydrocarbons such as propane (C_3H_8) and methane (CH_4) are also under investigation for use as refrigerants.

Thermodynamic property data for ammonia, propane, and Refrigerants 22 and 134a are included in the appendix tables. These data allow us to study refrigeration and heat pump systems in common use and to investigate some of the effects on refrigeration cycles of using alternative working fluids.

The industrial and heavy-commercial sectors were very satisfied with *ammonia*, and still are, although ammonia is toxic. The advantages of ammonia over other refrigerants are its low cost, higher COPs (and thus lower exergy cost), more favorable thermodynamic and transport properties and thus higher heat transfer coefficients (requires smaller and lower-cost heat exchangers), greater detect ability in the event of a leak, and no effect on the ozone layer. The major drawback of ammonia is its toxicity, which makes it unsuitable for domestic use. Ammonia is predominantly used in food refrigeration facilities such as the cooling of fresh fruits, vegetables, meat, and fish; refrigeration of beverages and dairy products such as beer, wine, milk, and cheese; freezing of ice cream and other foods; ice production; and low-temperature refrigeration in the pharmaceutical and other process industries.

R – 11 is used primarily in large-capacity water chillers serving air conditioning systems in buildings. R – 12 is used in domestic refrigerators and freezers, as well as automotive air conditioners. R – 22 is used in window air conditioners, heat pumps, air conditioners of commercial buildings, and large industrial refrigeration systems, and offers strong competition to ammonia. R – 502 (a blend of R – 115 and R – 22) is the dominant refrigerant used in commercial refrigeration systems such as those in supermarkets because it allows low temperatures at evaporators while operating at single stage compression.

Two important parameters that need to be considered in the selection of a refrigerant are the temperatures of the two media (the refrigerated space and the environment) with which the refrigerant exchanges heat.

The temperatures of the refrigerant in the evaporator and condenser are governed by the temperatures of the cold and warm regions, respectively, with which the system interacts thermally. This, in turn, determines the operating pressures in the evaporator and condenser. Consequently, the selection of a refrigerant is based partly on the suitability of its pressure – temperature relationship in the range of the particular application. It is generally desirable to avoid excessively low pressures in the evaporator and excessively high pressures in the condenser. Other considerations in refrigerant selection include chemical stability, toxicity, corrosiveness, and cost. The type of compressor also affects the choice of refrigerant. Centrifugal compressors are best suited for low evaporator pressures and refrigerants with large specific volumes at low pressure. Reciprocating compressors perform better over large pressure ranges and are better able to handle low specificvolume refrigerants.

Other desirable characteristics of a refrigerant include being nontoxic, noncorrosive, nonflammable, and chemically stable; having a high enthalpy of vaporization (minimizes the mass flow rate); and, of course, being available at low cost. In the case of heat pumps, the minimum temperature (and pressure) for the refrigerant may be considerably higher since heat is usually extracted from media that are well above the temperatures encountered in refrigeration systems.

Summary

Real gases exhibit ideal-gas behavior at relatively low pressures and high temperatures. The deviation from ideal-gas behavior can be properly accounted for by using the *compressibility factor* Z, The Z factor is approximately the same for all gases at the same *reduced temperature* and *reduced pressure*. We introduced and discussed several important property diagrams and tables. Phase-change processes of pure substance and saturated properties are also explained, including saturation temperature and saturation pressure. At last, how to choose the right refrigerant is analyzed.

Problems

5 – 1 A mass of 200 g of saturated liquid water is completely vaporized at a constant pressure of 100 kPa. Determine (a) the volume change and (b) the amount of exergy transferred to the water.

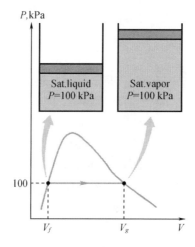

Fig. P5 – 1

5 – 2 An 80 L vessel contains 4 kg of refrigerant – 134a at a pressure of 160 kPa. Determine (a) the temperature, (b) the quality, (c) the enthalpy of the refrigerant, and (d) the volume occupied by the vapor phase.

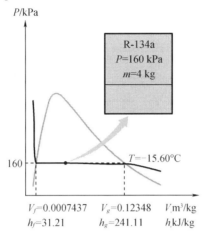

Fig. P5 – 2

5 – 3 Determine the temperature of water at a state of $P = 0.5$ MPa and $h = 2\,890$ kJ/kg.

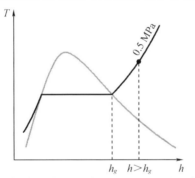

Fig. P5 – 3

5 – 4 A rigid container that is filled with R – 134a is heated. The initial pressure and final pressure are 300 kPa and 600 kPa, respectively. The temperature and total enthalpy are to be determined at the initial and final states.

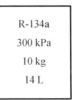

Fig. P5 – 4

Chapter 6　Atmospheric Air

At temperatures below the critical temperature, the gas phase of a substance is frequently referred to as a *vapor*. The term *vapor* implies a gaseous state that is close to the saturation region of the substance, raising the possibility of condensation during a process. We discussed mixtures of gases that are usually above their critical temperatures. Therefore, we were not concerned about any of the gases condensing during a process. Not having to deal with two phases greatly simplified the analysis. When we are dealing with a gas-vapor mixture, however, the vapor may condense out of the mixture during a process, forming a two-phase mixture. This may complicate the analysis considerably. Therefore, a gas-vapor mixture needs to be treated differently from an ordinary gas mixture. Several gas-vapor mixtures are encountered in engineering. In this chapter, we consider the *atmospheric air*, which is the most commonly encountered gas-vapor mixture in practice.

6 – 1　Atmospheric Air

Air is a mixture of nitrogen, oxygen, and small amounts of some other gases. Air in the atmosphere normally contains some water vapor (or *moisture*) and is referred to as atmospheric air. By contrast, air that contains no water vapor is called dry air. It is often convenient to treat air as a mixture of water vapor and dry air since the composition of dry air remains relatively constant, but the amount of water vapor changes as a result of condensation and evaporation from oceans, lakes, rivers, showers, and even the human body. Although the amount of water vapor in the air is small, it plays a major role in human comfort. Therefore, it is an important consideration in air-conditioning applications.

Taking 0 ℃ as the reference temperature, the enthalpy and enthalpy change of dry air can be determined from

$$h_{\text{dry air}} = c_p T = (1.005 \text{ kJ/kg} \cdot \text{℃}) T \quad (\text{kJ/kg}) \qquad (6-1)$$

and

$$\Delta h_{\text{dry air}} = c_p \Delta T = (1.005 \text{ kJ/kg} \cdot \text{℃}) \Delta T \quad (\text{kJ/kg}) \qquad (6-2)$$

where T is the air temperature in ℃ and T is the change in temperature. In air-conditioning processes we are concerned with the *changes* in enthalpy h, which is independent of the reference point selected.

It certainly would be very convenient to also treat the water vapor in the air as an ideal gas and you would probably be willing to sacrifice some accuracy for such convenience. Well, it turns out that we can have the convenience without much sacrifice. At 50 ℃, the saturation pressure of water is 12.3 kPa. At pressures below this value, water vapor can be treated as an ideal gas with negligible error (under 0.2 percent), even when it is a saturated vapor. Therefore, water vapor in air behaves as if it existed alone and obeys the ideal-gas relation $Pv = RT$. Then the atmospheric air can be treated as an ideal-gas mixture whose pressure is the sum of the partial pressure of dry air. Throughout this chapter, the subscript a and v denotes dry air and the water vapor. respectively P_a and that of water vapor P_v:

$$P = P_a + P_v \quad (\text{kPa}) \tag{6-3}$$

The partial pressure of water vapor is usually referred to as the vapor pressure. It is the pressure water vapor would exert if it existed alone at the temperature and volume of atmospheric air.

Since water vapor is an ideal gas, the enthalpy of water vapor is a function of temperature only, that is, $h = h(T)$. This can also be observed from the $T-s$ diagram of water given in Fig. B-2 and Fig. 6-1 where the constant enthalpy lines coincide with constant-temperature lines at temperatures below 50 ℃. Therefore, *the enthalpy of water vapor in air can be taken to be equal to the enthalpy of saturated vapor at the same temperature.* That is,

$$h_v(T, \text{low } P) \cong h_g(T) \tag{6-4}$$

The enthalpy of water vapor at 0 ℃ is 2 500.9 kJ/kg. The average c_p value of water vapor in the temperature range $-10 \sim 50$ ℃ can be taken to be 1.82 kJ/kg · ℃. Then the enthalpy of water vapor can be determined approximately from

$$h_g(T) \cong 2500.9 + 1.82T \quad (\text{kJ/kg}) \quad T \text{ in } ℃ \tag{6-5}$$

or

$$h_g(T) \cong 1060.9 + 0.435T \quad (\text{Btu/lbm}) \quad T \text{ in } ℉ \tag{6-6}$$

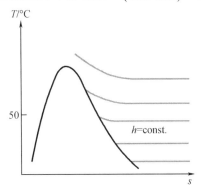

Fig. 6-1 At temperatures below 50 ℃, the h = constant lines coincide with the T = constant lines in the superheated vapor region of water

6-2 Parameters of Atmospheric Air

The amount of water vapor in the air can be specified in various ways. Probably the most logical way is to specify directly the mass of water vapor present in a unit mass of dry air. This is called absolute or specific humidity (also called *humidity ratio*) and is denoted by ω:

$$\omega = \frac{m_v}{m_a} \quad \text{(kg water vapor/kg dry air)} \tag{6-7}$$

The specific humidity can also be expressed as

$$\omega = \frac{m_v}{m_a} = \frac{P_v V/R_v T}{P_a V/R_a T} = \frac{P_v/R_v}{P_a/R_a} = 0.622 \frac{P_v}{P_a} \tag{6-8}$$

or

$$\omega = 0.622 \frac{P_v}{P - P_v} \quad \text{(kg water vapor/kg dry air)} \tag{6-9}$$

where P is the total pressure.

Consider 1 kg of dry air. By definition, dry air contains no water vapor, and thus its specific humidity is zero. If we add some water vapor to this dry air. the specific humidity will increase. As more vapor or moisture is added, the specific humidity will keep increasing until the air can hold no more moisture. At this point, the air is said to be saturated with moisture, and it is called saturated air. Any moisture introduced into saturated air will condense. The amount of water vapor in saturated air at a specified temperature and pressure can be determined from Eq. (6-9) by replacing P_v by P_g, the saturation pressure of water at that temperature.

The amount of moisture in the air has a definite effect on how comfortable we feel in an environment. However, the comfort level depends more on the amount of moisture the air holds (m_v) relative to the maximum amount of moisture the air can hold at the same temperature (m_g). The ratio of these two quantities is called the relative humidity φ

$$\varphi = \frac{m_v}{m_g} = \frac{P_v V/R_v T}{P_g V/R_v T} = \frac{P_v}{P_g} \tag{6-10}$$

where

$$P_g = P_{sat\,@\,T} \tag{6-11}$$

Combining Eqs. (6-9) and (6-10), we can also express the relative humidity as

$$\varphi = \frac{\omega P}{(0.622 + \omega) P_g} \quad \text{and} \quad \omega = \frac{0.622 \varphi P_g}{P - \varphi P_g} \tag{6-12a,b}$$

The relative humidity ranges from 0 for dry air to 1 for saturated air. Note that the amount of moisture air can hold depends on its temperature. Therefore, the relative humidity of air changes with temperature even when its specific humidity remains constant.

Atmospheric air is a mixture of dry air and water vapor, and thus the enthalpy of air is expressed in terms of the enthalpies of the dry air and the water vapor. In most practical applications, the amount of dry air in the air-water-vapor mixture remains constant, but the amount of water vapor changes. Therefore, the enthalpy of atmospheric air is expressed *per unit mass of dry air* instead of per unit mass of the air – water vapor mixture.

The total enthalpy (an extensive property) of atmospheric air is the sum of the enthalpies of dry air and the water vapor:

$$H = H_a + H_v = m_a h_a + m_v h_v$$

Dividing by m_a gives

$$h = \frac{H}{m_a} = h_a + \frac{m_v}{m_a} h_v = h_a + \omega h_v$$

or
$$h = h_a + \omega h_g \quad (\text{kJ/kg dry air}) \qquad (6-13)$$

since $h_v \cong h_g$.

The term *dry-bulb temperature* refers simply to the temperature that would be measured by a thermometer placed in the mixture.

If you live in a humid area, you are probably used to waking up most summer mornings and finding the grass wet. You know it did not rain the night before. So what happened? Well, the excess moisture in the air simply condensed on the cool surfaces, forming what we call *dew*. In summer, a considerable amount of water vaporizes during the day. As the temperature falls during the night, so does the "moisture capacity" of air, which is the maximum amount of moisture air can hold. (What happens to the relative humidity during this process?) After a while, the moisture capacity of air equals its moisture content. At this point, air is saturated, and its relative humidity is 100 percent. Any further drop in temperature results in the condensation of some of the moisture, and this is the beginning of dew formation.

The dew-point temperature T_{dp} is defined as *the temperature at which condensation begins when the air is cooled at constant pressure*. In other words, T_{dp} is the saturation temperature of water corresponding to the vapor pressure:

$$T_{dp} = T_{sat \, @ \, P_v} \qquad (6-14)$$

This is also illustrated in Fig. 6-2. As the air cools at constant pressure, the vapor pressure P_v remains constant. Therefore, the vapor in the air (state 1) undergoes a constant-pressure cooling process until it strikes the saturated vapor line (state 2). The temperature at this point is T_{dp}, and if the temperature drops any further, some vapor condenses out. As a result, the amount of vapor in the air decreases, which results in a decrease in P_v. The air remains saturated during the condensation process and thus follows a path of 100 percent relative humidity (the saturated vapor line). The ordinary temperature and the dew-point temperature of saturated air are identical.

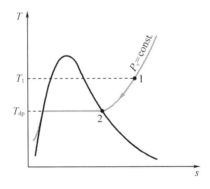

Fig. 6-2 Constant-pressure cooling of moist air and the dew-point temperature on the $T-s$ diagram of water

You have probably noticed that when you buy a cold canned drink from a vending machine on a hot and humid day, dew forms on the can. The formation of dew on the can indicates that the temperature of the drink is below the dew – point temperature of the surrounding air. The dew – point temperature of room air can be determined easily by cooling some water in a metal cup by adding small amounts of ice and stirring. The temperature of the outer surface of the cup when dew starts to form on the surface is the dew – point temperature of the air.

EXAMPLE 6 – 1 Fogging of the Windows in a House

In cold weather, condensation frequently occurs on the inner surfaces of the windows due to the lower air temperatures near the window surface. Consider a house, shown in Fig. 6 – 3, that contains air at 20 ℃ and 75 percent relative humidity. At what window temperature will the moisture in the air start condensing on the inner surfaces of the windows?

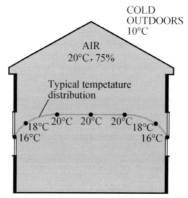

Fig. 6 – 3   Schematic for Example 6.1

Solution The interior of a house is maintained at a specified temperature and humidity. The window temperature at which fogging starts is to be determined.

Properties The saturation pressure of water at 20 ℃ is $P_{sat} = 2.339\ 2$ kPa (Table A – 4).

Analysis The temperature distribution in a house, in general, is not uniform. When the outdoor temperature drops in winter, so does the indoor temperature near the walls and the windows. Therefore, the air near the walls and the windows remains at a lower temperature than at the inner parts of a house even though the total pressure and the vapor pressure remain constant throughout the house. As a result, the air near the walls and the windows undergoes a $P_v =$ *constant* cooling process until the moisture in the air starts condensing. This happens when the air reaches its dew – point temperature T_{dp}, which is determined from Eq. (6 – 14) to be

$$T_{dp} = T_{sat\ @\ P_v}$$

where

$$P_v = \varphi P_{sat\ @\ 20℃} = (0.75)(2.339\ 2\ \text{kPa}) = 1.754\ \text{kPa}$$

Thus,

$$T_{dp} = T_{sat\ @\ 1.754\ \text{kPa}} = 15.4\ ℃$$

Discussion Note that the inner surface of the window should be maintained above 15.4 ℃ if condensation on the window surfaces is to be avoided.

Relative humidity and specific humidity are frequently used in engineering and atmospheric sciences, and it is desirable to relate them to easily measurable quantities such as temperature and pressure. One way of determining the relative humidity is to determine the dew-point temperature of air, as discussed in the last section. Knowing the dew-point temperature, we can determine the vapor pressure P_v and thus the relative humidity. However, this approach is simple, but not quite practical.

Another way of determining the absolute or relative humidity is related to an *adiabatic saturation process*, shown schematically and on a $T-s$ diagram in Fig. 6-4. The system consists of a long insulated channel that contains a pool of water. A steady stream of unsaturated air that has a specific humidity of ω_1 (unknown) and a temperature of T_1 is passed through this channel. As the air flows over the water, some water evaporates and mixes with the airstream. The moisture content of air increases during this process, and its temperature decreases, since part of the latent heat of vaporization of the water that evaporates comes from the air. If the channel is long enough, the air stream exits as saturated air ($\phi = 100$ percent) at temperature T_2, which is called the adiabatic saturation temperature (can also called wet-bulb temperature).

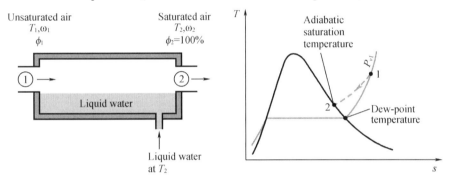

Fig. 6-4 The adiabatic saturation process and its representation on a $T-s$ diagram of water

If makeup water is supplied to the channel at the rate of evaporation at temperature T_2, the adiabatic saturation process described above can be analyzed as a steady-flow process. The process involves no heat or work interactions, and the kinetic and potential exergy changes can be neglected. Then the conservation of mass and conservation of exergy relations for this two inlet, one-exit steady-flow system reduces to the following:

Mass balance:

$$\dot{m}_{a_1} = \dot{m}_{a_2} = \dot{m}_a \begin{pmatrix} \text{The mass flow rate of dry air} \\ \text{remains constant} \end{pmatrix}$$

$$\dot{m}_{w_1} + \dot{m}_f = \dot{m}_{w_2} \begin{pmatrix} \text{The mass flow rate of vapor in the} \\ \text{air increases by an amount equal to} \\ \text{the rate of evaporation } \dot{m}_f \end{pmatrix}$$

or

$$\dot{m}_a \omega_1 + \dot{m}_f = \dot{m}_a \omega_2$$

Thus,

$$\dot{m}_f = \dot{m}_a (\omega_2 - \omega_1)$$

Exergy balance:

$$\dot{E}_{in} = \dot{E}_{out} \quad (\text{since } \dot{Q} = 0 \quad \text{and} \quad \dot{W} = 0)$$

$$\dot{m}_a h_1 + \dot{m}_a(\omega_2 - \omega_1) h_{f2} = \dot{m}_a h_2$$

or

$$\dot{m}_a h_1 + \dot{m}_a(\omega_2 - \omega_1) h_{f2} = \dot{m}_a h_2$$

Dividing by $\dot{m}_a$ gives

$$h_1 + (\omega_2 - \omega_1) h_{f2} = h_2$$

or

$$(c_p T_1 + \omega_1 h_{g_1}) + (\omega_2 - \omega_1) h_{f2} = (c_p T_2 + \omega_2 h_{g_2})$$

which yields

$$\omega_1 = \frac{c_p(T_2 - T_1) + \omega_2 h_{fg_2}}{h_{g_1} - h_{f2}} \qquad (6-15)$$

where, from Eq. (6 – 12b),

$$\omega_2 = \frac{0.622 P_{g_2}}{P_2 - P_{g_2}} \qquad (6-16)$$

since $\omega_2 = 100$ percent. Thus we conclude that the specific humidity (and relative humidity) of air can be determined from Eqs. (6 – 15) and (6 – 16) by measuring the pressure and temperature of air at the inlet and the exit of an adiabatic saturator.

If the air entering the channel is already saturated, then the adiabatic saturation temperature T_2 will be identical to the inlet temperature T_1, in which case Eq. (6 – 15) yields $\omega_1 = \omega_2$. In general, the adiabatic saturation temperature is between the inlet and dew – point temperatures.

The adiabatic saturation process discussed above provides a means of determining the absolute or relative humidity of air, but it requires a long channel or a spray mechanism to achieve saturation conditions at the exit. A more practical approach is to use a thermometer whose bulb is covered with a cotton wick saturated with water and to blow air over the wick, as shown in Fig. (6 – 5). The wet – bulb *temperature* T_{wb} is read from a wet – bulb thermometer, which is an ordinary liquid – in – glass thermometer whose bulb is enclosed by a wick moistened with water.

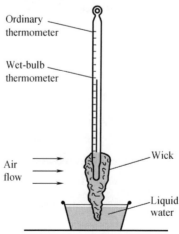

Fig. 6 – 5 A simple arrangement to measure the wet bulb temperature

The basic principle involved is similar to that in adiabatic saturation. When unsaturated air passes over the wet wick, some of the water in the wick evaporates. Therefore, the temperature of the water drops, creating a temperature difference between the air and the water. The temperature difference can be treated as the driving force for heat transfer After a while, the heat loss from the water by evaporation equals the heat gain from the air, and the water temperature stabilizes. The thermometer reading at this point is the wet-bulb temperature. The wet bulb temperature can also be measured by placing the wet-wicked thermometer in a holder attached to a handle and rotating the holder rapidly, that is, by moving the thermometer instead of the air. A device that works on this principle is called a *sling psychrometer* which is shown in Fig. 6 – 6. Usually a dry-bulb thermometer is also mounted on the frame of this device so that both the wet – and dry-bulb temperatures can be read simultaneously.

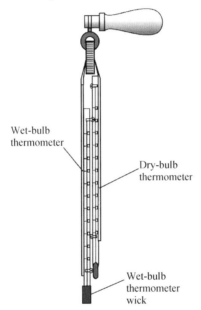

Fig. 6 – 6　Sling psychrometer

Advances in electronics made it possible to measure humidity directly in a fast and reliable way. It appears that sling psychrometer and wet-wicked thermometers are about to become things of the past. Today, hand-held electronic humidity measurement devices based on the capacitance change in a thin polymer film as it absorbs water vapor are capable of sensing and digitally displaying the relative humidity within 1 percent accuracy in a matter of seconds.

In general, the adiabatic saturation temperature and the wet-bulb temperature are not the same. However, for air – water vapor mixtures at atmospheric pressure, the wet-bulb temperature happens to be approximately equal to the adiabatic saturation temperature. Therefore, the wet – bulb temperature T_{wb} can be used in Eq. (6 – 15) in place of T_2 to determine the specific humidity of air.

6-3 The Psychrometric Chart

The state of the atmospheric air at a specified pressure is completely specified by two independent intensive properties. The rest of the properties can be calculated easily from the previous relations. The sizing of a typical airconditioning system involves numerous such calculations, which may eventually get on the nerves of even the most patient engineers. Therefore, there is clear motivation to computerize calculations or to do these calculations once and to present the data in the form of easily readable charts. Graphical representations of several important properties of moist air are provided by psychrometric charts. This chart is constructed for a mixture pressure of 1 atm, but charts for other mixture pressures are also available. A psychrometric chart for a pressure of 1 atm (101.325 kPa or 14.696 psia) is given in Fig. A-16 in SI units and in Fig. A-17 in English units. Psychrometric charts at other pressures (for use at considerably higher elevations than sea level) are also available.

The basic features of the psychrometric chart are illustrated in Fig. 6-7. The dry-bulb temperatures are shown on the horizontal axis, and the specific humidity is shown on the vertical axis. (Some charts also show the vapor pressure on the vertical axis since at a fixed total pressure P there is a one-to-one correspondence between the specific humidity v and the vapor pressure P_v, as can be seen from Eq. (6-9).) On the left end of the chart, there is a curve (called the *saturation line*) instead of a straight line. All the saturated air states are located on this curve. Therefore, it is also the curve of 100 percent relative humidity. Other constant relative-humidity curves have the same general shape.

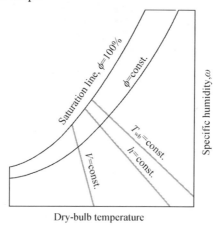

Fig. 6-7 **Schematic for a psychrometric chart**

Lines of constant wet-bulb temperature have a downhill appearance to the right. Lines of constant specific volume (in m^3/kg dry air) look similar, except they are steeper. Lines of constant enthalpy (in kJ/kg dry air) lie very nearly parallel to the lines of constant wet-bulb temperature. Therefore, the constantwet-bulb-temperature lines are used as constant-enthalpy

Chapter 6 Atmospheric Air

lines in some charts.

For saturated air, the dry-bulb, wet-bulb, and dew-point temperatures are identical. Therefore, the dew-point temperature of atmospheric air at any point on the chart can be determined by drawing a horizontal line (a line of ω = constant or P_v = constant) from the point to the saturated curve.

The temperature value at the intersection point is the dew – point temperature. The psychrometric chart also serves as a valuable aid in visualizing the air-conditioning processes. An ordinary heating or cooling process, for example, appears as a horizontal line on this chart if no humidification or dehumidification is involved (that is, ω = constant). Any deviation from a horizontal line indicates that moisture is added or removed from the air during the process.

EXAMPLE 6 – 2 The Use of the Psychrometric Chart

Consider a room that contains air at 1 atm, 35 ℃, and 40 percent relative humidity. Using the psychrometric chart, determine (a) the specific humidity, (b) the enthalpy, (c) the wet – bulb temperature, (d) the dew – point temperature, and (e) the specific volume of the air.

Solution The relative humidity of air in a room is given. The specific humidity, enthalpy, wet-bulb temperature, dew – point temperature, and specific volume of the air are to be determined using the psychrometric chart.

Analysis At a given total pressure, the state of atmospheric air is completely specified by two independent properties such as the dry-bulb temperature and the relative humidity. Other properties are determined by directly reading their values at the specified state.

(a) The specific humidity is determined by drawing a horizontal line from the specified state to the right until it intersects with the ω axis, as shown in Fig. 6 – 8. At the intersection point we read

$$\omega = 0.014\ 2 \text{ kg H}_2\text{O/kg dry air}$$

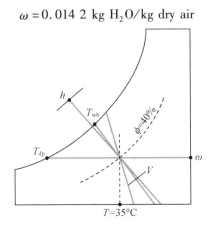

Fig. 6 – 8 Schematic for Example 6 – 2

(b) The enthalpy of air per unit mass of dry air is determined by drawing a line parallel to the h = constant lines from the specific state until it intersects the enthalpy scale, giving

$$h = 71.5 \text{ kJ/kg dry air}$$

(c) The wet-bulb temperature is determined by drawing a line parallel to the T_{wb} = constant

lines from the specified state until it intersects the saturation line, giving
$$T_{wb} = 24 \ °C$$
(d) The dew-point temperature is determined by drawing a horizontal line from the specified state to the left until it intersects the saturation line, giving
$$T_{dp} = 19.4 \ °C$$
(e) The specific volume per unit mass of dry air is determined by noting the distances between the specified state and the v = constant lines on both sides of the point. The specific volume is determined by visual interpolation to be
$$v = 0.893 \ m^3/kg \ dry \ air$$

Discussion Values read from the psychrometric chart inevitably involve reading errors, and thus are of limited accuracy.

6 – 4 Thermodynamic Processes of Atmospheric Air

Maintaining a living space or an industrial facility at the desired temperature and humidity requires some processes called air-conditioning processes. These processes include *simple heating* (raising the temperature), *simple cooling* (lowering the temperature), *humidifying* (adding moisture), and *dehumidifying* (removing moisture). Sometimes two or more of these processes are needed to bring the air to a desired temperature and humidity level.

Various air-conditioning processes are illustrated on the psychrometric chart in Fig. 6 – 9. Notice that simple heating and cooling processes appear as horizontal lines on this chart since the moisture content of the air remains constant (ω = constant) during these processes. Air is commonly heated and humidified in winter and cooled and dehumidified in summer. Notice how these processes appear on the psychrometric chart.

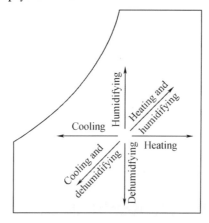

Fig. 6 – 9 Various air – conditioning processes

Most air-conditioning processes can be modeled as steady – flow processes, and thus the *mass balance* relation $\dot{m}_{in} = \dot{m}_{out}$ can be expressed for *dry air* and *water* as

Mass balance for dry air:

Chapter 6 Atmospheric Air

$$\sum \dot{m}_{a\,in} = \sum \dot{m}_{a\,out} \quad (\text{kg/s}) \qquad (6-17)$$

Mass balance for water:

$$\sum \dot{m}_{w\,in} = \sum \dot{m}_{w\,out} \quad \text{or} \quad \sum \dot{m}_a \omega_{in} = \sum \dot{m}_a \omega_{out} \qquad (6-18)$$

Disregarding the kinetic and potential exergy changes, the *steady-flow exergy balance* relation $\dot{E}_{in} = \dot{E}_{out}$ can be expressed in this case as

$$\dot{Q}_{in} + \dot{W}_{in} + \sum \dot{m} h_{in} = \dot{Q}_{out} + \dot{W}_{out} + \sum \dot{m} h_{out} \qquad (6-19)$$

The work term usually consists of the *fan work input*, which is small relative to the other terms in the exergy balance relation. Next we examine some commonly encountered processes in air-conditioning.

SIMPLE HEATING AND COOLING (ω = CONSTANT)

Many residential heating systems consist of a stove, a heat pump, or an electric resistance heater. The air in these systems is heated by circulating it through a duct that contains the tubing for the hot gases or the electric resistance wires, as shown in Fig. 6-10. The amount of moisture in the air remains constant during this process since no moisture is added to or removed from the air. That is, the specific humidity of the air remains constant (ω = constant) during a heating (or cooling) process with no humidification or dehumidification. Such a heating process proceeds in the direction of increasing dry-bulb temperature following a line of constant specific humidity on the psychrometric chart, which appears as a horizontal line.

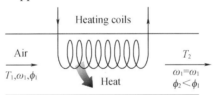

Fig. 6-10 During simple heating, specific humidity remains constant, but relative humidity decreases

Notice that the relative humidity of air decreases during a heating process even if the specific humidity v remains constant. This is because the relative humidity is the ratio of the moisture content to the moisture capacity of air at the same temperature, and moisture capacity increases with temperature. Therefore, the relative humidity of heated air may be well below comfortable levels, causing dry skin, respiratory difficulties, and an increase in static electricity.

A cooling process at constant specific humidity is similar to the heating process discussed above, except the dry bulb temperature decreases and the relative humidity increases during such a process, as shown in Fig. 6-11. Cooling can be accomplished by passing the air over some coils through which a refrigerant or chilled water flows.

The conservation of mass equations for a heating or cooling process that involves no humidification or dehumidification reduce to $\dot{m}_{a_1} = \dot{m}_{a_2} = \dot{m}_a$ for dry air and $\omega_1 = \omega_2$ for water. Neglecting any fan work that may be present, the conservation of exergy equation in this case reduces to

Thermodynamics

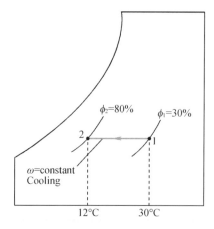

Fig. 6–11 During simple cooling, specific humidity remains constant, but relative humidity increases

$$\dot{Q} = \dot{m}_a(h_2 - h_1) \quad \text{or} \quad q = h_2 - h_1$$

where h_1 and h_2 are enthalpies per unit mass of dry air at the inlet and the exit of the heating or cooling section, respectively.

HEATING WITH HUMIDIFICATION

It is often necessary to increase the moisture content of the air circulated through occupied spaces. One way to accomplish this is to inject steam. Alternatively, liquid water can be sprayed into the air. This is accomplished by passing the air first through a heating section (process 1–2) and then through a humidifying section (process 2–3), as shown in Fig. 6–12.

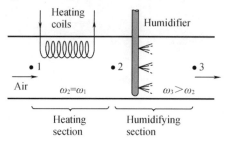

Fig. 6–12 Heating with humidification

The location of state 3 depends on how the humidification is accomplished. If steam is introduced in the humidification section, this will result in humidification with additional heating ($T_3 > T_2$). If humidification is accomplished by spraying water into the air stream instead, part of the latent heat of vaporization comes from the air, which results in the cooling of the heated airstream ($T_3 > T_2$). Air should be heated to a higher temperature in the heating section in this case to make up for the cooling effect during the humidification process.

COOLING WITH DEHUMIDITICATION

When a moist air stream is cooled at constant mixture pressure to a temperature below its dew point temperature, some condensation of the water vapor initially present would occur. Fig. 6-13 shows the schematic of a dehumidifier using this principle, and in conjunction with Example 6-3.

Hot and moist air enters the cooling section at state 1. As it passes through the cooling coils, its temperature declines and its relative humidity increases at constant specific humidity. If the cooling section is sufficiently long, air reaches its dew point (state x, saturated air). Further cooling of air results in the condensation of part of the moisture in the air. Air remains saturated during the entire condensation process, which follows a line of 100 percent relative humidity until the final state (state 2) is reached. The water vapor that condenses out of the air during this process is removed from the cooling section through a separate channel. The condensate is usually assumed to leave the cooling section at T_2.

The cool and saturated air at state 2 is usually routed directly to the room, where it mixes with the room air. In some cases, however, the air at state 2 may be at the right specific humidity but at a very low temperature. In such cases, air is passed through a heating section where its temperature is raised to a more comfortable level before it is routed to the room.

EXAMPLE 6-3 Cooling and Dehumidification of Air

Air enters a window air conditioner at 1 atm, 30 ℃, and 80 percent relative humidity at a rate of 10 m³/min, and it leaves as saturated air at 14 ℃. Part of the moisture in the air that condenses during the process is also removed at 14 ℃. Determine the rates of heat and moisture removal from the air.

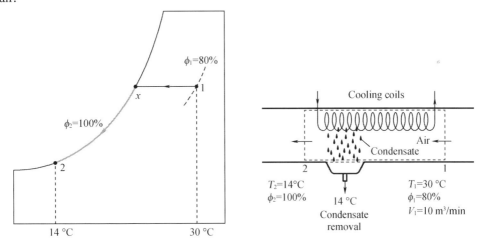

Fig. 6-13 Schematic and psychrometric chart for Example 6-3

Solution Air is cooled and dehumidified by a window air conditioner. The rates of heat and moisture removal are to be determined.

Assumptions (1) This is a steady-flow process and thus the mass flow rate of dry air remains constant during the entire process. (2) Dry air and the water vapor are ideal gases. (3) The kinetic and potential exergy changes are negligible.

Properties The enthalpy of saturated liquid water at 14 ℃ is 58.8 kJ/kg (Table A-4). Also, the inlet and the exit states of the air are completely specified, and the total pressure is 1 atm. Therefore, we can determine the properties of the air at both states from the psychrometric chart to be

$$h_1 = 85.4 \text{ kJ/kg dry air}$$
$$h_2 = 39.3 \text{ kJ/kg dry air}$$
$$\omega_1 = 0.0216 \text{ kg H}_2\text{O/kg dry air}$$
$$\omega_2 = 0.0100 \text{ kg H}_2\text{O/kg dry air}$$
$$v_1 = 0.889 \text{ m}^3/\text{kg dry air}$$

Analysis: We take the *cooling section* to be the system. The schematic of the system and the psychrometric chart of the process are shown in Fig. 6 – 13. We note that the amount of water vapor in the air decreases during the process ($\omega_2 < \omega_1$) due to dehumidification. Applying the mass and exergy balances on the cooling and dehumidification section gives

Dry air mass balance:
$$\dot{m}_{a_1} = \dot{m}_{a_2} = \dot{m}_a$$

Water mass balance:
$$\dot{m}_{a_1}\omega_1 = \dot{m}_{a_2}\omega_2 + \dot{m}_w \rightarrow \dot{m}_w = \dot{m}_a(\omega_2 - \omega_1)$$

Exergy balance:
$$\sum \dot{m}h_{in} = \sum \dot{m}h_{out} + \dot{Q}_{out} \rightarrow \dot{Q}_{out} = \dot{m}(h_1 - h_2) - \dot{m}_w h_w$$

Then,
$$\dot{m}_a = \frac{\dot{V}_1}{v_1} = \frac{10 \text{ m}^3/\text{min}}{0.889 \text{ m}^3/\text{kg dry air}} = 11.25 \text{ kg/min}$$

$$\dot{m}_w = (11.25 \text{ kg/min})(0.0216 - 0.01) = 0.131 \text{ kg/min}$$

$$\dot{Q}_{out} = (11.25 \text{ kg/min})[(85.4 - 39.3)\text{kJ/kg}] - (0.131 \text{ kg/min})(58.8 \text{ kJ/kg}) = 511 \text{ kJ/min}$$

Therefore, this air-conditioning unit removes moisture and heat from the air at rates of 0.131 kg/min and 511 kJ/min, respectively.

ADIABATIC MIXING OF AIRSTREAMS

common process in air – conditioning systems is the mixing of moist air streams, as shown in Fig. 6 – 14. The objective of the thermodynamic analysis of such a process is normally to fix the flow rate and state of the exiting stream for specified flow rates and states of each of the two inlet streams.

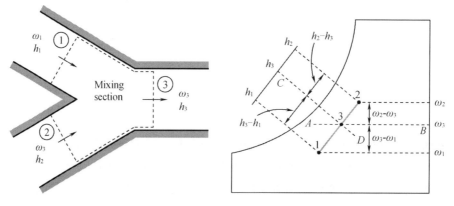

Fig. 6 – 14 When two air streams at states 1 and 2 are mixed adiabatically, the state of the mixture lies on the straight line connecting the two states

The heat transfer with the surroundings is usually small, and thus the mixing processes can be assumed to be adiabatic. Mixing processes normally involve no work interactions, and the changes in kinetic and potential energies, if any, are negligible. Then the mass and exergy balances for the adiabatic mixing of two airstreams reduce to

$$\dot{m}_{a_1} + \dot{m}_{a_2} = \dot{m}_{a_3} \ (\text{dry air}) \qquad (6-20)$$

$$\omega_1 \dot{m}_{a_1} + \omega_2 \dot{m}_{a_2} = \omega_3 \dot{m}_{a_3} \ (\text{water}) \qquad (6-21)$$

$$h_1 \dot{m}_{a_1} + h_2 \dot{m}_{a_2} = h_3 \dot{m}_{a_3} \qquad (6-22)$$

Eliminating $\dot{m}_{a_3}$ from the relations above, we obtain

$$\frac{\dot{m}_{a_1}}{\dot{m}_{a_2}} = \frac{\omega_2 - \omega_3}{\omega_3 - \omega_1} = \frac{h_2 - h_3}{h_3 - h_1} \qquad (6-23)$$

This equation has an instructive geometric interpretation on the psychrometric chart. It shows that the ratio of ($\omega_2 - \omega_3$) to ($\omega_3 - \omega_1$) is equal to the ratio of $\dot{m}_{a_1}$ to $\dot{m}_{a_2}$. The states that satisfy this condition are indicated by the dashed line *AB*. The ratio of ($h_2 - h_3$) to ($h_3 - h_1$) is also equal to the ratio of $\dot{m}_{a_1}$ to $\dot{m}_{a_2}$, and the states that satisfy this condition are indicated by the dashed line *CD*. The only state that satisfies both conditions is the intersection point of these two dashed lines, which is located on the straight line connecting states 1 and 2. Thus we conclude that *when two air streams at two different states (states 1 and 2) are mixed adiabatically, the state of the mixture (state 3) lies on the straight line connecting states 1 and 2 on the psychrometric chart, and the ratio of the distances (2−3) and (3−1) is equal to the ratio of mass flow rates $\dot{m}_{a_1}$ and $\dot{m}_{a_2}$*.

The concave nature of the saturation curve and the conclusion above lead to an interesting possibility. When states 1 and 2 are located close to the saturation curve, the straight line connecting the two states will cross the saturation curve, and state 3 may lie to the left of the saturation curve. In this case, some water will inevitably condense during the mixing process.

EVAPORATIVE COOLING

Cooling in hot, relatively dry climates can be accomplished by evaporative cooling. This involves either spraying liquid water into air or forcing air through a soaked pad that is kept replenished with water. The evaporative cooling process is shown schematically and on a psychrometric chart in Fig. 6-15. Owing to the low humidity of the moist air entering at state 1, part of the injected water evaporates. The exergy for evaporation is provided by the airstream, which is reduced in temperature and exits at state 2 with a lower temperature than the entering stream. Because the incoming air is relatively dry, the additional moisture carried by the exiting moist air stream is normally beneficial.

The evaporative cooling process is essentially identical to the adiabatic saturation process since the heat transfer between the air stream and the surroundings is usually negligible. Therefore, the evaporative cooling process follows aline of constant wet-bulb temperature on the psychrometric chart. We should note that this will not exactly be the case if the liquid water is supplied at a temperature different from the exit temperature of the airstream. Since the constantwet bulb-temperature lines almost coincide with the constant – enthalpy lines, the enthalpy of the airstream can also be assumed to remain constant. That is,

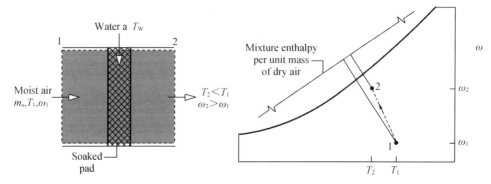

Fig. 6 – 15   Evaporative cooling

$$T_{wb} \cong const \qquad (6-24)$$

and

$$h \cong const \qquad (6-25)$$

during an evaporative cooling process. This is a reasonably accurate approximation, and it is commonly used in air-conditioning calculations.

WET COOLING TOWERS

Power plants invariably discharge considerable exergy to their surroundings by heat transfer. Although water drawn from a nearby river or lake can be employed to carry away this exergy, cooling towers provide an alternative in locations where sufficient cooling water cannot be obtained from natural sources or where concerns for the environment place a limit on the temperature at which cooling water can be returned to the surroundings. Cooling towers also are frequently employed to provide chilled water for applications other than those involving power plants.

Cooling towers can operate by *natural* or *forced* convection. Also they may be *counterflow*, *cross – flow*, or a combination of these. A schematic diagram of a forced – convection, counterflow cooling tower is shown in Fig. 6 – 16. The warm water to be cooled enters at 1 and is sprayed from the top of the tower. The falling water usually passes through a series of baffles intended to keep it broken up into fine drops to promote evaporation. Atmospheric air drawn in at 3 by the fan flows upward, counter to the direction of the falling water droplets.

As the two streams interact, a small fraction of the water stream evaporates into the moist air, which exits at 4 with a greater humidity ratio than the incoming moist air at 3. The exergy required for evaporation is provided mainly by the portion of the incoming water stream that does not evaporate, with the result that the water exiting at 2 is at a lower temperature than the water entering at 1. Since some of the incoming water is evaporated into the mois tair stream, an equivalent amount of make up water is added at 5 so that the return mass flowrate of the cool water equals the mass flow rate of the warm water entering at 1.

For operation at steady state, mass balances for the dry air and water and an exergy balance on the overall cooling tower provide information about cooling tower performance. In applying the exergy balance, heat transfer with the surroundings is usually neglected. The power input to the fan of forced – convection towers also may be negligible relative to other exergy rates involved. The example to follow illustrates the analysis of a cooling tower using conservation of mass and exergy together with property data for the dry air and water.

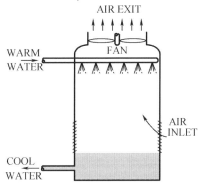

Fig. 6 – 16 An induced – draft counter flow cooling tower

Summary

In this chapter we discussed the air – water – vapor mixture, which is the most commonly encountered gas – vapor mixture in practice. The air in the atmosphere normally contains some water vapor, and it is referred to as *atmospheric air*. By contrast, air that contains no water vapor is called *dry air*. In the temperature range encountered in air-conditioning applications, both the dry air and the water vapor can be treated as ideal gases. We introduce several parameters of atmospheric air and how to use the psychrometric charts. At last, six classic thermodynamic processes of atmospheric air are analyzed and discussed, including heating and cooling, heating with humidification, cooling with dehumidification, adiabatic mixing, evaporative cooling and wet cooling towers.

Problems

6 – 1   A 5 m × 5 m × 3 m room shown in Fig. P6 – 1 contains air at 25 ℃ and 100 kPa at a relative humidity of 75 percent. Determine (a) the partial pressure of dry air, (b) the specific humidity, (c) the enthalpy per unit mass of the dry air, and (d) the masses of the dry air and water vapor in the room.

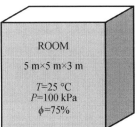

Fig. P6 – 1

6 – 2 The dry – and the wet – bulb temperatures of atmospheric air at 1 atm (101.325 kPa) pressure are measured with a sling psychrometer and determined to be 25 ℃ and 15 ℃, respectively. Determine (a) the specific humidity, (b) the relative humidity, and (c) the enthalpy of the air.

6 – 3 An air-conditioning system is to take in outdoor air at 10 ℃ and 30 percent relative humidity at a steady rate of 45 m³/min and to condition it to 25 ℃ and 60 percent relative humidity. The outdoor air is first heated to 22 ℃ in the heating section and then humidified by the injection of hot steam in the humidifying section. Assuming the entire process takes place at a pressure of 100 kPa, determine (a) the rate of heat supply in the heating section and (b) the mass flow rate of the steam required in the humidifying section.

6 – 4 Air enters an evaporative (or swamp) cooler at 14.7 psi, 95 ℉, and 20 percent relative humidity, and it exits at 80 percent relative humidity. Determine (a) the exit temperature of the air and (b) the lowest temperature to which the air can be cooled by this evaporative cooler.

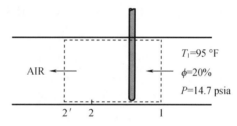

Fig. P6 – 4

Chapter 7 Flow and Compression of Gas and Steam

We start this chapter by introducing the concepts of *stagnation state*, *speed of sound*, and *Mach number* for compressible flows. The relationships between the static and stagnation fluid properties are developed for isentropic flows of ideal gases, and they are expressed as functions of specific heat ratios and the Mach number. The effects of area changes for one-dimensional isentropic subsonic and supersonic flows are discussed. These effects are illustrated by considering the isentropic flow through *converging* and *converging – diverging nozzles*.

7 – 1 Steady-Flow Characteristics of Gas and Steam

In this chapter, we discuss the steady-flow characteristics of gas and steam, and the gas that metioned is ideal gas and the steam is water vapor. As described in Chapter 2, the term *steady* implies *no change with time*. The opposite of steady is *unsteady*, or *transient*. The steady-flow process can be defined as a *process during which a fluid flows through a control volume steadily*. That is, the fluid properties can change from point to point within the control volume, but at any fixed point they remain the same during the entire process. Therefore, the volume V, the mass m, and the total exergy content E of the control volume remain constant during a steady flow process.

MACH NUMBER

sound wave is a small pressure disturbance that propagates through a gas, liquid, or solid at a velocity c that depends on the properties of the medium. In this section, we obtain an expression that relates the *velocity of sound*, or sonic velocity, to other properties. The velocity of sound is an important property in the study of compressible flows.

To obtain a relation for the speed of sound in a medium, consider a pipe that is filled with a fluid at rest. A piston fitted in the pipe is now moved to the right with a constant incremental velocity dV, creating a sonic wave. The wave front moves to the right through the fluid at the speed of sound c and separates the moving fluid adjacent to the piston from the fluid still at rest. The fluid to the left of the wave front experiences an incremental change in its thermodynamic properties, while the fluid on the right of the wave front maintains its original thermodynamic properties.

To simplify the analysis, consider a control volume that encloses the wave front and moves with it, as shown in Fig. 7 – 1. To an observer traveling with the wave front, the fluid to the right will appear to be moving toward the wave front with a speed of c and the fluid to the left to be moving away from the wave front with a speed of $c - \mathrm{d}V$. Of course, the observer will think the control volume that encloses the wave front is stationary, and the observer will be witnessing a steady-flow process. The mass balance for this single-stream, steady-flow process can be expressed as

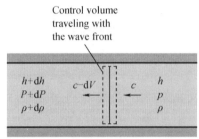

Fig. 7−1 Control volume moving with the small pressure wave along a duct

$$\dot{m}_{\text{right}} = \dot{m}_{\text{left}}$$

or

$$\rho A c = (\rho + \mathrm{d}\rho) A (c - \mathrm{d}V)$$

By canceling the cross − sectional (or flow) area A and neglecting the higher order terms, this equation reduces to

$$c\mathrm{d}\rho - \rho \mathrm{d}V = 0 \tag{7-1}$$

No heat or work crosses the boundaries of the control volume during this steady-flow process, and the potential exergy change, if any, can be neglected. Then the steady-flow exergy balance $e_{\text{in}} = e_{\text{out}}$ becomes

$$h + \frac{c^2}{2} = h + \mathrm{d}h + \frac{(c - \mathrm{d}V)^2}{2}$$

which yields

$$\mathrm{d}h - c\mathrm{d}V = 0 \tag{7-2}$$

where we have neglected the second-order term $\mathrm{d}V^2$. The amplitude of the ordinary sonic wave is very small and does not cause any appreciable change in the pressure and temperature of the fluid. Therefore, the propagation of a sonic wave is not only adiabatic but also very nearly is entropic.

$$T\mathrm{d}s \rightarrow 0 = \mathrm{d}h - \frac{\mathrm{d}P}{\rho}$$

or

$$\mathrm{d}h = \frac{\mathrm{d}P}{\rho} \tag{7-3}$$

Combining Eqs. (7−1), (7−2) and (7−3) yields the desired expression for the speed of sound as

$$c^2 = \frac{\mathrm{d}P}{\mathrm{d}\rho} \quad \text{at } s = \text{constant}$$

or

$$c^2 = \left(\frac{\partial P}{\partial \rho}\right)_s \tag{7-4}$$

It is left as an exercise for the reader to show, by using thermodynamic property relations that

Chapter 7 Flow and Compression of Gas and Steam

Eq. (7-4) can also be written as

$$c^2 = k\left(\frac{\partial P}{\partial \rho}\right)_T \tag{7-5}$$

where k is the specific heat ratio of the fluid. Note that the speed of sound in a fluid is a function of the thermodynamic properties of that fluid. When the fluid is an ideal gas ($P = rRT$), the differentiation in Eq. (7-5) can easily be performed to yield

$$c^2 = k\left(\frac{\partial P}{\partial \rho}\right)_T = k\left(\frac{\partial(\rho RT)}{\partial \rho}\right)_T = kRT$$

or

$$c = \sqrt{kRT} \tag{7-6}$$

Noting that the gas constant R has a fixed value for a specified ideal gas and the specific heat ratio k of an ideal gas is, at most, a function of temperature, we see that the speed of sound in a specified ideal gas is a function of temperature alone.

In subsequent discussions, the ratio of the velocity V at a state in a flowing fluid to the value of the sonic velocity c at the same state plays an important role. This ratio is called the Mach number, *Ma*.

$$Ma = \frac{V}{c} \tag{7-7}$$

When $Ma > 1$, the flow is said to be *supersonic*; when $Ma < 1$ the flow is *subsonic*; and when $Ma = 1$, the flow is *sonic*. The term *hypersonic* is used for flows with Mach numbers much greater than one, and the term *transonic* refers to flows where the Mach number is close to unity.

EXAMPLE 7-1 Mach Number of Air Entering a Diffuser

Air enters a diffuser shown in Fig. 7-2 with a velocity of 200 m/s. Determine (a) the speed of sound and (b) the Mach number at the diffuser inlet when the air temperature is 30 ℃.

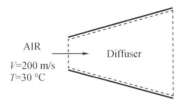

Fig. 7-2 Schematic for Example 7-1

Solution Air enters a diffuser with a high velocity. The speed of sound and the Mach number are to be determined at the diffuser inlet.

Assumptions Air at specified conditions behaves as an ideal gas.

Properties The gas constant of air is $R = 0.287$ kJ/kg · K, and its specific heat ratio at 30 ℃ is 1.4 (Table A-10a).

Analysis We note that the speed of sound in a gas varies with temperature, which is given to be 30 ℃.

(a) The speed of sound in air at 30 ℃ is determined from Eq. (7-6) to be

$$c = \sqrt{kRT} = \sqrt{(1.4)(0.287 \text{ kJ/kg} \cdot \text{K})[(273+30)\text{K}]\left(\frac{1\,000 \text{ m}^2/\text{s}^2}{1 \text{ kJ/kg}}\right)} = 349 \text{ m/s}$$

(b) Then the Mach number becomes

$$Ma = \frac{V}{c} = \frac{200 \text{ m/s}}{349 \text{ m/s}} = 0.573 < 1$$

Discussion The flow at the diffuser inlet is subsonic since $Ma < 1$.

ISENTROPIC FLOW OF GASES THROUGH NOZZLES

The objective of the present discussion is to establish criteria for determining whether a nozzle or diffuser should have a converging, diverging, or converging – diverging shape. This is accomplished using differential equations relating the principal variables that are obtained using mass and exergy balances together with property relations, as considered next.

Let us begin by considering a control volume enclosing a nozzle or diffuser. At steady state, the mass flow rate is constant, so

$$\dot{m} = \rho A V = const.$$

Differentiating and dividing the resultant equation by the mass flow rate, we obtain

$$\frac{d\rho}{\rho} + \frac{dA}{A} + \frac{dV}{V} = 0 \qquad (7-8)$$

Neglecting the potential exergy, the exergy balance for an isentropic flow with no work interactions can be expressed in the differential form as

Based on the conservation of exergy, for steady flow, we can figure out that $w = 0$, $q = 0$ and $\Delta e_p = 0$, then

$$h_1 + \frac{V_1^2}{2} = h_2 + \frac{V_2^2}{2} \quad \text{or} \quad h + \frac{V^2}{2} = const.$$

Differentiate,

$$dh + VdV = 0$$

Also,

$$Tds \to 0 \,(isentropic) = dh - vdP$$

$$dh = vdP = \frac{1}{\rho}dP$$

Substitute,

$$\frac{dP}{\rho} + VdV = 0 \qquad (7-9)$$

This relation is also the differential form of Bernoulli's equation when changes in potential exergy are negligible, which is a form of the conservation of momentum principle for steady-flow control volumes. Combining Eqs. (7-8) and (7-9) gives

$$\frac{dA}{A} = \frac{dP}{\rho}\left(\frac{1}{V^2} - \frac{d\rho}{dP}\right) \qquad (7-10)$$

Rearranging Eq. (7-4) as $\left(\frac{\partial \rho}{\partial P}\right)_s = \frac{1}{c^2}$ and substituting into Eq. (7-10) yield

$$\frac{dA}{A} = \frac{dP}{\rho V^2}(1 - Ma^2) \qquad (7-11)$$

This is an important relation for isentropic flow in ducts since it describes the variation of pressure with flow area. We note that A, r, and V are positive quantities. For *subsonic* flow ($Ma < 1$), the term $1 - Ma^2$ is positive; and thus dA and dP must have the same sign. That is, the pressure of the fluid must increase as the flow area of the duct increases and must decrease as the flow area of the duct decreases. Thus, at subsonic velocities, the pressure decreases in converging ducts (subsonic nozzles) and increases in diverging ducts (subsonic diffusers). In *supersonic* flow ($Ma > 1$), the term $1 - Ma^2$ is negative, and thus dA and dP must have opposite signs. That is, the pressure of the fluid must increase as the flow area of the duct decreases and must decrease as the flow area of the duct increases. Thus, at supersonic velocities, the pressure decreases in diverging ducts (supersonic nozzles) and increases in converging ducts (supersonic diffusers).

Another important relation for the isentropic flow of a fluid is obtained by substituting $\rho V = - dP/dV$ from Eq. (7-9) into Eq. (7-11):

$$\frac{dA}{A} = -\frac{dV}{V}(1 - Ma^2) \qquad (7-12)$$

This equation governs the shape of a nozzle or a diffuser in subsonic or supersonic isentropic flow. Noting that A and V are positive quantities, we conclude the following:

For subsonic flow ($Ma < 1$):

$$\frac{dA}{dV} < 0$$

For sonic flow ($Ma = 1$):

$$\frac{dA}{dV} = 0$$

For supersonic flow ($Ma > 0$):

$$\frac{dA}{dV} > 0$$

There fore, the proper shape of a nozzle depends on the highest velocity desired relative to the sonic velocity. To accelerate a fluid, we must use a converging nozzle at subsonic velocities and a diverging nozzle at supersonic velocities. The velocities encountered in most familiar applications are well below the sonic velocity, and thus it is natural that we visualize a nozzle as a converging duct. However, the highest velocity we can achieve by a converging nozzle is the sonic velocity, which occurs at the exit of the nozzle. If we extend the converging nozzle by further decreasing the flow area, in hopes of accelerating the fluid to supersonic velocities, as shown in Fig. 7-3, we are up for disappointment. Now the sonic velocity will occur at the exit of the converging extension, instead of the exit of the original nozzle, and the mass flow rate through the nozzle will decrease because of the reduced exit area.

Based on Eq. (7-11), which is an expression of the conservation of mass and exergy principles, we must add a diverging section to a converging nozzle to accelerate a fluid to supersonic velocities. The result is a converging-diverging nozzle. The fluid first passes through a subsonic (converging) section, where the Mach number increases as the flow area of the nozzle decreases, and then reaches the value of unity at the nozzle throat. The fluid continues to accelerate as it passes through a supersonic (diverging) section. Noting that $\dot{m} = \rho AV$ for steady flow, we see that the large decrease in density makes acceleration in the diverging section possible. An example of this type of flow is the flow of hot combustion gases through a nozzle in a gas turbine.

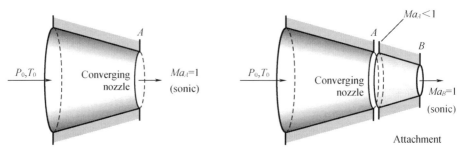

Fig. 7 – 3 We cannot obtain supersonic velocities by attaching a converging section to a converging nozzle. Doing so will only move the sonic cross section farther down stream and decrease the mass flow rate

The opposite process occurs in the engine inlet of a supersonic aircraft. The fluid is decelerated by passing it first through a supersonic diffuser, which has a flow area that decreases in the flow direction. Ideally, the flow reaches a Mach number of unity at the diffuser throat. The fluid is further decelerated in a subsonic diffuser, which has a flow area that increases in the flow direction, as shown in Fig. 7 – 4.

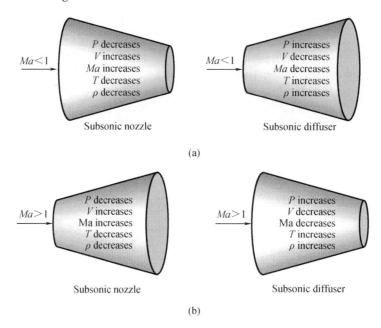

Fig. 7 – 4 Variation of flow properties in subsonic and supersonic nozzles and diffusers
(a) Subsonic flow; (b) Supersonic flow

7 − 2 Stagnation and Critical Properties of Gases

STAGNATION PROPERTIES

When dealing with compressible flows, it is often convenient to work with properties evaluated at a reference state known as the *stagnation state* h_0. The stagnation state is the state a flowing fluid would attain if it were decelerated to zero velocity is entropically, defined per unit mass as

$$h_0 = h + \frac{V^2}{2} \quad (\text{kJ/kg}) \qquad (7-13)$$

When the potential exergy of the fluid is negligible, the stagnation enthalpy represents the *total exergy of a flowing fluid stream* per unit mass. Thus its implifies the thermodynamic analysis of high − speed flows.

Throughout this chapter the ordinary enthalpy h is referred to as the static enthalpy, whenever necessary, to distinguish it from the stagnation enthalpy. Notice that the stagnation enthalpy is a combination property of a fluid, just like the static enthalpy, and these two enthalpies become identical when the kinetic exergy of the fluid is negligible.

Consider the steady flow of a fluid through a duct such as a nozzle, diffuser, or some other flow passage where the flow takes place adiabatically and with no shaft or electrical work. Assuming the fluid experiences little or no change in its elevation and its potential exergy, the exergy balance relation ($E_{in} = E_{out}$) for this single-stream steady-flow system reduces to

$$h_1 + \frac{V_1^2}{2} = h_2 + \frac{V_2^2}{2} \qquad (7-14)$$

or

$$h_{01} = h_{02} \qquad (7-15)$$

That is, in the absence of any heat and work interactions and any changes in potential exergy, the stagnation enthalpy of a fluid remains constant during a steady-flow process. Flows through nozzles and diffusers usually satisfy these conditions, and any increase in fluid velocity in these devices creates an equivalent decrease in the static enthalpy of the fluid. If the fluid were brought to a complete stop, then the velocity at state 2 would be zero and Eq. (7 − 14) would become

$$h_1 + \frac{V_1^2}{2} = h_2 = h_{02}$$

Thus the *stagnation enthalpy* represents the *enthalpy of a fluid when it is brought to rest adiabatically*. During a stagnation process, the kinetic exergy of a fluid is converted to enthalpy (internal exergy plus flow exergy), which results in an increase in the fluid temperature and pressure. The properties of a fluid at the stagnation state are called stagnation properties (stagnation temperature, stagnation pressure, stagnation density, etc.). The stagnation state and the stagnation properties are indicated by the subscript 0.

The stagnation state is called the isentropic stagnation state when the stagnation process is reversible as well as adiabatic (i. e., isentropic). The entropy of a fluid remains constant during an isentropic stagnation process. The actual (irreversible) and isentropic stagnation processes are shown on the $h-s$ diagram in Fig. 7 – 5. Notice that the stagnation enthalpy of the fluid (and the stagnation temperature if the fluid is an ideal gas) is the same for both cases. However, the actual stagnation pressure is lower than the isentropic stagnation pressure since entropy increases during the actual stagnation process as a result of fluid friction. The stagnation processes are often approximated to be isentropic, and the isentropic stagnation properties are simply referred to as stagnation properties.

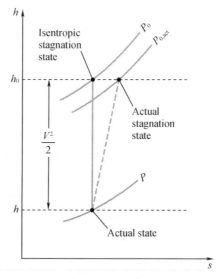

Fig. 7 – 5 The actual state, actual stagnation state, and isentropic stagnation state of a fluid on an $h-s$ diagram

When the fluid is approximated as an *ideal gas* with constant specific heats, its enthalpy can be replaced by $c_p T$ and Eq. (7 – 13) can be expressed as

$$c_p T_0 = c_p T + \frac{V^2}{2}$$

or

$$T_0 = T + \frac{V^2}{2c_p} \qquad (7-16)$$

Here T_0 is called the stagnation (or total) temperature, and it represents *the temperature an ideal gas attains when it is brought to rest adiabatically*. The term $V^2/2c_p$ corresponds to the temperature rise during such a process and is called the dynamic temperature. (Fig. 7 – 6) Note that for low-speed flows, the stagnation and static (or ordinary) temperatures are practically the same. But for high-speed flows, the temperature measured by a stationary probe placed in the fluid (the stagnation temperature) may be significantly higher than the static temperature of the fluid.

Chapter 7 Flow and Compression of Gas and Steam

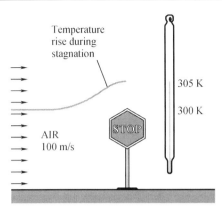

Fig. 7-6 The temperature of an ideal gas flowing at a velocity V rises by $V^2/2c_p$ when it is brought to a complete stop

The pressure a fluid attains when brought to rest is entropically is called the stagnation pressure P_0. For ideal gases with constant specific heats, P_0 is related to the static pressure of the fluid by

$$\frac{P_0}{P} = \left(\frac{T_0}{T}\right)^{k/(k-1)} \tag{7-17}$$

By noting that $\rho = 1/v$ and using the isentropic relation, the ratio of the stagnation density to static density can be expressed as

$$\frac{\rho_0}{\rho} = \left(\frac{T_0}{T}\right)^{1/(k-1)} \tag{7-18}$$

When stagnation enthalpies are used, there is no need to refer explicitly to kinetic exergy. Then the exergy balance for a single-stream, steady-flow device can be expressed as

$$q_{in} + w_{in} + (h_{01} + gz_1) = q_{out} + w_{out} + (h_{02} + gz_2) \tag{7-19}$$

where h_{01} and h_{02} are the stagnation enthalpies at states 1 and 2, respectively. When the fluid is an ideal gas with constant specific heats, Eq. (7-19) becomes

$$(q_{in} - q_{out}) + (w_{in} - w_{out}) = c_p(T_{02} - T_{01}) + g(z_2 - z_1) \tag{7-20}$$

where T_{01} and T_{02} are the stagnation temperatures.

Notice that kinetic exergy terms do not explicitly appear in Eqs. (7-19) and (7-20), but the stagnation enthalpy terms account for their contribution.

CRITICAL PROPERTIES

Next we develop relations between the static properties and stagnation properties of an ideal gas in terms of the specific heat ratio k and the Mach number Ma. We assume the flow is isentropic and the gas has constant specific heats.

The temperature T of an ideal gas anywhere in the flow is related to the stagnation temperature T_0 through Eq. (7-16):

$$T_0 = T + \frac{V^2}{2c_p}$$

or

$$\frac{T_0}{T} = 1 + \frac{V^2}{2c_p T}$$

Noting that $c_p = kR/(k-1)$, $c^2 = kRT$, and $Ma = V/c$, we see that

$$\frac{V^2}{2c_p T} = \frac{V^2}{2[kR/(k-1)]T} = \left(\frac{k-1}{2}\right)\frac{V^2}{c^2} = \left(\frac{k-1}{2}\right)Ma^2$$

Substituting yields

$$\frac{T_0}{T} = 1 + \left(\frac{k-1}{2}\right)Ma^2 \tag{7-21}$$

which is the desired relation between T_0 and T.

The ratio of the stagnation to static pressure is obtained by substituting Eq. (7-21) into Eq. (7-17):

$$\frac{P_0}{P} = \left[1 + \left(\frac{k-1}{2}\right)Ma^2\right]^{k/(k-1)} \tag{7-22}$$

The ratio of the stagnation to static density is obtained by substituting Eq. (7-21) into Eq. (7-18):

$$\frac{\rho_0}{\rho} = \left[1 + \left(\frac{k-1}{2}\right)Ma^2\right]^{1/(k-1)} \tag{7-23}$$

Numerical values of T/T_0, P/P_0, and ρ/ρ_0 are listed versus the Mach number in Table A-13 for $k = 1.4$, which are very useful for practical compressible flow calculations involving air.

The properties of a fluid at a location where the Mach number is unity (the throat) are called critical properties, and the ratios in Eqs. (7-21) through (7-23) are called critical ratios (Fig. 7-7). It is common practice in the analysis of compressible flow to let the superscript asterisk ($*$) represent the critical values. Setting $Ma = 1$ in Eqs. (7-21) through (7-23) yields

$$\frac{T^*}{T_0} = \frac{2}{k+1} \tag{7-24}$$

$$\frac{P^*}{P_0} = \left(\frac{2}{k+1}\right)^{k/(k-1)} \tag{7-25}$$

$$\frac{\rho^*}{\rho_0} = \left(\frac{2}{k+1}\right)^{1/(k-1)} \tag{7-26}$$

These ratios are evaluated for various values of k and are listed in Table 7-1. The critical properties of compressible flow should not be confused with the properties of substances at the critical point (such as the critical temperature T_c and critical pressure P_c).

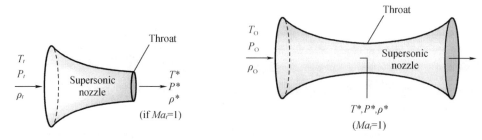

Fig. 7-7 When $Ma_t = 1$, the properties at the nozzle throat become the critical properties

Chapter 7 Flow and Compression of Gas and Steam

TABLE 7-1 The critical-pressure, critical-temperature, and critical-density ratios for isentropic flow of some ideal gases

	Superheated steam, $k=1.3$	Hot products of combustion, $k=1.33$	Air, $k=1.4$	Monatomic gases, $k=1.667$
$\dfrac{P^*}{P_0}$	0.545 7	0.540 4	0.528 3	0.487 1
$\dfrac{T^*}{T_0}$	0.869 6	0.858 4	0.833 3	0.749 9
$\dfrac{\rho^*}{\rho_0}$	0.627 6	0.629 5	0.634 0	0.649 5

During fluid flow through many devices such as nozzles, diffusers, and turbine blade passages, flow quantities vary primarily in the flow direction only, and the flow can be approximated as one-dimensional isentropic flow with good accuracy. Therefore, it merits special consideration. Before presenting a formal discussion of one-dimensional isentropic flow, we illustrate some important aspects of it with an example.

From Fig. 7-8, we see that a converging diffuser is required to decelerate a fluid flowing supersonically, but once $Ma = 1$ is achieved, further deceleration can occur only in a diverging diffuser. These findings suggest that a Mach number of unity can occur only at the location in a nozzle or diffuser where the cross-sectional area is a minimum. This location of minimum area is called the *throat*. Note that the velocity of the fluid keeps increasing after passing the throat although the flow area increases rapidly in that region. This increase in velocity past the throat is due to the rapid decrease in the fluid density. The flow area of the duct considered in this example first decreases and then increases. Such ducts are called converging-diverging nozzles. These nozzles are used to accelerate gases to supersonic speeds and should not be confused with *Venturi nozzles*, which are used strictly for incompressible flow. The first use of such a nozzle occurred in 1893 in a steam turbine designed by a Swedish engineer, Carl G. B. de Laval (1845-1913). As a consequence, therefore converging-diverging nozzles are often called *Laval nozzles*.

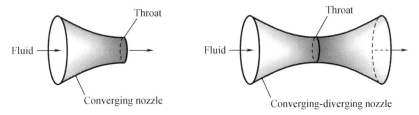

Fig. 7-8 The cross section of a nozzle at the smallest flow area is called the throat

7-3 Nozzles

In the present discussion we consider the effect of varying the *back pressure* on the rate of mass flow through nozzles. The *back pressure* is the pressure in the exhaust region outside the nozzle. The case of converging nozzles is taken up first and then converging – diverging nozzles are considered. Flow through the nozzle is steady, one – dimensional, and is entropic.

CONVERGING NOZZLES

In the present discussion we consider the effect of varying the *back pressure* on the rate of mass flow through nozzles. The *back pressure* is the pressure in the exhaust region outside the nozzle. The case of converging nozzles is taken up first and then converging – diverging nozzles are considered.

Fig. 7-9 shows a converging duct with stagnation conditions at the inlet, discharging into a region in which the back pressure p_B can be varied. For the series of cases labeled a through e, let us consider how the mass flow rate $\dot{m}$ and nozzle exit pressure p_E vary as the back pressure is decreased while keeping the inlet conditions fixed.

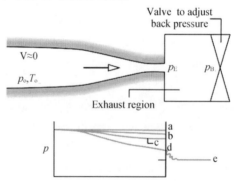

Fig. 7-9 The effect of back pressure P_B on the pressure distribution along a converging nozzle

When $p_B = p_E = p_0$, there is no flow, so This corresponds to case a of Fig. 7-9. If the back pressure p_B is decreased, as in cases b and c, there will be flow through the nozzle. As long as the flow is subsonic at the exit, information about changing conditions in the exhaust region can be transmitted upstream. Decreases in back pressure thus result in greater mass flow rates and new pressure variations within the nozzle. In each instance, the velocity is subsonic throughout the nozzle and the exit pressure equals the back pressure. The exit Mach number increases as p_B decreases, however, and eventually a Mach number of unity will be attained at the nozzle exit. The corresponding pressure is denoted by p^*, called the critical pressure. This case is represented by d on Fig. 7-9.

Recalling that the Mach number cannot increase beyond unity in a converging section, let us consider next what happens when the back pressure is reduced further to a value less than p^*, such as represented by case e. Since the velocity at the exit equals the velocity of sound, information about changing conditions in the exhaust region no longer can be transmitted upstream past the exit plane. Accordingly, reductions in p_B below p^* have no effect on flow conditions in the nozzle. Neither the pressure variation within the nozzle nor the mass flowrate is affected. Under these conditions, the nozzle is said to be choked. When a nozzle is choked, the mass flow rate is the *maximum possible for the given stagnation conditions*. For p_B less than p^*, the flow expands outside the nozzle to match the lower back pressure, as shown by case e of Fig. 7-8. The pressure variation outside the nozzle cannot be predicted using the one-dimensional flow model.

Under steady-flow conditions, the mass flow rate through the nozzle is constant and can be expressed as

$$\dot{m} = \rho A V = \left(\frac{P}{RT}\right) A (Ma \sqrt{kRT}) = PAMa \sqrt{\frac{k}{RT}}$$

Solving for T from Eq. (7-21) and for P from Eq. (7-22) and substituting,

$$\dot{m} = \frac{AMaP_0 \sqrt{k/(RT_0)}}{[1 + (k-1)Ma^2/2]^{(k+1)/(2(k-1))}} \quad (7-27)$$

Thus the mass flow rate of a particular fluid through a nozzle is a function of the stagnation properties of the fluid, the flow area, and the Mach number. Eq. 7-27 is valid at any cross section, and thus $\dot{m}$ can be evaluated at any location along the length of the nozzle.

For a specified flow area A and stagnation properties T_0 and P_0, the maximum mass flow rate can be determined by differentiating Eq. (7-27) with respect to Ma and setting the result equal to zero. It yields $Ma = 1$. Since the only location in a nozzle where the Mach number can be unity is the location of minimum flow area (the throat), the mass flow rate through a nozzle is a maximum when $Ma = 1$ at the throat. Denoting this area by A^*, we obtain an expression for the maximum mass flow rate by substituting $Ma = 1$ in Eq. (7-27):

$$\dot{m}_{max} = A^* P_0 \sqrt{k/(RT_0)} \left(\frac{2}{k+1}\right)^{(k+1)/(2(k-1))} \quad (7-28)$$

Thus, for a particular ideal gas, the maximum mass flow rate through a nozzle with a given throat area is fixed by the stagnation pressure and temperature of the inlet flow. The flow rate can be controlled by changing the stagnation pressure or temperature, and thus a converging nozzle can be used as a flowmeter. The flow rate can also be controlled, of course, by varying the throat area. This principle is vitally important for chemical processes, medical devices, flow meters, and anywhere the mass flux of a gas must be known and controlled.

A plot of $\dot{m}$ versus P_B/P_0 for a converging nozzle is shown in Fig. 7-10. Notice that the mass flow rate increases with decreasing P_B/P_0, reaches a maximum at $P_B = P^*$, and remains constant for P_B/P_0 values less than this critical ratio. Also illustrated on this figure is the effect of back pressure on the nozzle exit pressure P_E. We observe that

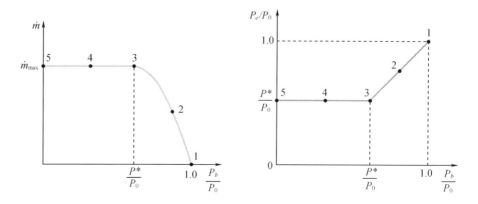

Fig. 7-10 The effect of back pressure P_b on the mass flow rate and the exit pressure P_e of a converging nozzle

$$P_E = \begin{cases} P_B & \text{for} \quad P_B \geq P^* \\ P^* & \text{for} \quad P_B < P^* \end{cases}$$

To summarize, for all back pressures lower than the critical pressure P^*, the pressure at the exit plane of the converging nozzle P_E is equal to P^*, the Mach number at the exit plane is unity, and the mass flow rate is the maximum (or choked) flow rate. Because the velocity of the flow is sonic at the throat for the maximum flow rate, a back pressure lower than the critical pressure cannot be sensed in the nozzle up stream flow and does not affect the flow rate.

The effects of the stagnation temperature T_0 and stagnation pressure P_0 on the mass flow rate through a converging nozzle are illustrated in Fig. 7-11 where the mass flow rate is plotted against the static-to-stagnation pressure ratio at the throat P_t/P_0. An increase in P_0 (or a decrease in T_0) will increase the mass flow rate through the converging nozzle; a decrease in P_0 (or an increase in T_0) will decrease it. We could also conclude this by carefully observing Eqs. (17-27) and (17-28).

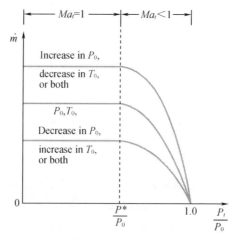

Fig. 7-11 The variation of the mass flow rate through a nozzle with inlet stagnation properties

A relation for the variation of flow area A through the nozzle relative to throat area A^* can be obtained by combining Eqs. (17-24) and (17-25) for the same mass flow rate and stagnation properties of a particular fluid. This yields

$$\frac{A}{A^*} = \frac{1}{Ma}\left[\left(\frac{2}{k+1}\right)\left(1 + \frac{k-1}{2}Ma^2\right)\right]^{(k+1)/(2(k-1))} \quad (7-29)$$

Table A-13 gives values of A/A^* as a function of the Mach number for air ($k = 1.4$). There is one value of A/A^* for each value of the Mach number, but there are two possible values of the Mach number for each value of A/A^*—one for subsonic flow and another for supersonic flow.

Another parameter sometimes used in the analysis of one-dimensional isentropic flow of ideal gases is Ma^*, which is the ratio of the local velocity to the speed of sound at the throat:

$$Ma^* = \frac{V}{c^*} \quad (7-30)$$

It can also be expressed as

$$Ma^* = \frac{V}{c}\frac{c}{c^*} = \frac{Mac}{c^*} = \frac{Ma\sqrt{kRT}}{\sqrt{kRT^*}} = Ma\sqrt{\frac{T}{T^*}}$$

where Ma is the local Mach number, T is the local temperature, and T^* is the critical temperature. Solving for T from Eq. 7-21 and for T^* from Eq. (7-24) and substituting, we get

$$Ma^* = Ma\sqrt{\frac{k+1}{2 + (k-1)Ma^2}} \quad (7-31)$$

Values of Ma^* are also listed in Table A-13 versus the Mach number for $k = 1.4$. Note that the parameter Ma^* differs from the Mach number Ma in that Ma^* is the local velocity nondimensionalized with respect to the sonic velocity at the *throat*, whereas Ma is the local velocity nondimensionalized with respect to the *local* sonic velocity. (Recall that the sonic velocity in a nozzle varies with temperature and thus with location.)

In fact, the friction can not be neglected when gases flow through nozzles, which leads to the decrease of exergy. Thus, the velocity of gas at exit is lower than that of in a reversible isentropic flow. A new quantity velocity coefficient φ, which is defined as

$$\varphi = \frac{V'_{exit}}{V_{exit}}$$

The exergy decrease coefficient ζ can be expressed as

$$\zeta = \frac{V_{exit}^2 - V'^2_{exit}}{V_{exit}^2} = 1 - \frac{V'^2_{exit}}{V_{exit}^2} = 1 - \varphi^2$$

CONVERGING – DIVERGING NOZZLES

When we think of nozzles, we ordinarily think of flow passages whose cross-sectional area decreases in the flow direction. However, the highest velocity to which a fluid can be accelerated in a converging nozzle is limited to the sonic velocity ($Ma = 1$), which occurs at the exit plane (throat) of the nozzle. Accelerating a fluid to supersonic velocities ($Ma > 1$) can be accomplished only by attaching a diverging flow section to the subsonic nozzle at the throat. The resulting combined flow section is a converging – diverging nozzle, which is standard equipment in supersonic aircraft and rocket propulsion.

Forcing a fluid through a converging – diverging nozzle is no guarantee that the fluid will be accelerated to a supersonic velocity. In fact, the fluid may find itself decelerating in the diverging section instead of accelerating if the back pressure is not in the right range. The state of the nozzle flow is determined by the overall pressure ratio P_b/P_0. Therefore, for given inlet conditions, the flow through a converging – diverging nozzle is governed by the back pressure P_b. A fluid enters the nozzle with a low velocity at stagnation pressure P_0. When $P_b = P_0$, there will be no flow through the nozzle. This is expected since the flow in a nozzle is driven by the pressure difference between the nozzle inlet and the exit.

To summarize, when the back pressure of a fluid is equal to the critical pressure or larger than the critical pressure ($P_b \geq P^*$), then the converging nozzle can be selected. Otherwise ($P_b < P^*$), the converging – diverging nozzle can be choose.

EXAMPLE 7 – 3 Gas Flow through a Converging-Diverging Duct

Carbon dioxide flows steadily through a varying cross-sectional-area duct such as a nozzle shown in Fig. 7 – 12 at a mass flow rate of 3 kg/s. The carbon dioxide enters the duct at a pressure of 1 400 kPa and 200 ℃ with a low velocity, and it expands in the nozzle to a pressure of 200 kPa. The duct is designed so that the flow can be approximated as isentropic. Determine the density, velocity, flow area, and Mach number at each location along the duct that corresponds to a pressure drop of 200 kPa.

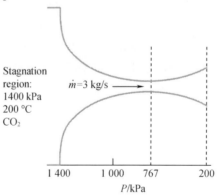

Fig. 7 – 12 Schematic for Example 7.3

Solution Carbon dioxide enters a varying cross – sectional – area duct at specified conditions. The flow properties are to be determined along the duct.

Assumptions 1. Carbon dioxide is an ideal gas with constant specific heat sat room temperature. 2. Flow through the duct is steady, one-dimensional, and isentropic.

Properties For simplicity we use $c_p = 0.846$ kJ/kg · K and $k = 1.289$ throughout the calculations, which are the constant-pressure specific heat and specific heat ratio values of carbon dioxide at room temperature. The gas constant of carbon dioxide is $R = 0.1889$ kJ/kg · K (Table A – 10a).

Analysis We note that the inlet temperature is nearly equal to the stagnation temperature since the inlet velocity is small. The flow is isentropic, and thus the stagnation temperature and pressure throughout the duct remain constant. Therefore,

$$T_0 \cong T_1 = 200 \ ℃ = 473 \ K$$

and

$$P_0 \cong P_1 = 1\ 400 \ kPa$$

To illustrate the solution procedure, we calculate the desired properties at the location where the pressure is 1 200 kPa, the first location that corresponds to a pressure drop of 200 kPa. From Eq. (7-17),

$$T = T_0 \left(\frac{P}{P_0}\right)^{(k-1)/k} = (473 \ K)\left(\frac{1\ 200 \ kPa}{1\ 400 \ kPa}\right)^{(1.289-1)/1.289} = 457 \ K$$

From Eq. (7-16),

$$V = \sqrt{2c_p(T_0 - T)}$$
$$= \sqrt{2(0.846 \ kJ/kg \cdot K)(473 \ K - 457 \ K)\left(\frac{1\ 000 \ m^2/s^2}{1 \ kJ/kg}\right)}$$
$$= 164.5 \ m/s$$

From the ideal-gas relation,

$$\rho = \frac{P}{RT} = \frac{1\ 200 \ kPa}{(0.188\ 9 \ kPa \cdot m^3/kg \cdot K)(457 \ K)} = 13.9 \ kg/m^3$$

From the mass flow rate relation,

$$A = \frac{\dot{m}}{\rho V} = \frac{3 \ kg/s}{(13.9 \ kg/m^3)(164.5 \ m/s)} = 13.1 \times 10^{-4} m^2 = 13.1 \ cm^2$$

From Eqs. (7-11) and (7-12),

$$c = \sqrt{kRT} = \sqrt{(1.289)(0.188\ 9 \ kJ/kg \cdot K)(457 \ K)\left(\frac{1\ 000 \ m^2/s^2}{1 \ kJ/kg}\right)} = 333.6 \ m/s$$

$$Ma = \frac{V}{c} = \frac{164.5 \ m/s}{333.6 \ m/s} = 0.493$$

Discussion: Note that as the pressure decreases, the temperature and speed of sound decrease while the fluid velocity and Mach number increase in the flow direction. The density decreases slowly at first and rapidly later as the fluid velocity increases.

7-4 Adiabatic Throttling Process

Throttling valves are *any kind of flow-restricting devices* that cause a significant pressure drop in the fluid. Some familiar examples are ordinary adjustable valves, capillary tubes, and porous plugs. Unlike turbines, they produce a pressure drop without involving any work. The pressure drop in the fluid is often accompanied by a *large drop in temperature*, and for that reason throttling devices are commonly used in refrigeration and air-conditioning applications. The magnitude of the temperature drop during a throttling process is governed by a property called the *Joule-*

Thomson coefficient.

Throttling valves are usually small devices, and the flow through them may be assumed to be adiabatic ($q \cong 0$) since there is neither sufficient time nor large enough area for any effective heat transfer to take place. Also, there is no work done ($w = 0$), and the change in potential exergy, if any, is very small ($\Delta e_p \cong 0$). Even though the exit velocity is often considerably higher than the inlet velocity, in many cases, the increase in kinetic exergy is insignificant ($\Delta e_k \cong 0$). Then the conservation of exergy equation for this single-stream steady-flow device reduces to

$$h_2 \cong h_1 \quad (\text{kJ/kg}) \tag{7-32}$$

That is, enthalpy values at the inlet and exit of a throttling valve are the same. For this reason, a throttling valve is sometimes called an *isenthalpic device*. Note, however, that for throttling devices with large exposed surface areas such as capillary tubes, heat transfer may be significant.

To gain some insight into how throttling affects fluid properties, let us express Eq. (7-32) as follows:

$$u_1 + P_1 v_1 = u_2 + P_2 v_2$$

or

Internal exergy + Flow exergy = Constant

Thus the final outcome of a throttling process depends on which of the two quantities increases during the process. If the flow exergy increases during the process ($P_2 v_2 > P_1 v_1$), it can do so at the expense of the internal exergy.

As a result, internal exergy decreases, which is usually accompanied by a drop in temperature. If the product Pv decreases, the internal exergy and the temperature of a fluid will increase during a throttling process. In the case of an ideal gas, $h = h(T)$, and thus the temperature has to remain constant during a throttling process.

When a fluid passes through a restriction such as a porous plug, a capillary tube, or an ordinary valve, its pressure decreases. The enthalpy of the fluid remains approximately constant during such a throttling process. You will remember that a fluid may experience a large drop in its temperature as a result of throttling, which forms the basis of operation for refrigerators and air conditioners. This is not always the case, however. The temperature of the fluid may remain unchanged, or it may even increase during a throttling process.

The temperature behavior of a fluid during a throttling (h = constant) process is described by the Joule – Thomson coefficient, defined as

$$\mu = \left(\frac{\partial T}{\partial P}\right)_h \tag{7-33}$$

Like other partial differential coefficients introduced in this section, the Joule – Thomson coefficient is defined in terms of thermodynamic properties only and thus is itself a property. The units of μ_{JT} are those of temperature divided by pressure. Notice that if

$$\mu_{JT} = \begin{cases} < 0 & \text{temperature, increases} \\ = 0 & \text{temperature, remains constant} \\ > 0 & \text{temperature, decreases} \end{cases}$$

during a throttling process.

A careful look at its defining equation reveals that the Joule – Thomson coefficient represents the slope of h = constant lines on a $T-P$ diagram. Such diagrams can be easily constructed from temperature and pressure measurements alone during throttling processes. A fluid at a fixed temperature and pressure T_1 and P_1 (thus fixed enthalpy) is forced to flow through a porous plug, and its temperature and pressure down stream (T_2 and P_2) are measured. The experiment is repeated for different sizes of porous plugs, each giving a different set of T_2 and P_2. Plotting the temperatures against the pressures gives us an h = constant line on a $T-P$ diagram, as shown in Fig. 7 – 13. Repeating the experiment for different sets of inlet pressure and temperature and plotting the results, we can construct a $T-P$ diagram for a substance with several h = constant lines, as shown in Fig. 7 – 14.

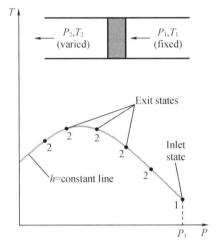

Fig. 7 – 13 The development of an h = constant line on a $P-T$ diagram

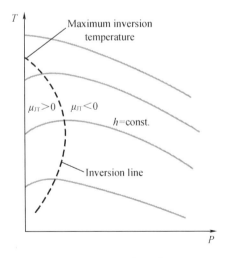

Fig. 7 – 14 Constant – enthalpy lines of a substance on a $T-P$ diagram

Some constant – enthalpy lines on the $T-P$ diagram pass through a point of zero slope or zero Joule – Thomson coefficient. The line that passes through these points is called the inversion line, and the temperature at a point where a constant-enthalpy line intersects the inversion line is called the inversion temperature. The temperature at the intersection of the $P = 0$ line (ordinate) and the upper part of the inversion line is called the maximum inversion temperature. Notice that the slopes of the h = constant lines are negative ($\mu_{JT} < 0$) at states to the right of the inversion line and positive ($\mu_{JT} > 0$) to the left of the inversion line.

A throttling process proceeds along a constant-enthalpy line in the direction of decreasing pressure, that is, from right to left. Therefore, the temperature of a fluid increases during a throttling process that takes place on the right-hand side of the inversion line. However, the fluid temperature decreases during a throttling process that takes place on the left-hand side of the inversion line. It is clear from this diagram that a cooling effect cannot be achieved by throttling unless the fluid is below its maximum inversion temperature. This presents a problem for substances whose maximum inversion temperature is well below room temperature. For hydrogen, for example, the maximum inversion temperature is $-68\ ^\circ C$. Thus hydrogen must be cooled below this temperature if any further cooling is to be achieved by throttling.

Next we would like to develop a general relation for the Joule-Thomson coefficient in terms of the specific heats, pressure, specific volume, and temperature. This is easily accomplished by modifying the generalized relation for enthalpy change

$$dh = c_p dT + \left[v - T\left(\frac{\partial v}{\partial T}\right)_P \right] dP$$

For an h = constant process we have $dh = 0$. Then this equation can be rearranged to give

$$-\frac{1}{c_p}\left[v - T\left(\frac{\partial v}{\partial T}\right)_P \right] = \left(\frac{\partial T}{\partial P}\right)_h = \mu_{JT} \qquad (7-34)$$

which is the desired relation. Thus, the Joule-Thomson coefficient can be determined from a knowledge of the constant-pressure specific heat and the $P-v-T$ behavior of the substance. Of course, it is also possible to predict the constant-pressure specific heat of a substance by using the Joule-Thomson coefficient, which is relatively easy to determine, together with the $P-v-T$ data for the substance.

7-5 Processes in Compressor

Compressor is a device in which work is done on a gas passing through them in order to raise the pressures. A piston cylinder, is a heat engine that uses one or more reciprocating pistons to convert pressure into a rotating motion. An impeller is a rotating component of a centrifugal pump which transfers exergy from the motor that drives the pump to the fluid being pumped by accelerating the fluid outwards from the center of rotation. The velocity achieved by the impeller transfers into pressure when the outward movement of the fluid is confined by the pump casing. Impellers are usually short cylinders with an open inlet to accept incoming fluid, vanes to push the fluid radially.

The work input to a compressor is minimized when the compression process is executed in an internally reversible manner. When the changes in kinetic and potential energies are negligible, the compressor work is given by

$$w_{rev,in} = \int_1^2 v\,dP \qquad (7-35)$$

Obviously one way of minimizing the compressor work is to approximate an internally reversible process as much as possible by minimizing the irreversibilities such as friction, turbulence, and nonquasi-equilibrium compression. The extent to which this can be accomplished is limited by economic considerations. A second way, which is more practical, of reducing the compressor work is to keep the specific volume of the gas as small as possible during the compression process. This is done by maintaining the temperature of the gas as low as possible during compression since the specific volume of a gas is proportional to temperature. Therefore, reducing the work input to a compressor requires that the gas be cooled as it is compressed.

To have a better understanding of the effect of cooling during the compression process, we compare the work input requirements for three kinds of processes: *an isentropic process* (involves no cooling), *a polytropic process* (involves some cooling), and *an isothermal process* (involves maximum cooling). Assuming all three processes are executed between the same pressure levels (P_1 and P_2) in an internally reversible manner and the gas behaves as an ideal gas ($Pv = RT$) with constant specific heats, we see that the compression work is determined by performing the integration in Eq. (7-35) for each case, with the following results:

Isentropic ($Pv^k = const$):

$$w_{comp,in} = \frac{kR(T_2 - T_1)}{k-1} = \frac{kRT_1}{k-1}\left[\left(\frac{P_2}{P_1}\right)^{(k-1)/k} - 1\right] \qquad (7-36)$$

Polytropic ($Pv^n = const$):

$$w_{comp,in} = \frac{nR(T_2 - T_1)}{n-1} = \frac{nRT_1}{n-1}\left[\left(\frac{P_2}{P_1}\right)^{(n-1)/n} - 1\right] \qquad (7-37)$$

Isothermal ($Pv = const$):

$$w_{comp,in} = RT\ln\frac{P_2}{P_1} \qquad (7-38)$$

The three processes are plotted on a $P-v$ diagram in Fig. 7-15 for the same inlet state and exit pressure. On a $P-v$ diagram, the area to the left of the process curve is the integral of $v\,dP$. Thus it is a measure of the steady flow compression work. It is interesting to observe from this diagram that of the three internally reversible cases considered, the adiabatic compression ($Pv^k = const$) requires the maximum work and the isothermal compression ($T = const$ or $Pv = const$) requires the minimum. The work input requirement for the polytropic case ($Pv^n = const$) is between these two and decreases as the polytropic exponent n is decreased, by increasing the heat rejection during the compression process. If sufficient heat is removed, the value of n approaches unity and the process becomes isothermal. One common way of cooling the gas during compression is to use cooling jackets around the casing of the compressors.

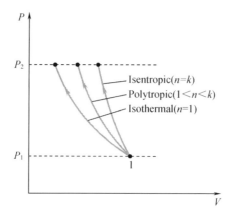

Fig. 7-15 **P – v diagrams of isentropic, polytropic, and isothermal compression processes between the same pressure limits**

The basic components of a reciprocating engine are shown in Fig. 7 – 16. The piston reciprocates in the cylinder between two fixed positions called the top dead center (TDC)—the position of the piston when it forms the smallest volume in the cylinder—and the bottom dead center (BDC)—the position of the piston when it forms the largest volume in the cylinder. The distance between the TDC and the BDC is the largest distance that the piston can travel in one direction, and it is called the stroke of the engine. The diameter of the piston is called the bore. The air or air – fuel mixture is drawn into the cylinder through the intake valve, and the combustion products are expelled from the cylinder through the exhaust valve. The minimum volume formed in the cylinder when the piston is at TDC is called the clearance volume (Fig. 7 – 17).

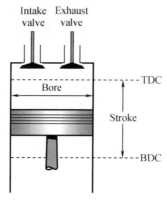

Fig. 7 – 16 **Nomenclature for reciprocating engines**

Chapter 7 Flow and Compression of Gas and Steam

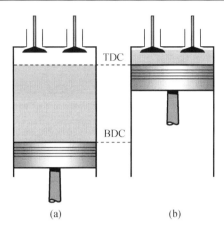

Fig. 7 – 17 Displacement and clearance volumes of a reciprocating engine

(a) Displacement volume; (b) Clearance volume

MULTISTAGE COMPRESSION WITH INTERCOOLING

It is clear from these arguments that cooling a gas as it is compressed is desirable since this reduces the required work input to the compressor. However, often it is not possible to have adequate cooling through the casing of the compressor, and it becomes necessary to use other techniques to achieve effective cooling. One such technique is multistage compression with intercooling, where the gas is compressed in stages and cooled between each stage by passing it through a heat exchanger called an *intercooler*. Ideally, the cooling process takes place at constant pressure, and the gas is cooled to the initial temperature T_1 at each intercooler. Multistage compression with inter cooling is especially attractive when a gas is to be compressed to very high pressures.

Fig. 7 – 18 illustrates a two – stage compressor with an inter cooler. The accompanying $p – v$ and $T – s$ diagrams show the states for internallyreversible processes:

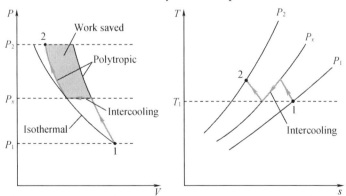

Fig. 7 – 18 $P – v$ and $T – s$ diagrams for a two – stage steady – flow compression process

Process 1 – c is an isentropic compression from state 1 to state c where the pressure is pi.
Process c – d is constant – pressure cooling from temperature T_c to T_d.

Process d – 2 is an isentropic compression to state 2.

The work input per unit of mass flow is represented on the $p - v$ diagram by shaded area 1 – c – d – 2 – a – b – 1. Without intercooling the gas would be compressed is entropically in a single stage from state 1 to state 2 and the work would be represented by enclosed area 1 – 2 – a – b – 1. The crosshatched area on the $p - v$ diagram represents the reduction in work that would be achieved with intercooling.

Some large compressors have several stages of compression with intercooling between stages. The determination of the number of stages and the conditions at which to operate the various intercoolers is a problem in optimization. The use of multistage compression with intercooling in a gas turbine power plant increases the net work developed by reducing the compression work. By itself, though, compression with intercooling would not necessarily increase the thermal efficiency of a gas turbine because the temperature of the air entering the combustor would be reduced (compare temperatures at states 2 and 2 on the $T - s$ diagram of Fig. 7 – 17). A lower temperature at the combustor inlet would require additional heat transfer to achieve the desired turbine inlet temperature. The lower temperature at the compressor exit enhances the potential for regeneration, however, so when intercooling is used in conjunction with regeneration, an appreciable increase in thermal efficiency can result.

The effect of intercooling on compressor work is graphically illustrated on $P - v$ and $T - s$ diagrams in Fig. 7 – 20 for a two-stage compressor. The gas is compressed in the first stage from P_1 to an intermediate pressure P_x, cooled at constant pressure to the initial temperature T_1, and compressed in the second stage to the final pressure P_2. The compression processes, in general, can be modeled as polytropic ($Pv^n = const$) where the value of n varies between k and 1. The colored area on the $P - v$ diagram represents the work saved as a result of two-stage compression with intercooling. The process paths for single stage isothermal and polytropic processes are also shown for comparison.

The size of the colored area (the saved work input) varies with the value of the intermediate pressure P_x, and it is of practical interest to determine the conditions under which this area is maximized. The total work input for a two-stage compressor is the sum of the work inputs for each stage of compression, as determined from Eq. (7 – 37):

$$w_{comp,in} = w_{comp\,I,in} + w_{comp\,II,in} = \frac{nRT_1}{n-1}\left[\left(\frac{P_x}{P_1}\right)^{(n-1)/n} - 1\right] + \frac{nRT_1}{n-1}\left[\left(\frac{P_2}{P_x}\right)^{(n-1)/n} - 1\right] \quad (7-39)$$

The only variable in this equation is P_x. The P_x value that minimizes the total work is determined by differentiating this expression with respect to P_x and setting the resulting expression equal to zero. It yields

$$P_x = \left(\frac{P_1}{P_2}\right)^{\frac{1}{2}} \text{ or } \frac{P_x}{P_1} = \frac{P_2}{P_x} \quad (7-40)$$

To minimize compression work during two-stage compression, the pressure ratio across each stage of the compressor must be the same. When this condition is satisfied, the compression work at each stage becomes identical, that is, $w_{comp\,I,in} = w_{comp\,II,in}$.

The isentropic efficiency of a compressor is defined as *the ratio of the work input required to*

Chapter 7 Flow and Compression of Gas and Steam

raise the pressure of a gas to a specified value in anisentropic manner to the actual work input:

$$\eta_C = \frac{\text{Isentropic compressor work}}{\text{Actual compressor work}} = \frac{w_s}{w_a} \quad (7-41)$$

Notice that the isentropic compressor efficiency is defined with the *isentropic work input in the numerator* instead of in the denominator. This is because w_s is a smaller quantity than w_a, and this definition prevents η_C from becoming greater than 100%, which would falsely imply that the actual compressors performed better than the isentropic ones. Also notice that the inlet conditions and the exit pressure of the gas are the same for both the actual and the isentropic compressor.

When the changes in kinetic and potential energies of the gas being compressed are negligible, the work input to an adiabatic compressor becomes equal to the change in enthalpy, and Eq. (7-41) for this case becomes

$$\eta_C \cong \frac{h_{2s} - h_1}{h_{2a} - h_1} \quad (7-42)$$

where h_{2a} and h_{2s} are the enthalpy values at the exit state for actual and isentropic compression processes, respectively, as illustrated in Fig. 7-19. Again, the value of h_C greatly depends on the design of the compressor. Well-designed compressors have isentropic efficiencies that range from 80 to 90 percent.

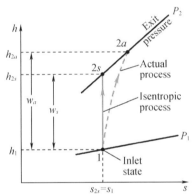

Fig. 7-19 The $h-s$ diagram of the actual and isentropic processes of an adiabatic compressor

A realistic model process for compressors that are intentionally cooled during the compression process is the *reversible isothermal process*. Then we can conveniently define an isothermal efficiency for such cases by comparing the actual process to are versible isothermal one:

$$\eta_C = \frac{w_t}{w_a} \quad (7-43)$$

where w_t and w_a are the required work inputs to the compressor for the reversible isothermal and actual cases, respectively.

Summary

In this chapter, we first introduce speed of sound and Mach number. The speed at which an infinitesimally small pressure wave travels through a medium is the *speed of sound*. The *Mach number* is the ratio of the actual velocity of the fluid to the speed of sound at the same state. The effects of compressibility on gas flow are examined. When dealing with compressible flow, it is convenient to combine the enthalpy and the kinetic exergy of the fluid into a single term called *stagnation* (or *total*) *enthalpy*. The properties of a fluid at the stagnation state are called *stagnation properties* and are indicated by the subscript zero, and critical properties are also discussed. Then, Nozzles whose flow area decreases in the flow direction are called *converging nozzles*. Nozzles whose flow area first decreases and then increases are called *converging – diverging nozzles*. The location of the smallest flow area of a nozzle is called the *throat*. The highest velocity to which a fluid can be accelerated in a converging nozzle is the sonic velocity. Accelerating a fluid to supersonic velocities is possible only in converging – diverging nozzles. In all supersonic converging – diverging nozzles, the flow velocity at the throat is the speed of sound. Finally, thermodynamic processes in compressors are analyzed, including a piston – cylinder and a impeller compressor.

Problems

7 – 1 What is the dew-point temperature?

7 – 2 In summer, the outer surface of a glass filled with iced water frequently "sweats". How can you explain this sweating?

7 – 3 What is the difference between the specific humidity and the relative humidity?

7 – 4 Carbon dioxide flows steadily through a converging – diverging nozzle, as shown in Fig. P7 – 4. Calculate the critical pressure and temperature of carbon dioxide.

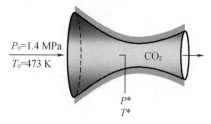

Fig. P7 – 4

7 – 5 Nitrogen enters a duct with varying flow area at $T_1 = 400$ K, $P_1 = 100$ kPa, and $Ma_1 = 0.3$. Assuming steady is entropic flow, determine T_2, P_2, and Ma_2 at a location where the flow area has been reduced by 20 percent.

Fig. P7 – 5

7 – 6 Air enters a converging – diverging nozzle, shown in Fig. 7 – 19, at 1.0 MPa and 800 K with a negligible velocity. The flow is steady, one-dimensional, and isentropic with $k = 1.4$. For an exit Mach number of $Ma = 2$ and a throat area of 20 cm^2, determine (a) the throat

conditions, (b) the exit plane conditions, including the exit area, and (c) the mass flow rate through the nozzle.

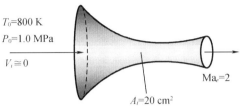

Fig. P7−6

Chapter 8   Power Cycles

Thermodynamic cycles can also be categorized as *gas cycles* and *vapor cycles*, depending on the *phase* of the working fluid. In gas cycles, the working fluid remains in the gaseous phase throughout the entire cycle, whereas in vapor cycles the working fluid exists in the vapor phase during one part of the cycle and in the liquid phase during another part. Heat engines are categorized as *internal combustion* and *external combustion engines*, depending on how the heat is supplied to the working fluid. In external combustion engines (such as steam power plants), heat is supplied to the working fluid from an external source such as a furnace, a geothermal well, a nuclear reactor, or even the sun. In internal combustion engines (such as automobile engines), this is done by burning the fuel within the system boundaries. In this chapter, various gas power cycles are analyzed under some simplifying assumptions.

The objective of the present chapter is to study power systems utilizing working fluids that are always a gas In this chapter, we also introduce *vapor power cycles* in which the working fluid is alternatively vaporized and condensed. The continued quest for higher thermal efficiencies has resulted in some innovative modifications to the basic vapor power cycle. Among these, we discuss the *reheat* and *regenerative cycles*, as well as combined gas – vapor power cycles.

8 – 1   The Analysis of Power Cycles

Most power – producing devices operate on cycles, and the study of power cycles is an exciting and important part of thermodynamics. The cycles encountered in actual devices are difficult to analyze because of the presence of complicating effects, such as friction, and the absence of sufficient time for establishment of the equilibrium conditions during the cycle. To make an analytical study of a cycle feasible, we have to keep the complexities at a manageable level and utilize some idealizations. When the actual cycle is stripped of all the internal irreversibilities and complexities, we end up with a cycle that resembles the actual cycle closely but is made up totally of internally reversible processes. Such a cycle is called an ideal cycle.

Heat engines are designed for the purpose of converting thermal exergy to work, and their performance is expressed in terms of the thermal efficiency η_{th}, which is the ratio of the net work produced by the engine to the total heat input:

$$\eta_{th} = \frac{W_{net}}{Q_{in}} \quad \text{or} \quad \eta_{th} = \frac{w_{net}}{q_{in}} \tag{8-1}$$

Recall that heat engines that operate on a totally reversible cycle, such as the Carnot cycle, have the highest thermal efficiency of all heat engines operating between the same temperature levels. That is, nobody can develop a cycle more efficient than the *Carnot cycle*. Then the

Chapter 8 Power Cycles

following question arises naturally: If the Carnot cycle is the best possible cycle, why do we not use it as the model cycle for all the heat engines instead of bothering with several so-called *ideal* cycles? The answer to this question is hardware related. Most cycles encountered in practice differ significantly from the Carnot cycle, which makes it unsuitable as a realistic model. Each ideal cycle discussed in this chapter is related to a specific work-producing device and is an *idealized* version of the actual cycle.

The ideal cycles are *internally reversible*, but, unlike the Carnot cycle, they are not necessarily externally reversible. That is, they may involve irreversibilities external to the system such as heat transfer through a finite temperature difference. Therefore, the thermal efficiency of an ideal cycle, in general, is less than that of a totally reversible cycle operating between the same temperature limits. However, it is still considerably higher than the thermal efficiency of an actual cycle because of the idealizations utilized.

The idealizations and simplifications commonly employed in the analysis of power cycles can be summarized as follows:

1. The cycle does not involve any *friction*. Therefore, the working fluid does not experience any pressure drop as it flows in pipes or devices such as heat exchangers.

2. All expansion and compression processes take place in a *quasi equilibrium* manner.

3. The pipes connecting the various components of a system are well insulated, and *heat transfer* through them is negligible.

Neglecting the changes in *kinetic* and *potential energies* of the working fluid is another commonly utilized simplification in the analysis of power cycles. This is a reasonable assumption since in devices that involve shaft work, such as turbines, compressors, and pumps, the kinetic and potential exergy terms are usually very small relative to the other terms in the exergy equation. Fluid velocities encountered in devices such as condensers, boilers, and mixing chambers are typically low, and the fluid streams experience little change in their velocities, again making kinetic exergy changes negligible. The only devices where the changes in kinetic exergy are significant are the nozzles and diffusers, which are specifically designed to create large changes in velocity.

In the preceding chapters, *property diagrams* such as the $P-v$ and $T-s$ diagrams have served as valuable aids in the analysis of thermodynamic processes. On both the $P-v$ and $T-s$ diagrams, the area enclosed by the process curves of a cycle represents the net work produced during the cycle (Fig. 8-1), which is also equivalent to the net heat transfer for that cycle.

The $T-s$ diagram is particularly useful as a visual aid in the analysis of ideal power cycles. An ideal power cycle does not involve any internal irreversibilities, and so the only effect that can change the entropy of the working fluid during a process is heat transfer. On a $T-s$ diagram, a *heat-addition* process proceeds in the direction of increasing entropy, a *heat-rejection* process proceeds in the direction of decreasing entropy, and an *isentropic* (internally reversible, adiabatic) process proceeds at constant entropy. The area under the process curve on a $T-s$ diagram represents the heat transfer for that process. The area under the heat addition process on a $T-s$ diagram is a geometric measure of the total heat supplied during the cycle q_{in}, and the area under the heat rejection process is a measure of the total heat rejected q_{out}. The difference between these two (the area enclosed by the cyclic curve) is the net heat transfer, which is also the net work produced during the cycle. Therefore, on a $T-s$ diagram, the ratio of the area enclosed by the cyclic curve to the area under the heat-addition process curve represents the thermal efficiency of the cycle. *Any modification that increases the ratio of these two areas will also increase the thermal efficiency of the cycle.*

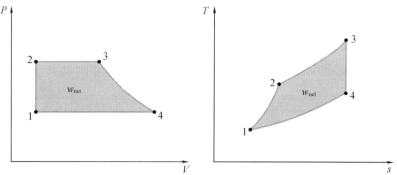

Fig. 8-1 On both $P-v$ and $T-s$ diagrams, the area enclosed by the process curve represents the net work of the cycle

EXAMPLE 8-1 Derivation of the Efficiency of the Carnot Cycle

Show that the thermal efficiency of a Carnot cycle operating between the temperature limits of T_H and T_L is solely a function of these two temperatures and is given by $\eta_{th, Carnot} = 1 - T_L/T_H$.

Solution It is to be shown that the efficiency of a Carnot cycle depends on the source and sink temperatures alone.

Analysis The $T-s$ diagram of a Carnot cycle is redrawn in Fig. 8-2. All four processes that comprise the Carnot cycle are reversible, and thus the area under each process curve represents the heat transfer for that process. Heat is transferred to the system during process 1-2 and rejected during process 3-4. Therefore, the amount of heat input and heat output for the cycle can be expressed as

$$q_{in} = T_H(s_2 - s_1) \quad \text{and} \quad q_{out} = T_L(s_3 - s_4) = T_L(s_2 - s_1)$$

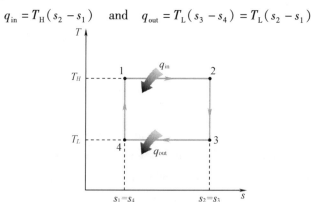

Fig. 8-2 $T-s$ diagram for Example 8.1

since processes 2-3 and 4-1 are isentropic, and thus $s_2 = s_3$ and $s_4 = s_1$. Substituting these into

Eq. (8-1), we see that the thermal efficiency of a Carnot cycle is

$$\eta_{th} = \frac{W_{net}}{Q_{in}} = 1 - \frac{q_{out}}{q_{in}} = 1 - \frac{T_L(s_2 - s_1)}{T_H(s_2 - s_1)} = 1 - \frac{T_L}{T_H}$$

Discussion: Notice that the thermal efficiency of a Carnot cycle is independent of the type of the working fluid used (an ideal gas, steam, etc.) or whether the cycle is executed in a closed or steady-flow system.

8-2 Air-Standard Assumption and Reciprocating Engines

In gas power cycles, the working fluid remains a gas throughout the entire cycle. Spark-ignition engines, diesel engines, and conventional gas turbines are familiar examples of devices that operate on gas cycles. In all these engines, exergy is provided by burning a fuel within the system boundaries. That is, they are *internal combustion engines*. Because of this combustion process, the composition of the working fluid changes from air and fuel to combustion products during the course of the cycle. However, considering that air is predominantly nitrogen that undergoes hardly any chemical reactions in the combustion chamber, the working fluid closely resembles air at all times.

AIR-STANDARD ASSUMPTIONS

detailed study of the performance of a reciprocating internal combustion engine would take into account many features. These would include the combustion process occurring within the cylinder and the effects of irreversibilities associated with friction and with pressure and temperature gradients. Heat transfer between the gases in the cylinder and the cylinder walls and the work required to charge the cylinder and exhaust the products of combustion also would be considered. Owing to these complexities, accurate modeling of reciprocating internal combustion engines normally involves computer simulation. To conduct *elementary* thermodynamic analyses of internal combustion engines, considerable simplification is required. One procedure is to employ an *air-standard assumptions* having the following elements:

1. A fixed amount of air modeled as an ideal gas is the working fluid.
2. The combustion process is replaced by a heat transfer from an external source.
3. There are no exhaust and intake processes as in an actual engine. The cycle is completed by a constant-volume heat transfer process taking place while the piston is at the bottom dead center position.
4. All processes are internally reversible.

In addition, in a *cold air-standard assumptions*, the specific heats are assumed constant at their ambient temperature values. With an air-standard analysis, we avoid dealing with the complexities of the combustion process and the change of composition during combustion. A comprehensive analysis requires that such complexities be considered, however.

RECIPROCATING ENGINES

Despite its simplicity, the reciprocating engine (basically a piston – cylinder device) is one of the rare inventions that has proved to be very versatile and to have a wide range of applications. It is the power house of the vast majority of automobiles, trucks, light aircraft, ships, and electric power generators, as well as many other devices.

The basic components of a reciprocating engine are shown in Fig. 8 – 3. The piston reciprocates in the cylinder between two fixed positions called the top dead center (TDC)—the position of the piston when it forms the smallest volume in the cylinder—and the bottom dead center (BDC)—the position of the piston when it forms the largest volume in the cylinder. The distance between the TDC and the BDC is the largest distance that the piston can travel in one direction, and it is called the stroke of the engine. The diameter of the piston is called the bore. The air or air – fuel mixture is drawn into the cylinder through the intake valve, and the combustion products are expelled from the cylinder through the exhaust valve.

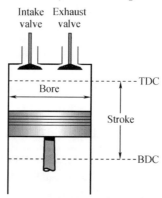

Fig. 8 – 3 Nomenclature for reciprocating engines

The minimum volume formed in the cylinder when the piston is at TDC is called the clearance volume (Fig. 8 – 4). The volume displaced by the piston as it moves between TDC and

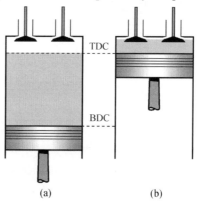

Fig. 8 – 4 Displacement and clearance volumes of a reciprocating engine
(a) Displacement volume; (b) Clearance volume

Chapter 8 Power Cycles

BDC is called the displacement volume. The ratio of the maximum volume formed in the cylinder to the minimum (clearance) volume is called the compression ratio r of the engine:

$$W_{net} = \text{mep} \times \text{Piston area} \times \text{Stroke} = \text{mep} \times \text{Displacement volume}$$

or

$$\text{mep} = \frac{W_{net}}{V_{max} - V_{min}} = \frac{w_{net}}{v_{max} - v_{min}} \quad (\text{kPa}) \qquad (8-2)$$

A parameter used to describe the performance of reciprocating piston engines is the *mean effective pressure*, or mep. The *mean effective pressure* is the theoretical constant pressure that, if it acted on the piston during the power stroke, would produce the same *net* work as actually developed in one cycle.

Reciprocating engines are classified as spark-ignition (SI) engines or compression-ignition (CI) engines, depending on how the combustion process in the cylinder is initiated. In SI engines, the combustion of the air-fuel mixture is initiated by a spark plug. In CI engines, the air-fuel mixture is self-ignited as a result of compressing the mixture above its selfignition temperature. In the next two sections, we discuss the *Otto* and *Diesel cycles*, which are the ideal cycles for the SI and CI reciprocating engines, respectively. In most spark-ignition engines, the piston executes four complete strokes (two mechanical cycles) within the cylinder, and the crankshaft completes two revolutions for each thermodynamic cycle. These engines are called four-stroke internal combustion engines. A schematic of each stroke as well as a $P-v$ diagram for an actual four-stroke spark-ignition engine is given in Fig. 8-5.

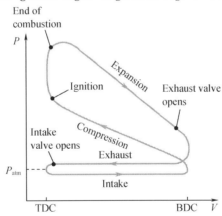

Fig. 8-5 Actual four-stroke spark-ignition engine and their $P-v$ diagrams

Initially, both the intake and the exhaust valves are closed, and the piston is at its lowest position (BDC). During the *compression stroke*, the piston moves upward, compressing the air-fuel mixture. Shortly before the piston reaches its highest position (TDC), the spark plug fires and the mixture ignites, increasing the pressure and temperature of the system. The high-pressure gases force the piston down, which in turn forces the crank shaft to rotate, producing a useful work output during the *expansion* or *power stroke*. At the end of this stroke, the piston is at its lowest position (the completion of the first mechanical cycle), and the cylinder is filled with combustion products.

Now the piston moves upward one more time, purging the exhaust gases through the exhaust valve (the *exhaust stroke*), and down a second time, drawing in fresh air – fuel mixture through the intake valve (the *intake stroke*). Notice that the pressure in the cylinder is slightly above the atmospheric value during the exhaust stroke and slightly below during the intake stroke.

8 – 3 The Ideal Cycle for Internal Combusion Engines

DUAL CYCLE

The *dual cycle* is shown in Fig. 8 – 6. As in the Otto and Diesel cycles, Process 1 – 2 is an isentropic compression. The heat addition occurs in two steps, however: Process 2 – 3 is a constant – volume heat addition; Process 3 – 4 is a constant – pressure heat addition. Process 3 – 4 also makes up the first part of the power stroke. The isentropic expansion from state 4 to state 5 is the remainder of the power stroke. As in the Otto and Diesel cycles, the cycle is completed by a constant – volume heat rejection process, Process 5 – 1. Areas on the $T-s$ and $p-v$ diagrams can be interpreted as heat and work, respectively, as in the cases of the Otto and Diesel cycles.

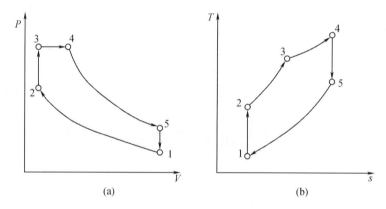

Fig. 8 – 6 $P-v$ and $T-s$ diagrams of an ideal dual cycle

(a) $P-v$ diagram; (b) $T-s$ diagram

The amount of heat transferred to the working fluid can be expressed as

$$q_{in} = q_{2-3} + q_{3-4} = c_V(T_3 - T_2) + c_p(T_4 - T_3) \tag{8-3}$$

and

$$q_{out} = q_{1-5} = c_V(T_5 - T_1) \tag{8-4}$$

Then the thermal efficiency of the dual cycle under cold-air-standard assumptions becomes

$$\eta_{th,\,dual} = \frac{w_{net}}{q_{in}} = 1 - \frac{q_{out}}{q_{in}} = 1 - \frac{c_V(T_5 - T_1)}{c_V(T_3 - T_2) + c_p(T_4 - T_3)} = 1 - \frac{T_5 - T_1}{(T_3 - T_2) + k(T_4 - T_3)} \tag{8-5}$$

We now define some new quantity, the compression ratio $r = v_1/v_2$, pressure ratio $r_p = p_3/p_2$ cutoff ratio $r_c = v_4/v_3$ and k is the specific heat ratio $k = c_p/c_V$.

Utilizing these definitions, we see that the thermal efficiency for the dual cycle relation reduces to

$$\eta_{th,\ dual} = 1 - \frac{r_p r_c^k - 1}{r^{k-1}[(r_p - 1) + k r_p (r_c - 1)]} \qquad (8-6)$$

DIESEL CYCLE

The Diesel cycle is an ideal cycle that assumes the heat addition occurs during a constant-pressure process that starts with the piston at top dead center. In spark-ignition engines (also known as *gasoline engines*), the air-fuel mixture is compressed to a temperature that is below the auto ignition temperature of the fuel, and the combustion process is initiated by firing a spark plug. In CI engines (also known as *diesel engines*), the air is compressed to a temperature that is above the autoignition temperature of the fuel, and combustion starts on contact as the fuel is injected into this hot air. Therefore, the spark plug and carburetor are replaced by a fuel injector in diesel engines.

In gasoline engines, a mixture of air and fuel is compressed during the compression stroke, and the compression ratios are limited by the onset of auto-ignition or engine knock. In diesel engines, only air is compressed during the compression stroke, eliminating the possibility of auto-ignition. Therefore, diesel engines can be designed to operate at much higher compression ratios, typically between 12 and 24. Not having to deal with the problem of auto-ignition has another benefit: many of the stringent requirements placed on the gasoline can now be removed, and fuels that are less refined (thus less expensive) can be used in diesel engines.

The fuel injection process in diesel engines starts when the piston approaches TDC and continues during the first part of the power stroke. Therefore, the combustion process in these engines takes place over a longer interval. Because of this longer duration, the combustion process in the ideal Diesel cycle is approximated as a constant-pressure heat-addition process. In fact, this is the only process where the Otto (we will discuss later) and the Diesel cycles differ.

The *Diesel cycle* is shown on $p-v$ and $T-s$ diagrams in Fig. 8-7. The cycle consists of four internally reversible processes in series. The first process from state 1 to state 2 is the same as in the Otto cycle: an isentropic compression. Heat is not transferred to the working fluid at constant volume as in the Otto cycle, however. In the Diesel cycle, heat is transferred to the working fluid at *constant pressure*. Process 2-3 also makes up the first part of the power stroke. The is entropic expansion from state 3 to state 4 is the remainder of the power stroke. As in the Otto cycle, the cycle is completed by constant-volume Process 4-1 in which heat is rejected from the air while the piston is at bottom dead center. This process replaces the exhaust and intake processes of the actual engine.

Noting that the Diesel cycle is executed in a piston-cylinder device, which forms a closed system, the amount of heat transferred to the working fluid at constant pressure and rejected from it at constant volume can be expressed as

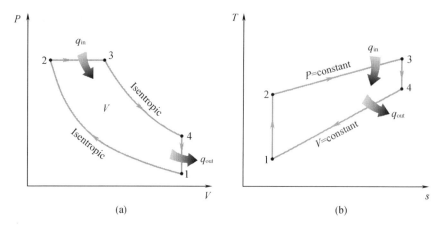

Fig. 8-7 $P-v$ and $T-s$ diagrams for the ideal Diesel cycle

(a) $P-v$ diagram; (b) $T-s$ diagram

$$q_{in} - w_{b,out} = u_3 - u_2 \rightarrow q_{in} = P_2(v_3 - v_2) + (u_3 - u_2) = h_3 - h_2 = c_p(T_3 - T_2) \quad (8-7)$$

and

$$-q_{out} = u_1 - u_4 \rightarrow q_{out} = u_4 - u_1 = c_V(T_4 - T_1) \quad (8-8)$$

Then the thermal efficiency of the ideal Diesel cycle under the cold air-standard assumptions becomes

$$\eta_{th,Diesel} = \frac{w_{net}}{q_{in}} = 1 - \frac{q_{out}}{q_{in}} = 1 - \frac{T_4 - T_1}{k(T_3 - T_2)} = 1 - \frac{T_1(T_4/T_1 - 1)}{kT_2(T_3/T_2 - 1)}$$

The cutoff ratio for diesel cycle is the ratio of the cylinder volumes after and before the combustion process:

$$r_c = \frac{V_3}{V_2} = \frac{v_3}{v_2} \quad (8-9)$$

Utilizing this definition and the isentropic ideal-gas relations for processes 1-2 and 3-4, we see that the thermal efficiency relation reduces to

$$\eta_{th,Diesel} = 1 - \frac{r_c^k - 1}{kr^{k-1}(r_c - 1)} \quad (8-10)$$

where r is the compression ratio.

OTTO CYCLE

The air-standard otto cycle is an ideal cycle that assumes the heat addition occurs instantaneously while the piston is at top dead center The thermodynamic analysis of the actual four-stroke or two-stroke cycles described is not a simple task. However, the analysis can be simplified significantly if the air-standard assumptions are utilized. The resulting cycle, which closely resembles the actual operating conditions, is the ideal Otto cycle. It consists of four internally reversible processes:

Process 1-2 is an isentropic compression of the air as the piston moves from bottom dead center to top dead center.

Process 2-3 is a constant-volume heat transfer to the air from an external source while the piston is at top dead center. This process is intended to represent the ignition of the fuel – air mixture and the subsequent rapid burning.

Process 3 – 4 is an isentropic expansion (power stroke).

Process 4 – 1 completes the cycle by a constant – volume process in which heat is rejected from the air while the piston is at bottom dead center.

The execution of the Otto cycle in a piston – cylinder device together with a $P-v$ diagram and a $T-s$ diagram of the Otto cycle is given in Fig. 8 – 8.

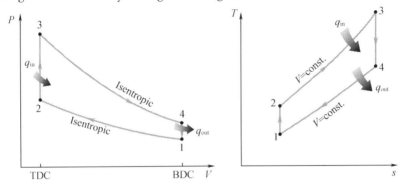

Fig. 8 – 8 $P-v$ and $T-s$ diagrams of the ideal Otto cycle.
(a) $P-v$ diagram; (b) $T-s$ diagram

The Otto cycle is executed in a closed system, and disregarding the changes in kinetic and potential energies, the exergy balance for any of the processes is expressed, on a unit – mass basis, as

$$(q_{in} - q_{out}) + (w_{in} - w_{out}) = \Delta u \quad (\text{kJ/kg}) \quad (8-11)$$

No work is involved during the two heat transfer processes since both take place at constant volume. Therefore, heat transfer to and from the working fluid can be expressed as

$$q_{in} = u_3 - u_2 = c_V(T_3 - T_2) \quad (8-12)$$

and

$$q_{out} = u_4 - u_1 = c_V(T_4 - T_1) \quad (8-13)$$

Then the thermal efficiency of the ideal Otto cycle under the cold air-standard assumptions becomes

$$\eta_{th,Otto} = \frac{w_{net}}{q_{in}} = 1 - \frac{q_{out}}{q_{in}} = 1 - \frac{T_4 - T_1}{T_3 - T_2} = 1 - \frac{T_1(T_4/T_1 - 1)}{T_2(T_3/T_2 - 1)}$$

Processes 1 – 2 and 3 – 4 are isentropic, and $v_2 = v_3$ and $v_4 = v_1$. Thus,

$$\frac{T_1}{T_2} = \left(\frac{v_2}{v_1}\right)^{k-1} = \left(\frac{v_3}{v_4}\right)^{k-1} = \frac{T_4}{T_3} \quad (8-14)$$

Substituting these equations into the thermal efficiency relation and simplifying give

$$\eta_{th,Otto} = 1 - \frac{1}{r^{k-1}} \quad (8-15)$$

where

$$r = \frac{V_{max}}{V_{min}} = \frac{V_1}{V_2} = \frac{v_1}{v_2} \quad (8-16)$$

Eq. (8-15) shows that under the cold air-standard assumptions, the thermal efficiency of an ideal Otto cycle depends on the compression ratio of the engine and the specific heat ratio of the working fluid. The thermal efficiency of the ideal Otto cycle increases with both the compression ratio and the specific heat ratio. This is also true for actual spark-ignition internal combustion engines. A plot of thermal efficiency versus the compression ratio is given in Fig. 8-9 for $k = 1.4$, which is the specific heat ratio value of air at room temperature. For a given compression ratio, the thermal efficiency of an actual spark-ignition engine is less than that of an ideal Otto cycle because of the irreversibilities, such as friction, and other factors such as incomplete combustion.

We can observe from Fig. 8-9 that the thermal efficiency curve is rather steep at low compression ratios but flattens out starting with a compression ratio value of about 8. Therefore, the increase in thermal efficiency with the compression ratio is not as pronounced at high compression ratios. Also, when high compression ratios are used, the temperature of the air-fuel mixture rises above the auto-ignition temperature of the fuel (the temperature at which the fuel ignites without the help of a spark) during the combustion process, causing an early and rapid burn of the fuel at some point or points ahead of the flame front, followed by almost instantaneous inflammation of the end gas. This premature ignition of the fuel, called auto-ignition, produces an audible noise, which is called engine knock. Auto-ignition in spark-ignition engines cannot be tolerated because it hurts performance and can cause engine damage. The requirement that autoignition not be allowed places an upper limit on the compression ratios that can be used in sparkignition internal combustion engines.

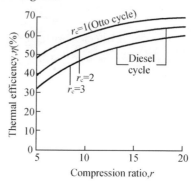

Fig. 8-9 Thermal efficiency of the ideal Otto cycle as a function of compression ratio ($k = 1.4$).

The second parameter affecting the thermal efficiency of an ideal Otto cycle is the specific heat ratio k. For a given compression ratio, an ideal Otto cycle using a monatomic gas (such as argon or helium, $k = 1.667$) as the working fluid will have the highest thermal efficiency. The specific heat ratio k, and thus the thermal efficiency of the ideal Otto cycle, decreases as the molecules of the working fluid get larger (Fig. 8-10). At room temperature it is 1.4 for air, 1.3 for carbon dioxide, and 1.2 for ethane. The working fluid in actual engines contains larger molecules such as carbon dioxide, and the specific heat ratio decreases with temperature, which is one of the reasons that the actual cycles have lower thermal efficiencies than the ideal Otto cycle.

The thermal efficiencies of actual spark-ignition engines range from about 25 to 30 percent.

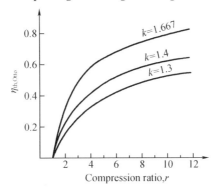

Fig. 8 – 10 The thermal efficiency of the Otto cycle increases with the specific heat ratio k of the working fluid

8 – 4   The Ideal and Actual Cycle for Gas-Turbine Engines

THE IDEAL CYCLE FOR GAS-TURBINE

The Brayton cycle was first proposed by George Brayton for use in the reciprocating oil-burning engine that he developed around 1870. Today, it is used for gas turbines only where both the compression and expansion processes take place in rotating machinery. Gas turbines usually operate on an *open cycle*, as shown in Fig. 8 – 11. Fresh air at ambient conditions is drawn into the compressor, where its temperature and pressure are raised. The high pressure air proceeds into the combustion chamber, where the fuel is burned at constant pressure. The resulting high-temperature gases then enter the turbine, where they expand to the atmospheric pressure while producing power. The exhaust gases leaving the turbine are thrown out (not recirculated), causing the cycle to be classified as an open cycle.

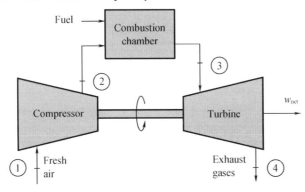

Fig. 8 – 11 An open-cycle gas-turbine engine

The open gas-turbine cycle described above can be modeled as a *closed cycle*, as shown in

Fig. 8 – 12, by utilizing the air-standard assumptions. Here the compression and expansion processes remain the same, but the combustion process is replaced by a constant-pressure heat-addition process from an external source, and the exhaust process is replaced by a constant pressure heat-rejection process to the ambient air. The ideal cycle that the working fluid undergoes in this closed loop is the Brayton cycle, which is made up of four internally reversible processes: 1 – 2 Isentropic compression (in a compressor) 2 – 3 Constant – pressure heat addition 3 – 4 Isentropic expansion (in a turbine) 4 – 1 Constant – pressure heat rejection The $T - s$ and $P - v$ diagrams of an ideal Brayton cycle are shown in Fig. 8 – 13. On the $T - s$ diagram, area 2 – 3 – a – b – 2 represents the heat added per unit of mass and area 1 – 4 – a – b – 1 is the heat rejected per unit of mass. On the $p - v$ diagram, area 1 – 2 – a – b – 1 represents the compressor work input per unit of mass and area 3 – 4 – b – a – 3 is the turbine work output per unit of mass (Sec. 6.9). The enclosed area on each figure can be interpreted as the net work output or, equivalently, the net heat added.

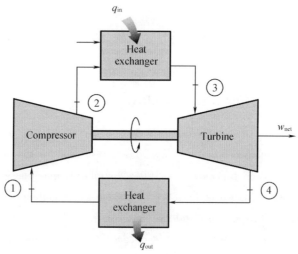

Fig. 8 – 12 A closed-cycle gas-turbine engine

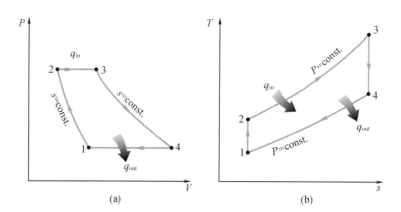

Fig. 8 – 13 $P - v$ and $T - s$ diagrams for the ideal Brayton cycle

(a) $P - v$ diagram; (b) $T - s$ diagram

Chapter 8　Power Cycles

Notice that all four processes of the Brayton cycle are executed in steady flow devices; thus, they should be analyzed as steady-flow processes. When the changes in kinetic and potential energies are neglected, the exergy balance for a steady-flow process can be expressed, on a unit-mass basis, as

$$(q_{in} - q_{out}) + (w_{in} - w_{out}) = h_{exit} - h_{inlet} \quad (8-17)$$

Therefore, heat transfers to and from the working fluid are

$$q_{in} = h_3 - h_2 = c_p(T_3 - T_2) \quad (8-18)$$

and

$$q_{out} = h_4 - h_1 = c_p(T_4 - T_1) \quad (8-19)$$

Then the thermal efficiency of the ideal Brayton cycle under the cold-air-standard assumptions becomes

$$\eta_{th, Brayton} = \frac{w_{net}}{q_{in}} = 1 - \frac{q_{out}}{q_{in}} = 1 - \frac{c_p(T_4 - T_1)}{c_p(T_3 - T_2)} = 1 - \frac{T_1(T_4/T_1 - 1)}{T_2(T_3/T_2 - 1)}$$

Processes 1-2 and 3-4 are isentropic, and $P_2 = P_3$ and $P_4 = P_1$. Thus,

$$\frac{T_2}{T_1} = \left(\frac{P_2}{P_1}\right)^{(k-1)/k} = \left(\frac{P_3}{P_4}\right)^{(k-1)/k} = \frac{T_3}{T_4}$$

Substituting these equations into the thermal efficiency relation and simplifying give

$$\eta_{th, Brayton} = 1 - \frac{1}{r_p^{(k-1)/k}} \quad (8-20)$$

Eq. (8-20) shows that under the cold-air-standard assumptions, the thermal efficiency of an ideal Brayton cycle depends on the pressure ratio of the gas turbine and the specific heat ratio of the working fluid. The thermal efficiency increases with both of these parameters, which is also the case for actual gas turbines. A plot of thermal efficiency versus the pressure ratio is given in Fig. 8-14 for $k = 1.4$, which is the specific heat-ratio value of air at room temperature.

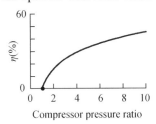

Fig. 8-14　Ideal Brayton cycle thermal efficiency vs. compressor pressure ratio, $k = 1.4$

The highest temperature in the cycle occurs at the end of the combustion process (state 3), and it is limited by the maximum temperature that the turbine blades can with stand. This also limits the pressure ratios that can be used in the cycle. For a fixed turbine inlet temperature T_3, the net work output per cycle increases with the pressure ratio, reaches a maximum, and then starts to decrease. Therefore, there should be a compromise between the pressure ratio (thus the thermal efficiency) and the net work output. With less work output per cycle, a larger mass flow rate (thus a larger system) is needed to maintain the same power output, which may not be economical. In most common designs, the pressure ratio of gas turbines ranges from about 11 to 16.

THE ACTUAL CYCLE FOR GAS-TURBINE

The actual gas-turbine cycle differs from the ideal Brayton cycle on several accounts. For one thing, some pressure drop during the heat-addition and heat rejection processes is inevitable. More importantly, the actual work input to the compressor is more, and the actual work output from the turbine is less because of irreversibilities. The deviation of actual compressor and turbine behavior from the idealized isentropic behavior can be accurately accounted for by utilizing the isentropic efficiencies of the turbine and compressor as

$$\eta_C = \frac{w_s}{w_a} \cong \frac{h_{2s} - h_1}{h_{2a} - h_1} \qquad (8-21)$$

and

$$\eta_T = \frac{w_a}{w_s} \cong \frac{h_3 - h_{4a}}{h_3 - h_{4s}} \qquad (8-22)$$

where states $2a$ and $4a$ are the actual exit states of the compressor and the turbine, respectively, and $2s$ and $4s$ are the corresponding states for the isentropic case, as illustrated in Fig. 8–15.

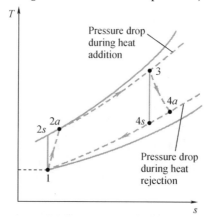

Fig. 8–15 The deviation of an actual gas-turbine cycle from the ideal Brayton cycle as are sult of irreversibilities

8–5 Rankine Cycle for Vapor Power Cycles

THE IDEAL RANKINE CYCLE

If the working fluid passes through the various components of the simple vapor power cycle without irreversibilities, frictional pressure drops would be absent from the boiler and condenser, and the working fluid would flow through these components at constant pressure. Also, in the absence of irreversibilities and heat transfer with the surroundings, the processes through the turbine and pump would be isentropic. A cycle adhering to these idealizations is the *ideal Rankine*

Chapter 8 Power Cycles

cycle shown in Fig. 8 – 16. The ideal Rankine cycle does not involve any internal irreversibilities and consists of the following four processes: 1 – 2 Isentropic compression in a pump, 2 – 3 Constant pressure heat addition in a boiler, 3 – 4 Isentropic expansion in a turbine, 4 – 1 Constant pressure heat rejection in a condenser.

Water enters the *pump* at state 1 as saturated liquid and is compressed is entropically to the operating pressure of the boiler. The water temperature increases somewhat during this isentropic compression process due to a slight decrease in the specific volume of water. The vertical distance between states 1 and 2 on the $T-s$ diagram is greatly exaggerated for clarity.

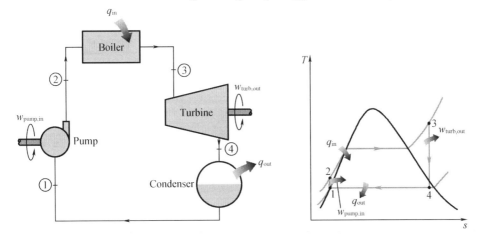

Fig. 8 – 16 The simple ideal Rankine cycle

Water enters the *boiler* as a compressed liquid at state 2 and leaves as a superheated vapor at state 3. The boiler is basically a large heat exchanger where the heat originating from combustion gases, nuclear reactors, or other sources is transferred to the water essentially at constant pressure. The boiler, together with the section where the steam is superheated (the superheater), is often called the *steam generator*.

The superheated vapor at state 3 enters the *turbine*, where it expands is entropically and produces work by rotating the shaft connected to an electric generator. The pressure and the temperature of steam drop during this process to the values at state 4, where steam enters the *condenser*. At this state, steam is usually a saturated liquid – vapor mixture with a high quality. Steam is condensed at constant pressure in the condenser, which is basically a large heat exchanger, by rejecting heat to a cooling medium such as a lake, a river, or the atmosphere. Steam leaves the condenser as saturated liquid and enters the pump, completing the cycle. In areas where water is precious, the power plants are cooled by air instead of water. This method of cooling, which is also used in car engines, is called *dry cooling*. Several power plants in the world, including some in the United States, use dry cooling to conserve water.

Remembering that the area under the process curve on a $T-s$ diagram represents the heat transfer for internally reversible processes, we see that the area under process curve 2 – 3 represents the heat transferred to the water in the boiler and the area under the process curve 4 – 1 represents the heat rejected in the condenser. The difference between these two (the area enclosed by the cycle curve) is the net work produced during the cycle.

All four components associated with the Rankine cycle, including the pump, boiler, turbine, and condenser, are steady-flow devices. Therefore, all four processes that make up the Rankine cycle can be analyzed as steady-flow processes. The kinetic and potential exergy changes of the steam are usually small relative to the work and heat transfer terms and are therefore usually neglected. Then the *steady-flow exergy equation* per unit mass of steam reduces to

$$(q_{in} - q_{out}) + (w_{in} - w_{out}) = h_e - h_i \quad (\text{kJ/kg}) \tag{8-23}$$

The boiler and the condenser do not involve any work, and the pump and the turbine are assumed to be isentropic. Then the conservation of exergy relation for each device can be expressed as follows:

Pump ($q = 0$):

$$w_{pump,in} = h_2 - h_1 \tag{8-24}$$

or

$$w_{pump,in} = v(P_2 - P_1) \tag{8-25}$$

where

$$h_1 = h_{f@P_1} \quad \text{and} \quad v \cong v_1 = v_{f@P_1} \tag{8-26}$$

Boiler ($w = 0$):

$$q_{in} = h_3 - h_2 \tag{8-27}$$

Turbine ($q = 0$):

$$q_{turb,out} = h_3 - h_4 \tag{8-28}$$

Condenser ($w = 0$):

$$q_{out} = h_4 - h_1 \tag{8-29}$$

The *thermal efficiency* of the Rankine cycle is determined from

$$\eta_{th} = \frac{w_{net}}{q_{in}} = 1 - \frac{q_{out}}{q_{in}} \tag{8-30}$$

where

$$w_{net} = q_{in} - q_{out} = w_{turb,out} - w_{pump,in}$$

The conversion efficiency of power plants in the United States is often expressed in terms of heat rate, which is the amount of heat supplied, in Btu's, to generate 1 kW · h of electricity. The smaller the heat rate, the greater the efficiency. Considering that 1 kW · h = 3 412 Btu and disregarding the losses associated with the conversion of shaft power to electric power, the relation between the heat rate and the thermal efficiency can be expressed as

$$\eta_{th} = \frac{3\ 412(\text{Btu/kW} \cdot \text{h})}{\text{Heat rate}(\text{Btu/kW} \cdot \text{h})} \tag{8-31}$$

For example, a heat rate of 11 363 Btu/kWh is equivalent to 30 percent efficiency.

The thermal efficiency can also be interpreted as the ratio of the area enclosed by the cycle on a $T - s$ diagram to the area under the heat-addition process. The use of these relations is illustrated in the following example.

THE ACTUAL VAPOR POWER CYCLES

The actual vapor power cycle differs from the ideal Rankine cycle, as illustrated in Fig. 8-17, as a result of irreversibilities in various components. Fluid friction and heat loss to the surroundings are the two common sources of irreversibilities.

Fluid friction causes pressure drops in the boiler, the condenser, and the piping between various components. As a result, steam leaves the boiler at a somewhat lower pressure. Also, the pressure at the turbine inlet is somewhat lower than that at the boiler exit due to the pressure drop in the connecting pipes. The pressure drop in the condenser is usually very small. To compensate for these pressure drops, the water must be pumped to a sufficiently higher pressure than the ideal cycle calls for. This requires a larger pump and larger work input to the pump.

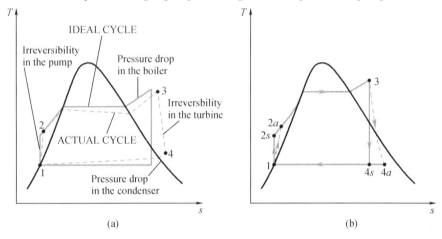

Fig. 8 – 17　(a) Deviation of actual vapor power cycle from the ideal Rankine cycle.
(b) The effect of pump and turbine irreversibilities on the ideal Rankine cycle

The other major source of irreversibility is the *heat loss* from the steam to the surroundings as the steam flows through various components. To maintain the same level of net work output, more heat needs to be transferred to the steam in the boiler to compensate for these undesired heat losses. As a consequence, cycle efficiency decreases.

Of particular importance are the irreversibilities occurring within the pump and the turbine. A pump requires a greater work input, and a turbine produces a smaller work output as a result of irreversibilities. Under ideal conditions, the flow through these devices is isentropic. The deviation of actual pumps and turbines from the isentropic ones can be accounted for by utilizing *isentropic efficiencies*, defined as

$$\eta_P = \frac{w_s}{w_a} = \frac{h_{2s} - h_1}{h_{2a} - h_1} \qquad (8-32)$$

and

$$\eta_T = \frac{w_a}{w_s} = \frac{h_3 - h_{4a}}{h_3 - h_{4s}} \qquad (8-33)$$

where states $2a$ and $4a$ are the actual exit states of the pump and the turbine, respectively, and $2s$ and $4s$ are the corresponding states for the isentropic case.

Other factors also need to be considered in the analysis of actual vapor power cycles. In actual condensers, for example, the liquid is usually subcooled to prevent the onset of *cavitation*, the rapid vaporization and condensation of the fluid at the low-pressure side of the pump impeller, which may damage it. Additional losses occur at the bearings between the moving parts as a result of friction. Steam that leaks out during the cycle and air that leaks into the condenser represent two other sources of loss. Finally, the power consumed by the auxiliary equipment such as fans that supply air to the furnace should also be considered in evaluating the overall performance of power plants.

The effect of irreversibilities on the thermal efficiency of a steam power cycle is illustrated below with an example.

EXAMPLE 8 – 2 An Actual Steam Power Cycle

A steam power plant operates on the cycle shown in Fig. 8 – 18. If the isentropic efficiency of the turbine is 87 percent and the isentropic efficiency of the pump is 85 percent, determine (a) the thermal efficiency of the cycle and (b) the net power output of the plant for a mass flow rate of 15 kg/s.

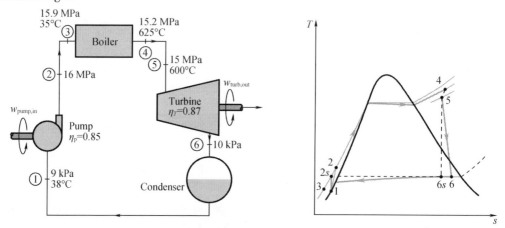

Fig. 8 – 18 Schematic and $T-s$ diagram for Example 8 – 2

Solution A steam power cycle with specified turbine and pump efficiency is considered. The thermal efficiency and the net power output are to be determined.

Assumptions ①Steady operating conditions exist. ②Kinetic and potential exergy changes are negligible.

Analysis The schematic of the power plant and the $T-s$ diagram of the cycle are shown in Fig. 8 – 18. The temperatures and pressures of steam at various points are also indicated on the figure. We note that the power plant involves steady-flow components and operates on the Rankine cycle, but the imperfections at various components are accounted for.

(a) The thermal efficiency of a cycle is the ratio of the net work output to the heat input, and it is determined as follows:

Pump work input:

$$w_{pump,in} = \frac{w_{s,pump,in}}{\eta_p} = \frac{v_1(P_2 - P_1)}{\eta_p}$$

$$= \frac{(0.001\ 009\ \text{m}^3/\text{kg})[(16\ 000 - 9)\text{kPa}]}{0.85}\left(\frac{1\ \text{kJ}}{1\ \text{kPa}\cdot\text{m}^3}\right)$$

$$= 19 \text{ kJ/kg}$$

Turbine work output:
$$w_{turb,out} = \eta_T w_{s,turb,out}$$
$$= \eta_T (h_5 - h_{6s}) = 0.87(3\,583.1 - 2\,115.3)\text{kJ/kg}$$
$$= 1\,277 \text{ kJ/kg}$$

Boiler heat input:
$$q_{in} = h_4 - h_3 = (3\,647.6 - 160.1)\text{kJ/kg} = 3\,487.5 \text{ kJ/kg}$$

Thus,
$$w_{net} = w_{turb,out} - w_{pump,in} = (1\,277 - 19)\text{kJ/kg} = 1\,258 \text{ kJ/kg}$$
$$\eta_{th} = \frac{w_{net}}{q_{in}} = \frac{1258 \text{ kJ/kg}}{3487.5 \text{ kJ/kg}} = 0.361$$

(b) The power produced by this power plant is
$$\dot{W}_{net} = \dot{m} w_{net} = (15 \text{ kg/s})(1258 \text{ kJ/kg}) = 18.9 \text{ MW}$$

Discussion: Without the irreversibilities, the thermal efficiency of this cycle would be 43.0 percent.

8-6 The Ideal Reheat Rankine Cycle

We noted in the last section that increasing the boiler pressure increases the thermal efficiency of the Rankine cycle, but it also increases the moisture content of the steam to unacceptable levels. Then it is natural to ask the following question:

How can we take advantage of the increased efficiencies at higher boiler pressures without facing the problem of excessive moisture at the final stages of the turbine?

Two possibilities come to mind:

1. Superheat the steam to very high temperatures before it enters the turbine. This would be the desirable solution since the average temperature at which heat is added would also increase, thus increasing the cycle efficiency. This is not a viable solution, however, since it requires raising the steam temperature to metallurgically unsafe levels.

2. Expand the steam in the turbine in two stages, and reheat it in between. In other words, modify the simple ideal Rankine cycle with a reheat process. With reheat, a power plant can take advantage of the increased efficiency that results with higher boiler pressures and yet avoid low-quality steam at the turbine exhaust.

The $T-s$ diagram of the ideal reheat Rankine cycle and the schematic of the power plant operating on this cycle are shown in Fig. 8-19. The ideal reheat Rankine cycle differs from the simple ideal Rankine cycle in that the expansion process takes place in two stages. In the first stage (the high pressure turbine), steam is expanded is entropically to an intermediate pressure and sent back to the boiler where it is reheated at constant pressure, usually to the inlet temperature of the first turbine stage. Steam then expands is entropically in the second stage (low-pressure turbine) to the condenser pressure. Thus the total heat input and the total turbine work output for are heat cycle become

$$q_{in} = q_{primary} + q_{reheat} = (h_3 - h_2) + (h_5 - h_4) \qquad (8-34)$$

and

$$w_{turb,out} = w_{turb,\,I} + w_{turb,\,II} = (h_3 - h_4) + (h_5 - h_6) \qquad (8-35)$$

The incorporation of the single reheat in a modern power plant improves the cycle efficiency by 4 to 5 percent by increasing the average temperature at which heat is transferred to the steam.

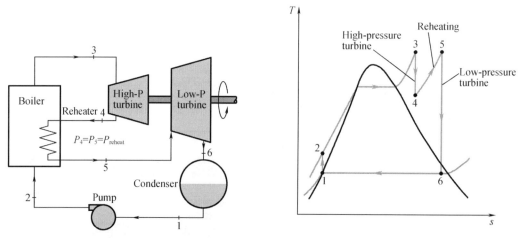

Fig. 8 – 19   The ideal reheat Rankine cycle

8 – 7   The Ideal Regenerative Rankine Cycle

A careful examination of the $T - s$ diagram of the Rankine cycle redrawn in Fig. 8 – 20 reveals that heat is transferred to the working fluid during process 2 – 2 at a relatively low temperature. This lowers the average heat addition temperature and thus the cycle efficiency. To remedy this shortcoming, we look for ways to raise the temperature of the liquid leaving the pump (called the *feedwater*) before it enters the boiler.

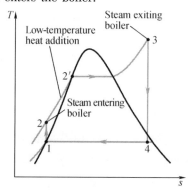

Fig. 8 – 20   The first part of the heat-addition process in the boiler takes place at trelatively low temperatures

Chapter 8 Power Cycles

One such possibility is to transfer heat to the feedwater from the expanding steam in a counter flow heat exchanger built into the turbine, that is, to use regeneration. This solution is also impractical because it is difficult to design such a heat exchanger and because it would increase the moisture content of the steam at the final stages of the turbine.

A practical regeneration process in steam power plants is accomplished by extracting, or "bleeding," steam from the turbine at various points. This steam, which could have produced more work by expanding further in the turbine, is used to heat the feedwater instead. The device where the feedwater is heated by regeneration is called a regenerator, or a feedwater heater (FWH).

Regeneration not only improves cycle efficiency, but also provides a convenient means of deaerating the feedwater (removing the air that leaks in at the condenser) to prevent corrosion in the boiler. It also helps control the large volume flow rate of the steam at the final stages of the turbine (due to the large specific volumes at low pressures). Therefore, regeneration has been used in all modern steam power plants since its introduction in the early 1920s.

A feedwater heater is basically a heat exchanger where heat is transferred from the steam to the feedwater either by mixing the two fluid streams (open feedwater heaters) or without mixing them (closed feedwater heaters). Regeneration with both types of feedwater heaters is discussed below.

An open (or direct-contact) feedwater heater is basically a *mixing chamber*, where the steam extracted from the turbine mixes with the feed water exiting the pump. Ideally, the mixture leaves the heater as a saturated liquid at the heater pressure. The schematic of a steam power plant with one open feed water heater (also called *single-stage regenerative cycle*) and the $T-s$ diagram of the cycle are shown in Fig. 8–21.

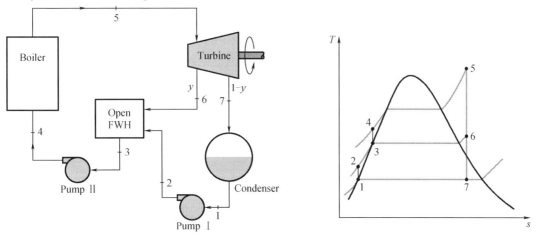

Fig. 8–21 The ideal regenerative Rankine cycle with an open feedwater heater

In an ideal regenerative Rankine cycle, steam enters the turbine at the boiler pressure (state 5) and expands is entropically to an intermediate pressure (state 6). Some steam is extracted at this state and routed to the feedwater heater, while the remaining steam continues to expand is entropically to the condenser pressure (state 7). This steam leaves the condenser as a saturated liquid at the condenser pressure (state 1). The condensed water, which is also called the *feedwater*, *then enters an isentropic pump*, *where it is compressed to the feedwater heater pressure* (state 2) and is routed to the feedwater heater, where it mixes with the steam extracted from the turbine. The fraction of the steam extracted is such that the mixture leaves the heater as a saturated liquid at the heater pressure (state 3). A second pump raises the pressure of the water to the boiler pressure (state 4). The cycle is completed by heating the water in the boiler to the turbine inlet state (state 5).

In the analysis of steam power plants, it is more convenient to work with quantities expressed per unit mass of the steam flowing through the boiler. For each 1 kg of steam leaving the boiler, y kg expands partially in the turbine and is extracted at state 6. The remaining $(1-y)$ kg expands completely to the condenser pressure. Therefore, the mass flow rates are different in different components. If the mass flow rate through the boiler is $\dot{m}$, for example, it is $(1-y)\dot{m}$ through the condenser. This aspect of the regenerative Rankine cycle should be considered in the analysis of the cycle as well as in the interpretation of the areas on the $T-s$ diagram. In light of Fig. 8-21, the heat and work interactions of a regenerative Rankine cycle with one feed water heater can be expressed per unit mass of steam flowing through the boiler as follows:

$$q_{in} = h_5 - h_4 \qquad (8-36)$$

$$q_{out} = (1-y)(h_7 - h_1) \qquad (8-37)$$

$$w_{turb,out} = (h_5 - h_6) + (1-y)(h_6 - h_7) \qquad (8-38)$$

$$w_{pump,in} = (1-y)w_{pump\,I,in} + w_{pump\,II,in} \qquad (8-39)$$

where

$$y = \frac{\dot{m}_6}{\dot{m}_5} \quad \text{(fraction of steam extracted)} \qquad (8-40)$$

$$w_{pump\,I,in} = v_1(P_2 - P_1) \qquad (8-41)$$

$$w_{pump\,II,in} = v_3(P_4 - P_3) \qquad (8-42)$$

The thermal efficiency of the Rankine cycle increases as a result of regeneration. This is because regeneration raises the average temperature at which heat is transferred to the steam in the boiler by raising the temperature of the water before it enters the boiler. The cycle efficiency increases further as the number of feed water heaters is increased. Many large plants in operation today use as many as eight feed water heaters. The optimum number of feedwater heaters is determined from economical considerations. The use of an additional feedwater heater cannot be justified unless it saves more from the fuel costs than its own cost.

Another type of feedwater heater frequently used in steam power plants is the closed feedwater heater, in which heat is transferred from the extracted steam to the feedwater without any mixing taking place. The two streams now can be at different pressures, since they do not mix. The schematic of a steam power plant with one closed feedwater heater and the $T-s$ diagram of the cycle are shown in Fig. 8-22.

In an ideal closed feedwater heater, the feedwater is heated to the exit temperature of the extracted steam, which ideally leaves the heater as a saturated liquid at the extraction pressure. In actual power plants, the feedwater leaves the heater below the exit temperature of the extracted

Chapter 8 Power Cycles

steam because a temperature difference of at least a few degrees is required for any effective heat transfer to take place. The condensed steam is then either pumped to the feedwater line or routed to another heater or to the condenser through a device called a trap. A trap allows the liquid to be throttled to a lower pressure region but *traps* the vapor. The enthalpy of steam remains constant during this throttling process.

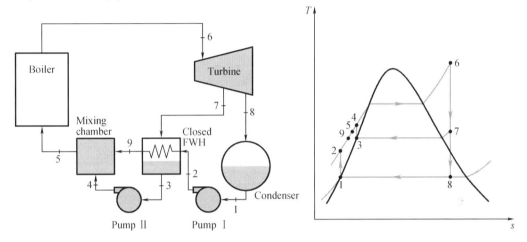

Fig. 8 – 22 The ideal regenerative Rankine cycle with a closed feedwater heater

Summary

A cycle during which a net amount of work is produced is called a *power cycle*, and a power cycle during which the working fluid remains a gas throughout is called a *gas power cycle*. The actual gas cycles are rather complex. The approximations used to simplify the analysis are known as the *air-standard assumptions*. Under these assumptions, all the processes are assumed to be internally reversible; the working fluid is assumed to be air, which behaves as an ideal gas; and the combustion and exhaust processes are replaced by heat-addition and heat-rejection processes, respectively. The air-standard assumptions are called *cold-air-standard assumptions* if air is also assumed to have constant specific heats at room temperature. In reciprocating engines, the *compression ratio r* and the *mean effective pressure* are defined. The ideal cycles for internal combustion engines are discussed, including the dual cycle, Otto cycle and Diesel cycle. We define some new quantities in this chapter, such as pressure ratio, compression ratio and cutoff ratio. Then, we also analyze the ideal and actual cycles for gas turbine in this chapter.

The *Rankine cycle* is a model cycle for vapor power cycles, which is composed of four internally reversible processes: constant-pressure heat addition in a boiler, isentropic expansion in a turbine, constant-pressure heat rejection in a condenser, and isentropic compression in a pump. To take advantage of the improved efficiencies at higher boiler pressures and lower condenser pressures, steam is usually *reheated* after expanding partially in the high-pressure turbine. Another way of increasing the thermal efficiency of the Rankine cycle is *regeneration*.

Problems

8 – 1 What is the difference between air-standard assumptions and the cold-air-standard assumptions?

8 – 2 What four processes make up the ideal Otto cycle?

8 – 3 What four processes make up the simple ideal Rankine cycle?

8 – 4 An ideal Otto cycle with air as the working fluid has a compression ratio of 9.5. The highest pressure and temperature in the cycle, the amount of heat transferred, the thermal efficiency, and the mean effective pressure are to be determined.

8 – 5 An ideal diesel cycle has a a cutoff ratio of 1.2. The power produced is to be determined.

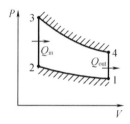 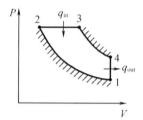

Fig. P8 – 4   Fig. P8 – 5

8 – 6 Consider a steam power plant operating on the ideal reheat Rankine cycle. Steam enters the high-pressure turbine at 15 MPa and 600 ℃ and is condensed in the condenser at a pressure of 10 kPa. If the moisture content of the steam at the exit of the low-pressure turbine is not to exceed 10.4 percent, determine (a) the pressure at which the steam should be reheated and (b) the thermal efficiency of the cycle. Assume the steam is reheated to the inlet temperature of the high – pressure turbine.

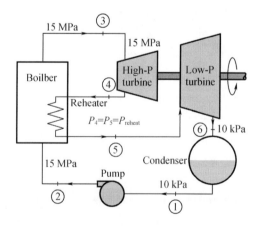

Fig. P8 – 6

Chapter 9　Solution Thermodynamics

In this chapter, we introduce concepts of solution and ideal solution. Phase equilibrium, the phase rule and the Raoult's law are discussed. A system is said to be in equilibrium if no changes occur within the system when it is isolated from its surroundings. The Raoult's law is applied for gases dissolved in liquids.

9 – 1　Phase Equilibrium

We know from experience that a wet T-shirt hanging in an open are a eventually dries, a small amount of water left in a glass evaporates, and the after shave in an open bottle quickly disappears. These examples suggest that there is a driving force between the two phases of a substance that forces the mass to transform from one phase to another. The magnitude of this force depends, among other things, on the relative concentrations of the two phases. A wet T-shirt dries much quicker in dry air than it does in humid air. In fact, it does not dry at all if the relative humidity of the environment is 100 percent. In this case, there is no transformation from the liquid phase to the vapor phase, and the two phases are in phase equilibrium. The conditions of phase equilibrium change, however, if the temperature or the pressure is changed. Therefore, we examine phase equilibrium at a specified temperature and pressure.

In chemistry, a solution is a homogeneous mixture composed of two or more substances. In such a mixture, a solute is a substance dissolved in another substance, known as a solvent. The concentration of a solute in a solution is the mass of that solute expressed as a percentage of the mass of the whole solution.

The situation is similar at *solid – liquid* interfaces. Again, at a given temperature, only a certain amount of solid can be dissolved in a liquid, and the solubility of the solid in the liquid is determined from the requirement that thermodynamic equilibrium exists between the solid and the solution at the interface. The solubility represents *the maximum amount of solid that can be dissolved in a liquid at a specified temperature* and is widely available in chemistry handbooks.

The mole fraction of a gas dissolved in the liquid (or solid) is usually expressed as a function of the partial pressure of the gas in the gas phase and the temperature. An approximate relation in this case for the *mole fractions* of a species on the *liquid* and *gas sides* of the interface is given by Raoult's law as

$$P_{i,\text{gas side}} = y_{i,\text{gas side}} P_{\text{total}} = y_{i,\text{liquid side}} P_{i,\text{sat}}(T) \qquad (9-1)$$

where $P_{i,\text{sat}}(T)$ is the *saturation pressure* of the species i at the interface temperature and P_{total} is the *total pressure* on the gas phase side. Tabular data are available in chemical handbooks for common solutions such as the ammonia – water solution that is widely used in absorption –

refrigeration systems.

Gases may also dissolve in *solids*, but the diffusion process in this case can be very complicated. The dissolution of a gas may be independent of the structure of the solid, or it may depend strongly on its porosity. Some dissolution processes (such as the dissolution of hydrogen in titanium, similar to the dissolution of CO_2 in water) are *reversible*, and thus maintaining the gas content in the solid requires constant contact of the solid with a reservoir of that gas. Some other dissolution processes are *irreversible*. For example, oxygen gas dissolving in titanium forms TiO_2 on the surface, and the process does not reverse itself. The molar density of the gas species i in the solid at the interface $\bar{\rho}_{i,\text{solid side}}$ is proportional to the *partial pressure* of the species i in the gas $P_{i,\text{gas side}}$ on the gas side of the interface and is expressed as

$$\bar{\rho}_{i,\text{solid side}} = \varphi P_{i,\text{gas side}} \quad (\text{kmol}/\text{m}^3) \qquad (9-2)$$

where φ is the solubility.

An ideal solution or ideal mixture is a solution with thermodynamic properties analogous to those of a mixture of ideal gases and the vapor pressure of the solution obeys Raoult's law.

9-2 The Phase Rule

Notice that a single-component two-phase system may exist in equilibrium at different temperatures (or pressures). However, once the temperature is fixed, the system is locked into an equilibrium state and all intensive properties of each phase (except their relative amounts) are fixed. Therefore, a single-component two-phase system has one independent property, which may be taken to be the temperature or the pressure.

In general, the number of independent variables associated with a multi-component, multiphase system is given by the Gibbs phase rule, expressed as

$$IV = C - PH + 2 \qquad (9-3)$$

where IV = the number of independent variables, C = the number of components, and PH the number of phases present in equilibrium. For the single-component ($C = 1$) two-phase ($PH = 2$) system discussed above, for example, one independent intensive property needs to be specified ($IV = 1$). At the triple point, however, $PH = 3$ and thus $IV = 0$. That is, none of the properties of a pure substance at the triple point can be varied. Also, based on this rule, a pure substance that exists in a single phase ($PH = 1$) has two independent variables. In other words, two independent intensive properties need to be specified to fix the equilibrium state of a pure-substance in a single phase.

9-3 Phase Equilibrium for a Multi-component System

Many multi-phase systems encountered in practice involve two or more components. A multicomponent multi-phase system at a specified temperature and pressure is in phase

Chapter 9   Solution Thermodynamics

equilibrium when there is no driving force between the different phases of each component.

In this section we examine the phase equilibrium of two-component systems that involve two phases (liquid and vapor) in equilibrium. For such systems, $C = 2$, $PH = 2$, and thus $IV = 2$. That is, a two-component, two-phase system has two independent variables, and such a system will not be in equilibrium unless two independent intensive properties are fixed.

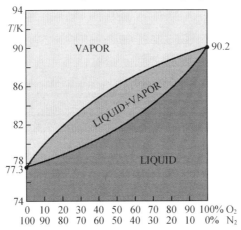

Fig. 9 – 1   Equilibrium diagram for the two-phase mixture of oxygen and nitrogen (0.1 MPa)

In general, the two phases of a two-component system do not have the same composition in each phase. That is, the mole fraction of a component is different in different phases. This is illustrated in Fig. 9 – 1 for the two phase mixture of oxygen and nitrogen at a pressure of 0.1 MPa. On this diagram, the vapor line represents the equilibrium composition of the vaporo phase at various temperatures, and the liquid line does the same for the liquid phase. At 84 K, for example, the mole fractions are 30 percent nitrogen and 70 percent oxygen in the liquid phase and 66 percent nitrogen and 34 percent oxygen in the vapor phase. Notice that

$$y_{f,N_2} + y_{f,O_2} = 0.3 + 0.7 = 1 \tag{9-4}$$

$$y_{f,N_2} + y_{f,O_2} = 0.66 + 0.34 = 1 \tag{9-5}$$

Therefore, once the temperature and pressure (two independent variables) of a two-component, two-phase mixture are specified, the equilibrium composition of each phase can be determined from the phase diagram, which is based on experimental measurements.

Summary

Two phases are said to be in *phase equilibrium* when there is no transformation from one phase to the other. Two phases of a pure substance are in equilibrium when each phase has the same value of specific Gibbs function. In general, the number of independent variables associated with a multi-component, multi-phase system is given by the *Gibbs phase rule*. When a gas is highly soluble in a liquid, the mole fractions of the species of a two-phase mixture in the liquid and gas phases are given approximately by Raoult's law.

Problems

9 – 1 Define the concept of phase equilibrium.

9 – 2 What is the phase rule?

9 – 3 What is ideal solution?

9 – 4 In absorption refrigeration systems, a two-phase equilibrium mixture of liquid ammonia (NH_3) and water (H_2O) is frequently used. Consider one such mixture at 40 ℃, shown in Fig. P9 – 4. If the composition of the liquid phase is 70 percent NH_3 and 30 percent H_2O by mole numbers, determine the composition of the vapor phase of this mixture.

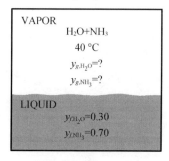

Fig. P9 – 4

Chapter 10 Refrigeration Cycles

A major application area of thermodynamics is refrigeration, which is the transfer of heat from a lower temperature region to a higher temperature one. Devices that produce refrigeration are called *refrigerators*, and the cycles on which they operate are called *refrigeration cycles*. The most frequently used refrigeration cycle is the *vapor-compression refrigeration cycle* in which the refrigerant is vaporized and condensed alternately and is compressed in the vapor phase. Other refrigeration cycles also discussed in this chapter are *absorption refrigeration* and *steam jet refrigeration*.

10 – 1 The Reversed Carnot Cycle

Recall from Chap. 4 that the Carnot cycle is a totally reversible cycle that consists of two reversible isothermal and two isentropic processes. It has the maximum thermal efficiency for given temperature limits, and it serves as a standard against which actual power cycles can be compared.

Since it is a reversible cycle, all four processes that comprise the Carnot cycle can be reversed. Reversing the cycle does also reverse the directions of any heat and work interactions. The result is a cycle that operates in the counter clock wise direction on a $T-s$ diagram, which is called the reversed Carnot cycle. A refrigerator or heat pump that operates on the reversed Carnot cycle is called a Carnot refrigerator or a Carnot heat pump.

Consider a reversed Carnot cycle executed within the saturation dome of a refrigerant, as shown in Fig. 10 – 1. The refrigerant absorbs heat isothermally from a low-temperature source at T_L in the amount of Q_L (process 1 – 2), is compressed isentropically to state 3 (temperature rises to T_H), rejects heat isothermally to a high-temperature sink at T_H in the amount of Q_H (process 3 – 4), and expands isentropically to state 1 (temperature drops to T_L). The refrigerant changes from a saturated vapor state to a saturated liquid state in the condenser during process 3 – 4.

The coefficients of performance of Carnot refrigerators and heat pumps are expressed in terms of temperatures as

$$\text{COP}_{R,\text{Carnot}} = \frac{1}{T_H/T_L - 1} \qquad (10-1)$$

and

$$\text{COP}_{HP,\text{Carnot}} = \frac{1}{1 - T_L/T_H} \qquad (10-2)$$

Notice that both COPs increase as the difference between the two temperatures decreases, that is, as T_L rises or T_H falls. The reversed Carnot cycle is the *most efficient* refrigeration cycle operating between two specified temperature levels. Therefore, it is natural to look at it first as a prospective ideal cycle for refrigerators and heat pumps. If we could, we certainly would adapt it as the ideal cycle. As explained below, however, the reversed Carnot cycle is not a suitable model for refrigeration cycles.

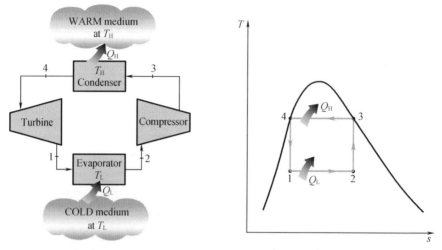

Fig. 10 – 1 Schematic of a Carnot refrigerator and $T-s$ diagram of the reversed Carnot cycle

The two isothermal heat transfer processes are not difficult to achieve in practice since maintaining a constant pressure automatically fixes the temperature of a two-phase mixture at the saturation value. Therefore, processes 1 – 2 and 3 – 4 can be approached closely in actual evaporators and condensers. However, processes 2 – 3 and 4 – 1 cannot be approximated closely in practice. This is because process 2 – 3 involves the compression of a liquid – vapor mixture, which requires a compressor that will handle two phases, and process 4 – 1 involves the expansion of high-moisture-content refrigerant in a turbine.

It seems as if these problems could be eliminated by executing the reversed Carnot cycle outside the saturation region. But in this case we have difficulty in maintaining isothermal conditions during the heat-absorption and heat-rejection processes. Therefore, we conclude that the reversed Carnot cycle cannot be approximated in actual devices and is not a realistic model for refrigeration cycles. However, the reversed Carnot cycle can serve as a standard against which actual refrigeration cycles are compared.

EXAMPLE 10 – 1 The Reversed Carnot Cycle for Heat Pump(Fig. 10 – 2)

A heat pump maintains a house at a specified temperature. The rate of heat loss of the house and the power consumption of the heat pump are given. It is to be determined if this heat pump can do the job.

Assumptions The heat pump operates steadily.

Analysis The power input to a heat pump will be a minimum when the heat pump operates in a reversible manner. The coefficient of performance of a reversible heat pump depends on the temperature limits in the cycle only, and is determined from

Chapter 10 Refrigeration Cycles

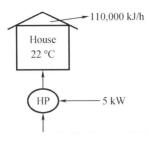

Fig. 10 – 2 Schematic for Example 10 – 1

$$COP_{HP,Carnot} = \frac{1}{1 - T_L/T_H} = \frac{1}{1 - (2+273 \text{ K})(22+273 \text{ K})} = 14.75$$

The required power input to this reversible heat pump is determined from the definition of the coefficient of performance to be

$$\dot{W}_{net,in,min} = \frac{\dot{Q}_H}{COP_{HP}} = \frac{110\,000 \text{ kJ/h}}{14.75}\left(\frac{1 \text{ h}}{3\,600 \text{ s}}\right) = 2.07 \text{ kW}$$

This heat pump is powerful enough since 5 kW > 2.07 kW.

10 – 2 Vapor-Compression Refrigeration Cycle

The most frequently used refrigeration cycle is the *vapor-compression refrigeration cycle*. The vapor-compression refrigeration cycle is the most widely used cycle for refrigerators, air-conditioning systems, and heat pumps. It consists of four processes: isentropic compression in a compressor, constant-pressure heat rejection in a condenser, throttling in an expansion device and constant-pressure heat absorption in an evaporator. The schematic, lg $p - h$ and $T - s$ diagrams are shown in Fig. 10 – 3 and 10 – 4, respectively.

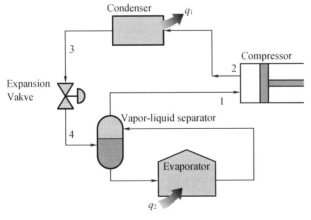

Fig. 10 – 3 Schematic for vapor-compression refrigeration cycles

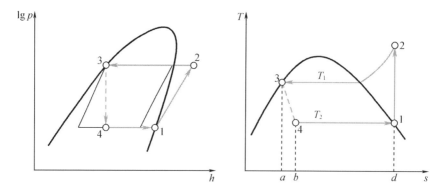

Fig. 10-4 lg $p-h$ and $T-s$ diagrams for vapor-compression refrigeration cycles

In a vapor-compression refrigeration cycle, the refrigerant enters the compressor at state 1 as saturated vapor and is compressed isentropically to the condenser pressure. The temperature of the refrigerant increases during this isentropic compression process to well above the temperature of the surrounding medium. The refrigerant then enters the condenser as superheated vapor at state 2 and leaves as saturated liquid at state 3 as a result of heat rejection to the surroundings. The temperature of the refrigerant at this state is still above the temperature of the surroundings.

The saturated liquid refrigerant at state 3 is throttled to the evaporator pressure by passing it through an expansion valve or capillary tube. The temperature of the refrigerant drops below the temperature of the refrigerated space during this process. The refrigerant enters the evaporator at state 4 as a low-quality saturated mixture, and it completely evaporates by absorbing heat from the refrigerated space. The refrigerant leaves the evaporator as saturated vapor and reenters the compressor, completing the cycle.

Remember that the area under the process curve on a $T-s$ diagram represents the heat transfer for internally reversible processes. The area under the process curve 4-1 represents the heat absorbed by the refrigerant in the evaporator, and the area under the process curve 2-3 represents the heat rejected in the condenser. A rule of thumb is that the *COP improves by 2 to 4 percent for each ℃ the evaporating temperature is raised or the condensing temperature is lowered.* Notice that the vapor-compression refrigeration cycle is not an internally reversible cycle since it involves an irreversible (throttling) process.

All four components associated with the vapor-compression refrigeration cycle are steady-flow devices, and thus all four processes that make up the cycle can be analyzed as steady-flow processes. The kinetic and potential exergy changes of the refrigerant are usually small relative to the work and heat transfer terms, and therefore they can be neglected. The condenser and the evaporator do not involve any work, and the compressor can be approximated as adiabatic. Then the COP of refrigerators operating on the vapor-compression refrigeration cycle can be expressed as

$$\text{COP}_R = \frac{q_2}{w_{net}} = \frac{q_2}{q_1 - q_2} = \frac{h_1 - h_4}{h_2 - h_3} \qquad (10-3)$$

Vapor-compression refrigeration dates back to 1834 when the Englishman Jacob Perkins received a patent for a closed-cycle ice machine using ether or other volatile fluids as refrigerants.

Chapter 10 Refrigeration Cycles

A working model of this machine was built, but it was never produced commercially. In 1850, Alexander Twining began to design and build vapor-compression ice machines using ethylether, which is a commercially used refrigerant in vapor-compression systems. Initially, vapor-compression refrigeration systems were large and were mainly used for ice making, brewing, and cold storage. They lacked automatic controls and were steam-engine driven. In the 1890s, electric motor driven smaller machines equipped with automatic controls started to replace the older units, and refrigeration systems began to appear in butcher shops and households. By 1930, the continued improvements made it possible to have vapor-compression refrigeration systems that were relatively efficient, reliable, small, and inexpensive.

EXAMPLE 10-2 The Vapor-Compression Refrigeration Cycle(Fig. 10-5)

A refrigerator uses refrigerant-134a as the working fluid and operates on an ideal vapor-compression refrigeration cycle between 0.14 and 0.8 MPa. If the mass flow rate of the refrigerant is 0.05 kg/s, determine (a) the rate of heat removal from the refrigerated space and the power input to the compressor, (b) the rate of heat rejection to the environment, and (c) the COP of the refrigerator.

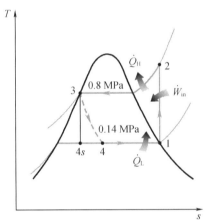

Fig. 10-5 $T-s$ **diagram of the vapor-compression refrigeration cycle described in Example 10-2**

Solution A refrigerator operates on an ideal vapor-compression refrigeration cycle between two specified pressure limits. The rate of refrigeration, the power input, the rate of heat rejection, and the COP are to be determined.

Assumptions ①Steady operating conditions exist. ②Kinetic and potential exergy changes are negligible.

Analysis The $T-s$ diagram of the refrigeration cycle is shown in Fig. 10-6. We note that this is an ideal vapor-compression refrigeration cycle, and thus the compressor is isentropic and the refrigerant leaves the condenser as a saturated liquid and enters the compressor as saturated vapor. From the refrigerant-134a tables, the enthalpies of the refrigerant at all four states are determined as follows:

$$P_1 = 0.14 \text{ MPa} \rightarrow h_1 = h_{g@0.14 \text{ MPa}} = 239.16 \text{ kJ/kg}$$
$$s_1 = s_{g@0.14 \text{ MPa}} = 0.944\,56 \text{ kJ/kg} \cdot \text{K}$$

$$\left.\begin{array}{l}P_2 = 0.8 \text{ MPa} \\ s_2 = s_1\end{array}\right\} h_2 = 275.39 \text{ kJ/kg}$$

$$P_3 = 0.8 \text{ MPa} \rightarrow h_3 = h_{f@0.8 \text{ MPa}} = 95.47 \text{ kJ/kg}$$

$$h_4 \cong h_3 (\text{throttling}) \rightarrow h_4 = 95.47 \text{ kJ/kg}$$

(a) The rate of heat removal from the refrigerated space and the power input to the compressor are determined from their definitions:

$$\dot{Q}_L = \dot{m}(h_1 - h_4) = (0.05 \text{ kg/s})[(239.16 - 95.47) \text{ kJ/kg}] = 7.18 \text{ kW}$$

and

$$\dot{W}_{in} = \dot{m}(h_2 - h_1) = (0.05 \text{ kg/s})[(275.39 - 239.16) \text{ kJ/kg}] = 1.81 \text{ kW}$$

(b) The rate of heat rejection from the refrigerant to the environment is

$$\dot{Q}_H = \dot{m}(h_2 - h_3) = (0.05 \text{ kg/s})[(275.39 - 95.47) \text{ kJ/kg}] = 9 \text{ kW}$$

It could also be determined from

$$\dot{Q}_H = \dot{Q}_L + \dot{W}_{in} = 7.18 + 1.81 = 8.99 \text{ kW}$$

(c) The coefficient of performance of the refrigerator is

$$\text{COP}_R = \frac{\dot{Q}_L}{\dot{W}_{in}} = \frac{7.18}{1.81} = 3.97$$

That is, this refrigerator removes about 4 units of thermal exergy from the refrigerated space for each unit of electric exergy it consumes.

Discussion It would be interesting to see what happens if the throttling valve were replaced by an isentropic turbine. The enthalpy at state 4s (the turbine exit with P_{4s} = 0.14 MPa, and s_{4s} = s_3 = 0.35404 kJ/kg · K) is 88.94 kJ/kg, and the turbine would produce 0.33 kW of power. This would decrease the power input to the refrigerator from 1.81 ~ 1.48 kW and increase the rate of heat removal from the refrigerated space from 7.18 ~ 7.51 kW. As a result, the COP of the refrigerator would increase from 3.97 to 5.07, an increase of 28 percent.

10-3 Absorption Refrigeration Cycle

Another form of refrigeration that becomes economically attractive when there is a source of inexpensive thermal exergy at a temperature of 100 ~ 200 ℃ is absorption refrigeration. These cycles have some features in common with the vapor-compression cycles considered previously but differ in two important respects:

One is the nature of the compression process. Instead of compressing a vapor between the evaporator and the condenser, the refrigerant of an absorption system is absorbed by a secondary substance, called an absorbent, to form a *liquid solution*. The liquid solution is then pumped to the higher pressure. Because the average specific volume of the liquid solution is much less than that of the refrigerant vapor, significantly less work is required. Accordingly, absorption refrigeration systems have the advantage of relatively small work input compared to vapor-

Chapter 10 Refrigeration Cycles

compression systems.

The other main difference between absorption and vapor-compression systems is that some means must be introduced in absorption systems to retrieve the refrigerant vapor from the liquid solution before the refrigerant enters the condenser. This involves heat transfer from a relatively high-temperature source. Steam or waste heat that otherwise would be discharged to the surroundings without use is particularly economical for this purpose. Natural gas or some other fuel can be burned to provide the heat source, and there have been practical applications of absorption refrigeration using alternative exergy sources such as solar and geothermal exergy.

As the name implies, absorption refrigeration systems involve the absorption of a *refrigerant* by a *transport medium*. The most widely used absorption refrigeration system is the ammonia-water system, where ammonia (NH_3) serves as the refrigerant and water (H_2O) as the transport medium. Other absorption refrigeration systems include water-lithium bromide and water-lithium chloride systems, where water serves as the refrigerant. The latter two systems are limited to applications such as air-conditioning where the minimum temperature is above the freezing point of water.

To understand the basic principles involved in absorption refrigeration, we examine the NH_3-H_2O system shown in Fig. 10-6. You will immediately notice from the figure that this system looks very much like the vapor-compression system, except that the compressor has been replaced by a complex absorption mechanism consisting of an absorber, a pump, a generator, a regenerator, a valve, and a rectifier. Once the pressure of NH_3 is raised by the components in the box (this is the only thing they are set up to do), it is cooled and condensed in the condenser by rejecting heat to the surroundings, is throttled to the evaporator pressure, and absorbs heat from the refrigerated space as it flows through the evaporator. So, there is nothing new there. Here is what happens in the box:

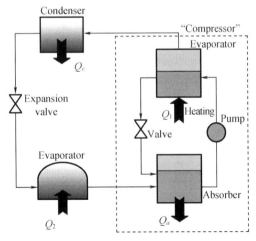

Fig. 10-6 Ammonia absorption refrigeration cycle

Ammonia vapor leaves the evaporator and enters the absorber, where it dissolves and reacts with water to form $NH_3 \cdot H_2O$. This is an exothermic reaction; thus heat is released during this process. The amount of NH_3 that can be dissolved in H_2O is inversely proportional to the temperature. Therefore, it is necessary to cool the absorber to maintain its temperature as low as possible, hence to maximize the amount of NH_3 dissolved in water. The liquid NH_3 and H_2O solution, which is rich in NH_3, is then pumped to the generator. Heat is transferred to the solution from a source to vaporize some of the solution. The vapor, which is rich in NH_3, passes through a rectifier, which separates the water and returns it to the generator. The high-pressure pure NH_3 vapor then continues its journey through the rest of the cycle. The hot $NH_3 \cdot H_2O$ solution, which is weak in NH_3, then passes through a regenerator, where it transfers some heat to the rich solution leaving the pump, and is throttled to the absorber pressure.

Compared with vapor-compression systems, absorption refrigeration systems have one major advantage: A liquid is compressed instead of a vapor. The steady-flow work is proportional to the specific volume, and thus the work input for absorption refrigeration systems is very small and often neglected in the cycle analysis. The operation of these systems is based on heat transfer from an external source. Therefore, absorption refrigeration systems are often classified as *heat-driven systems*.

The absorption refrigeration systems are much more expensive than the vapor-compression refrigeration systems. They are more complex and occupy more space, they are much less efficient thus requiring much larger cooling towers to reject the waste heat, and they are more difficult to service since they are less common. Therefore, absorption refrigeration systems should be considered only when the unit cost of thermal exergy is low and is projected to remain low relative to electricity. Absorption refrigeration systems are primarily used in large commercial and industrial installations.

The COP of absorption refrigeration systems is defined as

$$\text{COP}_{absorption} = \frac{\text{Desired output}}{\text{Required input}} = \frac{Q_2}{Q_1} \qquad (10-4)$$

Another type of absorption system uses *lithium bromide* as the absorbent and *water* as the refrigerant. The basic principle of operation is the same as for ammonia – water systems. To achieve refrigeration at lower temperatures than are possible with water as the refrigerant, a lithium bromide-water absorption system may be combined with another cycle using a refrigerant with good low-temperature characteristics, such as ammonia, to form a cascaderefrigeration system.

10 – 4   Steam Jet Refrigeration Cycle

Steam jet refrigeration uses a high-pressure jet of steam to cool water or other fluid media. Typical uses include industrial sites, where a suitable steam supply already exists for other purposes or, historically, for air conditioning on passenger trains which use steam for heating. Steam jet refrigeration experienced a wave of popularity during the early 1930s for air conditioning large buildings. Steam jet refrigeration cycles were later supplanted by systems using mechanical compressors. The schematic and $T-s$ diagram for steam jet refrigeration cycle are shown in Fig.

Chapter 10   Refrigeration Cycles

10 – 7.

The utilization coefficient of exergy for steam jet refrigeration cycle is defined as

$$\xi_{\text{steam jet}} = \frac{Q_2}{Q} \qquad (10-5)$$

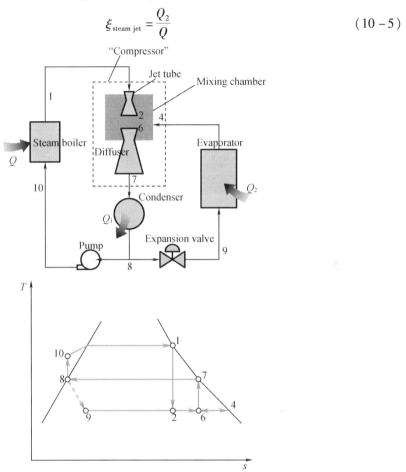

Fig. 10 – 7   Schematic and $T-s$ diagram for steam jet refrigeration cycle

Summary

The transfer of heat from lower temperature regions to higher temperature ones is called *refrigeration*. Devices that produce refrigeration are called *refrigerators*, and the cycles on which they operate are called *refrigeration cycles*. The working fluids used in refrigerators are called *refrigerants*. Refrigerators used for the purpose of heating a space by transferring heat from a cooler medium are called *heat pumps*. The performance of refrigerators and heat pumps is expressed in terms of *coefficient of performance* (COP). The standard of comparison for refrigeration cycles is the *reversed Carnot cycle*. A refrigerator or heat pump that operates on the reversed Carnot cycle is called a *Carnot refrigerator* or a *Carnot heat pump* The most widely used refrigeration cycle is the *vapor-compression refrigeration cycle*. Some other refrigeration cycles, including absorption refrigeration cycle and steam jet refrigeration are introduced.

Problems

10 – 1 Why is the reversed Carnot cycle executed within the saturation dome not a realistic model for refrigeration cycles?

10 – 2 Does the ideal vapor-compression refrigeration cycle involve any internal irreversibilities?

10 – 3 An inventor claims to have developed a heat engine. The inventor reports temperature, heat transfer, and work output measurements. The claim is to be evaluated.

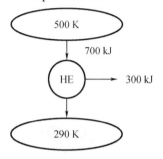

Fig. P10 – 3

10 – 4 A reversible heat pump with specified reservoir temperatures is considered. The entropy change of two reservoirs is to be calculated and it is to be determined if this heat pump satisfies the increase in entropy principle.

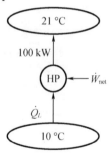

Fig. P10 – 4

ANSWERS

Chapter 1

1 – 1 A barometer is used to measure the height of a building by recording reading at the bottom and at the top of the building. The height of the building is to be determined.

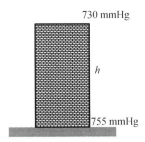

Fig. P1 – 1

Assumptions The variation of air density with altitude is negligible.

Properties The density of air is given to be $\rho = 1.18 \text{ kg/m}^3$. The density of mercury is 13 600 kg/m³.

Analysis Atmospheric pressures at the top and at the bottom of the building are

$$P_{top} = (\rho g h)_{top}$$
$$= (13\ 600 \text{ kg/m}^3)(9.807 \text{ m/s}^2)(0.730 \text{ m})\left(\frac{1 \text{ N}}{1 \text{ kg} \cdot \text{m/s}^2}\right)\left(\frac{1 \text{ kPa}}{1\ 000 \text{ N/m}^2}\right)$$
$$= 97.36 \text{ kPa}$$

$$P_{bottom} = (\rho g h)_{bottom}$$
$$= (13\ 600 \text{ kg/m}^3)(9.807 \text{ m/s}^2)(0.755 \text{ m})\left(\frac{1 \text{ N}}{1 \text{ kg} \cdot \text{m/s}^2}\right)\left(\frac{1 \text{ kPa}}{1\ 000 \text{ N/m}^2}\right)$$
$$= 100.70 \text{ kPa}$$

Taking an air column between the top and the bottom of the building and writing a force balance per unit base area, we obtain

$$W_{air}/A = P_{bottom} - P_{top}$$
$$(\rho g h)_{air} = P_{bottom} - P_{top}$$
$$(1.18 \text{ kg/m}^3)(9.807 \text{ m/s}^2)(h)\left(\frac{1 \text{ N}}{1 \text{ kg} \cdot \text{m/s}^2}\right)\left(\frac{1 \text{ kPa}}{1000 \text{ N/m}^2}\right) = (100.70 - 97.36) \text{ kPa}$$

It yields $h = 288.6 \text{ m}$

which is also the height of the building.

1 – 2 The air pressure in a duct is measured by an inclined manometer. For a given vertical level difference, the gage pressure in the duct and the length of the differential fluid column are to

be determined.

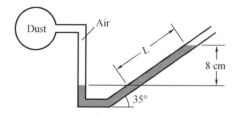

Fig. P1 – 2

Assumptions The manometer fluid is an incompressible substance.
Properties The density of the liquid is given to be
$$\rho = 0.81 \text{ kg/L} = 810 \text{ kg/m}^3$$
Analysis The gage pressure in the duct is determined from
$$P_{gage} = P_{abs} - P_{atm} = \rho g h$$
$$= (810 \text{ kg/m}^3)(9.81 \text{ m/s}^2)(0.08 \text{ m})\left(\frac{1 \text{ N}}{1 \text{ kg} \cdot \text{m/s}^2}\right)\left(\frac{1 \text{ Pa}}{1 \text{ N/m}^2}\right)$$
$$= 636 \text{ Pa}$$
The length of the differential fluid column is
$$L = h/\sin \theta = (8 \text{ cm})/\sin 35° = 13.9 \text{ cm}$$

Discussion Note that the length of the differential fluid column is extended considerably by inclining the manometer arm for better readability.

1 – 3 The water in a tank is pressurized by air, and the pressure is measured by a multifluid manometer as shown in Fig. P1 – 3. The tank is located on a mountain at an altitude of 1 400 m where the atmospheric pressure is 85.6 kPa. Determine the air pressure in the tank if $h_1 = 0.1$ m, $h_2 = 0.2$ m, and $h_3 = 0.35$ m. Take the densities of water, oil, and mercury to be 1 000 kg/m^3, 850 kg/m^3, and 13,600 kg/m^3, respectively.

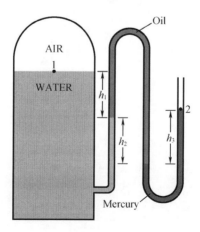

Fig. P1 – 3

Assumption The air pressure in the tank is uniform (i. e., its variation with elevation is

ANSWERS

negligible due to its low density), and thus we can determine the pressure at the air – water interface.

Properties The densities of water, oil, and mercury are given to be 1 000 kg/m³, 850 kg/m³, and 13 600 kg/m³, respectively.

Analysis Starting with the pressure at point 1 at the air – water interface, moving along the tube by adding or subtracting the ρgh terms until we reach point 2, and setting the result equal to P_{atm} since the tube is open to the atmosphere gives

$$P_1 + \rho_{water}gh_1 + \rho_{oil}gh_2 - \rho_{mercury}gh_3 = P_{atm}$$

Solving for P_1 and substituting,

$$\begin{aligned} P_1 &= P_{atm} - \rho_{water}gh_1 - \rho_{oil}gh_2 + \rho_{mercury}gh_3 \\ &= P_{atm} + g(\rho_{mercury}h_3 - \rho_{water}h_1 - \rho_{oil}h_2) \\ &= 85.6 \text{ kPa} + (9.81 \text{ m/s}^2)[(13\,600 \text{ kg/m}^3)(0.35 \text{ m}^2) - (1\,000 \text{ kg/m}^3)(0.1 \text{ m}^2) - \\ &\quad (850 \text{ kg/m}^3)(0.2 \text{ m})](\frac{1 \text{ N}}{1 \text{ kg} \cdot \text{m/s}^2})(\frac{1 \text{ kPa}}{1\,000 \text{ N/m}^2}) \\ &= 130 \text{ kPa} \end{aligned}$$

Discussion Note that jumping horizontally from one tube to the next and realizing that pressure remains the same in the same fluid simplifies the analysis considerably. Also note that mercury is a toxic fluid, and mercury manometers and thermometers are being replaced by ones with safer fluids because of the risk of exposure to mercury vapor during an accident.

1 – 4 The gas is slowly expanded in the cylinder, volume of gas increases from 0.1 m³ to 0.25 m³, and relationship between pressure and volume is $\{p\}_{MPa} = 0.24 - 0.4\{V\}_{m^3}$, in the process. Friction force between cylinder and piston is 1 200 N, the local atmospheric pressure is 0.1 MPa, area of cylinder is 0.1 m². Determine:

(a) expansion work of gas;
(b) useful work of the system output;
(c) useful work of the system output if no friction.

Assumption The system will include gas, piston and cylinder boundary, and the gas is uniform in this system.

Properties The initial volume of gas is 0.1 m³, the final volume is 0.25 m³, friction force between cylinder and piston is 1 200 N, the local atmospheric pressure is 0.1 MPa, and area of cylinder is 0.1 m².

Analysis The movement distance is determined by volume change and area of cylinder. The expansion work is determined by equation pdV, the work against atmosphere W_r is determined by atmosphere pressure and volume change and the work by friction is determined by friction force and movement distance.

$$L = \frac{V_2 - V_1}{A} = \frac{0.25 \text{ m}^3 - 0.1 \text{ m}^3}{0.1 \text{ m}^2} = 1.5 \text{ m}$$

(a) expansion work of gas W

$$W = \int_1^2 pdV = \int_1^2 (0.24 - 0.4V)dV = 0.24(V_2 - V_1) - 0.2(V_2^2 - V_1^2)$$

$$= 0.24 \times (0.25 - 0.1)\,\text{m} - 0.2 \times (0.25^2 - 0.1^2)\,\text{m}^2$$
$$= 0.025\ 5 \times 10^6\,\text{J}$$

(b) work of gas against atmosphere W_r
$$W_r = p_0(V_2 - V_1) = 0.1\ \text{MPa} \times (0.25 - 0.1)\,\text{m}^3 = 0.015 \times 10^6\,\text{J}$$

Dissipated work by friction W_1
$$W_1 = FL = 1\ 200\ \text{N} \times 1.5\ \text{m} = 1\ 800\ \text{J}$$

Thus, useful work of the system output is
$$W_u = W - W_r - W_1 = 0.025\ 5 \times 10^6\,\text{J} - 0.015 \times 10^6\,\text{J} - 1\ 800\ \text{J} = 8\ 700\ \text{J}$$

(c) useful work of the system output if no friction is
$$W_u = W - W_r = 0.025\ 5 \times 10^6\,\text{J} - 0.015 \times 10^6\,\text{J} = 10\ 500\ \text{J}$$

Discussion Note that expansion work of gas including useful work, work against atmosphere and dissipated work by friction. The expansion work from point $1-2$ in a process is calculated by equation $W = \int_1^2 p\,dV$.

Chapter 2

2-1 A well-insulated electric oven is being heated through its heating element. If the entire oven, including the heating element, is taken to be the system, determine whether this is a heat or work interaction if the system is taken as only the air in the oven without the heating element.

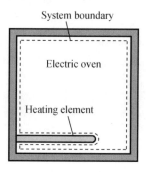

Fig. P2-1

Solution A well-insulated electric oven is being heated by its heating element. It is to be determined whether this is a heat or work interaction by taking the system to be only the air in the oven.

Analysis This time, the system boundary will include the outer surface of the heating element and will not cut through it. Therefore, no electrons will be crossing the system boundary at any point. Instead, the exergy generated in the interior of the heating element will be transferred to the air around it as a result of the temperature difference between the heating element and the air in the oven. Therefore, this is a heat transfer process.

Discussion For this case and Example 2-2, the amount of exergy transfer to the air is the same. The problem show that an exergy transfer can be heat or work, depending on how the system is selected.

ANSWERS

2-2 A classroom is to be air-conditioned using window air-conditioning units. The cooling load is due to people, lights, and heat transfer through the walls and the windows. The value of each people cooling load is 360 kJ/h and that of a bulb is 100 W. The number of bulbs and people in this room are 10 and 40, respectively. The number of 5 kW window air conditioning units required is to be determined.

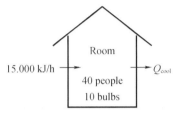

Fig. P2-2

Assumptions There are no heat dissipating equipment (such as computers, TVs, or ranges) in the room.

Analysis The total cooling load of the room is determined from

$$\dot{Q}_{cooling} = \dot{Q}_{lights} + \dot{Q}_{people} + \dot{Q}_{heat\ gain}$$

where

$$\dot{Q}_{lights} = 10 \times 100 \text{ W} = 1 \text{ kW}$$

$$\dot{Q}_{people} = 40 \times 360 \text{ kJ/h} = 4 \text{ kW}$$

$$\dot{Q}_{heat\ gain} = 15\ 000 \text{ kJ/h} = 4.17 \text{ kW}$$

Substituting,

$$\dot{Q}_{cooling} = 1 + 4 + 4.17 = 9.17 \text{ kW}$$

Thus the number of air-conditioning units required is

$$\frac{9.17 \text{ kW}}{5 \text{ kW/unit}} = 1.83 \rightarrow 2 \text{ units}$$

2-3 A rigid tank contains a hot fluid that is cooled while being stirred by a paddle wheel. Initially, the internal exergy of the fluid is 800 kJ. During the cooling process, the fluid loses 500 kJ of heat, and the paddle wheel does 100 kJ of work on the fluid. Determine the final internal exergy of the fluid. Neglect the exergy stored in the paddle wheel.

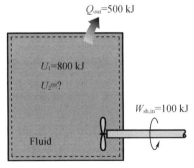

Fig. P2-3

Solution A fluid in a rigid tank looses heat while being stirred. The final internal exergy of the fluid is to be determined.

Assumptions ①The tank is stationary and thus the kinetic and potential exergy changes are zero, $\Delta KE = \Delta PE = 0$. Therefore, $\Delta E = \Delta U$ and internal exergy is the only form of the system's exergy that may change during this process. ②Exergy stored in the paddle wheel is negligible.

Analysis Take the contents of the tank as the *system* (Fig. P2-3). This is a *closed system* since no mass crosses the boundary during the process. We observe that the volume of a rigid tank is constant, and thus there is no moving boundary work. Also, heat is lost from the system and shaft work is done on the system. Applying the exergy balance on the system gives

$$\underbrace{\dot{E}_{in} - \dot{E}_{out}}_{\substack{\text{Net exergy transfer} \\ \text{by heat, work, and mass}}} = \underbrace{\Delta E_{system}}_{\substack{\text{Change in internal, kinetic,} \\ \text{potential, etc., energies}}}$$

$$W_{sh,in} - Q_{out} = \Delta U = U_2 - U_1$$

$$100 \text{ kJ} - 500 \text{ kJ} = U_2 - 800 \text{ kJ}$$

$$U_2 = 400 \text{ kJ}$$

Therefore, the final internal exergy of the system is 400 kJ.

2-4 The available head, flow rate, and efficiency of a hydro electric turbine are given. The height is 120 m and volume flow rate is 100 m³/s. The electric power output is to be determined.

Assumptions ①The flow is steady. ②Water levels at the reservoir and the discharge site remain constant. ③Frictional losses in piping are negligible.

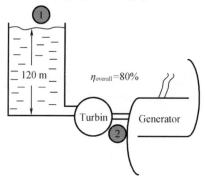

Fig. P2-4

Properties We take the density of water to be

$$\rho = 1\ 000 \text{ kg/m}^3 = 1 \text{ kg/L}$$

Analysis The total mechanical exergy the water in a dam possesses is equivalent to the potential exergy of water at the free surface of the dam (relative to free surface of discharge water), and it can be converted to work entirely. Therefore, the power potential of water is its potential exergy, which is gz per unit mass, and for a given mass flow rate.

$$e_{mech} = pe = gz = (9.81 \text{ m/s}^2)(120 \text{ m})\left(\frac{1 \text{ kJ/kg}}{1\ 000 \text{ m}^2/\text{s}^2}\right) = 1.177 \text{ kJ/kg}$$

The mass flow rate is

$$\dot{m} = \rho \dot{V} = (1\ 000 \text{ kg/m}^3)(100 \text{ m}^3/\text{s}) = 100\ 000 \text{ kg/s}$$

ANSWERS

Then the maximum and actual electric power generation become

$$\dot{W}_{max} = \dot{E}_{mech} = \dot{m}e_{mech} = (100\ 000\ \text{kg/s})(1.177\ \text{kJ/kg})\left(\frac{1\ \text{MW}}{1\ 000\ \text{kJ/s}}\right) = 117.7\ \text{MW}$$

$$\dot{W}_{electric} = \eta_{overall}\dot{W}_{max} = 0.80(117.7\ \text{MW}) = 94.2\ \text{MW}$$

Discussion Note that the power generation would increase by more than 1 MW for each percentage point improvement in the efficiency of the turbine-generator unit.

2 – 5 Air at 150 kPa and 300 K is compressed steadily to 700 kPa and 550 K. The mass flow rate of the air is 0.03 kg/s, and a heat loss of 20 kJ/kg occurs during the process. Assuming the changes in kinetic and potential energies are negligible, determine the necessary power input to the compressor.

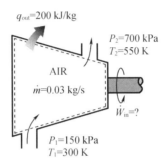

Fig. P2 – 5

Solution Air is compressed steadily by a compressor to a specified temperature and pressure. The power input to the compressor is to be determined.

Assumptions

1 This is a steady-flow process since there is no change with time at any point and thus $\Delta m_{CV} = 0$ and $\Delta E_{CV} = 0$.

2 Air is an ideal gas since it is at a high temperature and low pressure relative to its critical-point values.

3 The kinetic and potential exergy changes are zero, $\Delta e_k = \Delta e_p = 0$.

Analysis We take the *compressor* as the system (Fig. P2 – 5). This is a *control volume* since mass crosses the system boundary during the process. We observe that there is only one inlet and one exit and thus $\dot{m}_1 = \dot{m}_2 = \dot{m}$. Also, heat is lost from the system and work is supplied to the system. Under stated assumptions and observations, the exergy balance for this steady-flow system can be expressed in the rate form as

$$\underbrace{\dot{E}_{in} = \dot{E}_{out}}_{\substack{\text{Rate of net exergy transfer}\\\text{by heat, work, and mass}}} = \underbrace{dE_{system}/dt}_{\substack{\text{Rate of change in internal, kinetic,}\\\text{potential, etc., energies}}}\!\!\!\!{}^{0(\text{steady})} = 0$$

$$\dot{E}_{in} = \dot{E}_{out}$$

$$\dot{W}_{in} + \dot{m}h_1 = \dot{Q}_{out} + \dot{m}h_2 \quad (\text{since } \Delta e_k = \Delta e_p \cong 0)$$

$$\dot{W}_{in} = \dot{m}q_{out} + \dot{m}(h_2 - h_1)$$

The enthalpy of an ideal gas depends on temperature only, and the enthalpies of the air at

the specified temperatures are determined from the air table to be
$$h_1 = h_{@300K} = 302.29 \text{ kJ/kg}$$
$$h_2 = h_{@550K} = 556.76 \text{ kJ/kg}$$

Substituting, the power input to the compressor is determined to be

$$\dot{W}_{in} = (0.03 \text{ kg/s})(20 \text{ kJ/kg}) + (0.03 \text{ kg/s})(556.76 - 302.29) \text{ kJ/kg}$$
$$= 8.23 \text{ kW}$$

Discussion Note that the mechanical exergy input to the compressor manifests itself as a rise in enthalpy of air and heat loss from the compressor.

Chapter 3

3-1 Helium gas is compressed by an adiabatic compressor from an initial state of 14 psia and 50 °F to a final temperature of 320 °F in a reversible manner. Determine the exit pressure of helium.

Solution Helium is compressed from a given state to a specified pressure isentropically. The exit pressure of helium is to be determined.

Assumptions At specified conditions, helium can be treated as an ideal gas. Therefore, the isentropic relations developed earlier for ideal gases are applicable.

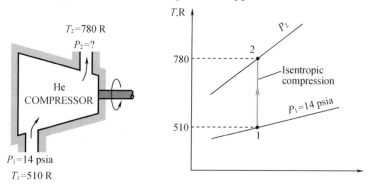

Fig. P3-1

Analysis A sketch of the system and the $T-s$ diagram for the process are given in Fig. P3-1. The specific heat ratio k of helium is 1.667 and is independent of temperature in the region where it behaves as an ideal gas. Thus the final pressure of helium can be determined:

$$P_2 = P_1 \left(\frac{T_2}{T_1}\right)^{k/(k-1)} = (14 \text{ psia})\left(\frac{780 \text{ R}}{510 \text{ R}}\right)^{1.667/0.667} = 40.5 \text{ psia}$$

3-2 A piston-cylinder device contains a liquid-vapor mixture of water at 300 K. During a constant-pressure process, 750 kJ of heat is transferred to the water. As a result, part of the liquid in the cylinder vaporizes. Determine the entropy change of the water during this process.

Solution Heat is transferred to a liquid-vapor mixture of water in a piston-cylinder device at constant pressure. The entropy change of water is to be determined.

Assumptions No irreversibilities occur within the system boundaries during the process.

Analysis We take the *entire water* (liquid-vapor) in the cylinder as the system (Fig. P3-2).

ANSWERS

This is a *closed system* since no mass crosses the system boundary during the process. We note that the temperature of the system remains constant at 300 K during this process since the temperature of a pure substance remains constant at the saturation value during a phase change process at constant pressure. The system undergoes an internally reversible, isothermal process, and thus its entropy change can be determined directly from $\Delta S = \dfrac{Q}{T_0}$ to be

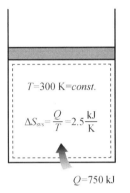

Fig. P3 – 2

$$\Delta S_{sys,\,isothermal} = \frac{Q}{T_{sys}} = \frac{750 \text{ kJ}}{300 \text{ K}} = 2.5 \text{ kJ/K}$$

Discussion Note that the entropy change of the system is positive, as expected, since heat transfer is *to* the system.

3 – 3 A rigid tank contains 2 kmol of N_2 and 6 kmol of CO_2 gases at 300 K and 15 MPa (Fig. P3 – 3). Estimate the volume of the tank on the basis of the ideal-gas equation of state.

| 2 kmol N_2 |
| 6 kmol CO_2 |
| 300K |
| 15 MPa |
| $V_m = ?$ |

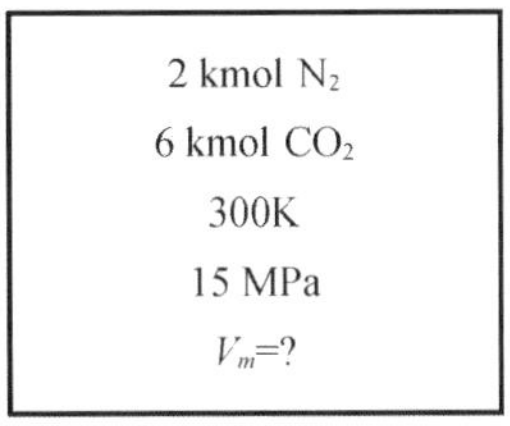

Fig. P3 – 3

Solution The composition of a mixture in a rigid tank is given. The volume of the tank is to be determined using the ideal-gas equation of state.

Assumptions Stated in each section.

Analysis When the mixture is assumed to behave as an ideal gas, the volume of the mixture is easily determined from the ideal-gas relation for the mixture:

Since
$$N_m = N_{N_2} + N_{CO_2} = 2 + 6 = 8 \text{ kmol}$$

thus
$$V_m = \frac{N_m R_u T_m}{P_m} = \frac{(8 \text{ kmol})(8.314 \text{ kJ/kmol} \cdot \text{K})(300 \text{ K})}{15\,000 \text{ kPa}} = 1.33 \text{ m}^3$$

3 – 4 An insulated rigid tank is divided into two compartments by a partition, as shown in Fig. P3 – 4. One compartment contains 7 kg of oxygen gas at 40 ℃ and 100 kPa, and the other compartment contains 4 kg of nitrogen gas at 20 ℃ and 150 kPa. Now the partition is removed, and the two gases are allowed to mix. Determine (a) the mixture temperature and (b) the mixture pressure after equilibrium has been established.

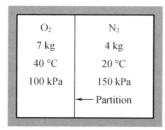

Fig. P3 – 4

Solution A rigid tank contains two gases separated by a partition. The pressure and temperature of the mixture are to be determined after the partition is removed.

Assumptions ①We assume both gases to be ideal gases, and their mixture to be an ideal-gas mixture. This assumption is reasonable since both the oxygen and nitrogen are well above their critical temperatures and well below their critical pressures. ②The tank is insulated and thus there is no heat transfer. ③There are no other forms of work involved.

Properties The constant-volume specific heats of N_2 and O_2 at room temperature are $c_{v,N2}$ = 0.743 kJ/kg · K and $c_{v,O2}$ = 0.658 kJ/kg · K (Table A – 10a).

Analysis We take the entire contents of the tank (both compartments) as the system. This is a *closed system* since no mass crosses the boundary during the process. We note that the volume of a rigid tank is constant and thus there is no boundary work done.

(a) Noting that there is no exergy transfer to or from the tank, the exergy balance for the system can be expressed as

$$E_{in} - E_{out} = \Delta E_{system}$$
$$0 = \Delta U = \Delta U_{N_2} + \Delta U_{O_2}$$
$$[mc_v(T_m - T_1)]_{N_2} + [mc_v(T_m - T_1)]_{O_2} = 0$$

By using c_v values at room temperature, the final temperature of the mixture is determined to be

$$(4 \text{ kg})(0.743 \text{ kJ/kg} \cdot \text{K})(T_m - 20 \text{ ℃}) + (7 \text{ kg})(0.658 \text{ kJ/kg} \cdot \text{K})(T_m - 40 \text{ ℃}) = 0$$
$$T_m = 32.2 \text{ ℃}$$

(b) The final pressure of the mixture is determined from the ideal-gas relation

$$P_m V_m = N_m R_u T_m$$

where

$$N_{O_2} = \frac{m_{O_2}}{M_{O_2}} = \frac{7 \text{ kg}}{32 \text{ kg/kmol}} = 0.219 \text{ kmol}$$

$$N_{N_2} = \frac{m_{N_2}}{M_{N_2}} = \frac{4 \text{ kg}}{28 \text{ kg/kmol}} = 0.143 \text{ kmol}$$

ANSWERS

$$N_m = N_{O_2} + N_{N_2} = 0.219 + 0.143 = 0.362 \text{ kmol}$$

and

$$V_{O_2} = \left(\frac{NR_uT_1}{P_1}\right)_{O_2} = \frac{(0.219 \text{ kmol})(8.314 \text{ kPa} \cdot \text{m}^3/\text{kmol} \cdot \text{K})(313 \text{ K})}{100 \text{ kPa}} = 5.7 \text{ m}^3$$

$$V_{N_2} = \left(\frac{NR_uT_1}{P_1}\right)_{N_2} = \frac{(0.143 \text{ kmol})(8.314 \text{ kPa} \cdot \text{m}^3/\text{kmol} \cdot \text{K})(293 \text{ K})}{150 \text{ kPa}} = 2.32 \text{ m}^3$$

$$V_m = V_{O_2} + V_{N_2} = 5.7 + 2.32 = 8.02 \text{ m}^3$$

Thus,

$$P_m = \frac{N_m R_u T_m}{V_m} = \frac{(0.362 \text{ kmol})(8.314 \text{ kPa} \cdot \text{m}^3/\text{kmol} \cdot \text{K})(305.2 \text{ K})}{8.02 \text{ m}^3} = 114.5 \text{ kPa}$$

Discussion We could also determine the mixture pressure by using $P_m V_m = m_m R_m T_m$, where R_m is the apparent gas constant of the mixture. This would require a knowledge of mixture composition in terms of mass or mole fractions.

3-5 Oxygen is heated to experience a specified temperature change. The heat transfer is to be determined for two cases (a) at constant volume process, (b) at constant pressure process.

Assumptions ①Oxygen is an ideal gas since it is at a high temperature and low pressure relative to its critical point values of 154.8 K and 5.08 MPa. ②The kinetic and potential exergy changes are negligible, $\Delta ke \cong \Delta pe \cong 0$. 3 Constant specific heats can be used for oxygen.

Properties The specific heats of oxygen at the average temperature of $(25+300)/2 = 162.5$ ($C = 436$ K are $c_p = 0.952$ kJ/kg·K and $c_v = 0.692$ kJ/kg·K (Table A-10b).

Analysis We take the oxygen as the system. This is a *closed system* since no mass crosses the boundaries of the system. The exergy balance for a constant-volume process can be expressed as

$$\underbrace{E_{in} - E_{out}}_{\substack{\text{Net exergy transfer} \\ \text{by heat, work, and mass}}} = \underbrace{\Delta E_{system}}_{\substack{\text{Change in internal, kinetic,} \\ \text{potential, etc. energies}}}$$

$$Q_{in} = \Delta U = mc_v(T_2 - T_1)$$

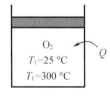

The exergy balance during a constant-pressure process (such as in a piston-cylinder device) can be expressed as

$$\underbrace{E_{in} - E_{out}}_{\substack{\text{Net exergy transfer} \\ \text{by heat, work, and mass}}} = \underbrace{\Delta E_{system}}_{\substack{\text{Change in internal, kinetic,} \\ \text{potential, etc. energies}}}$$

$$Q_{in} - W_{b,out} = \Delta U$$

$$Q_{in} = W_{b,out} + \Delta U$$

$$Q_{in} = \Delta H = mc_p(T_2 - T_1)$$

since $\Delta U + W_b = \Delta H$ during a constant pressure quasi-equilibrium process. Substituting for both cases,

$$Q_{in, V=const} = mc_v(T_2 - T_1) = (1 \text{ kg})(0.692 \text{ kJ/kg} \cdot \text{K})(300-25) \text{ K} = 190.3 \text{ kJ}$$

$$Q_{in, P=const} = mc_p(T_2 - T_1) = (1 \text{ kg})(0.952 \text{ kJ/kg} \cdot \text{K})(300-25) \text{ K} = 261.8 \text{ kJ}$$

Chapter 4

4-1 A Carnot heat engine, shown in Fig. P4-1, receives 500 kJ of heat per cycle from a high – temperature source at 652 ℃ and rejects heat to a low-temperature sink at 30℃. Determine (a) the thermal efficiency of this Carnot engine and (b) the amount of heat rejected to the sink per cycle.

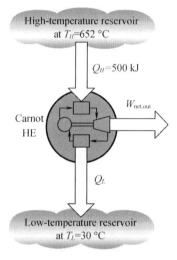

Fig. P4 – 1

Solution The heat supplied to a Carnot heat engine is given. The thermal efficiency and the heat rejected are to be determined.

Analysis (a) The Carnot heat engine is a reversible heat engine, and so its efficiency can be determined

$$\eta_{th,C} = \eta_{th,rev} = 1 - \frac{T_L}{T_H} = 1 - \frac{(30+273)\,K}{(652+273)\,K} = 0.672$$

That is, this Carnot heat engine converts 67.2 percent of the heat it receives to work.

(b) The amount of heat rejected Q_L by this reversible heat engine is easily determined to be

$$Q_{L,rev} = \frac{T_L}{T_H} Q_{H,rev} = \frac{(30+273)\,K}{(652+273)\,K}(500\text{ kJ}) = 164\text{ kJ}$$

Discussion Note that this Carnot heat engine rejects to a low-temperature sink 164 kJ of the 500 kJ of heat it receives during each cycle.

4 – 2 An inventor claims to have developed a refrigerator that maintains the refrigerated space at 35 ℉ while operating in a room where the temperature is 75 ℉ and that has a COP of 13.5. Is this claim reasonable?

Solution An extraordinary claim made for the performance of a refrigerator is to be evaluated.

Assumptions Steady operating conditions exist.

Analysis The performance of this refrigerator (shown in Fig. P4 – 2) can be evaluated by comparing it with a reversible refrigerator operating between the same temperature limits:

$$COP_{R,max} = COP_{R,rev} = \frac{1}{T_H/T_L - 1} = \frac{1}{(75+460\text{ R})/(35+460\text{ R}) - 1} = 12.4$$

ANSWERS

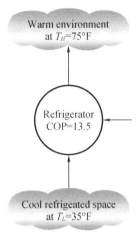

Fig. P4-2

Discussion This is the highest COP a refrigerator can have when absorbing heat from a cool medium at 35 °F and rejecting it to a warmer medium at 75 °F. Since the COP claimed by the inventor is above this maximum value, the claim is *false*.

4-3 A 50 kg block of iron casting at 500 K is thrown into a large lake that is at a temperature of 285 K. The iron block eventually reaches thermal equilibrium with the lake water. Assuming an average specific heat of 0.45 kJ/(kg · K) for the iron, determine (a) the entropy change of the iron block, (b) the entropy change of the lake water, and (c) the entropy generated during this process.

Fig. P4-3

Solution A hot iron block is thrown into a lake, and cools to the lake temperature. The entropy changes of the iron and of the lake as well as the entropy generated during this process are to be determined.

Assumptions ①Both the water and the iron block are incompressible substances. ②Constant specific heats can be used for the water and the iron. ③The kinetic and potential exergy changes of the iron are negligible, $ke = pe = 0$ and thus $E = U$.

Properties The specific heat of the iron is 0.45 kJ/(kg · K) (Table A-3).

Analysis We take the *iron casting* as the system (Fig. P4-3). This is a *closed system* since no mass crosses the system boundary during the process. To determine the entropy change for the iron block and for the lake, first we need to know the final equilibrium temperature. Given that the thermal exergy capacity of the lake is very large relative to that of the iron block, the lake will absorb all the heat rejected by the iron block without experiencing any change in its temperature. Therefore, the iron block will cool to 285 K during this process while the lake temperature remains constant at 285 K.

(a) The entropy change of the iron block can be determined from

$$\Delta S_{icon} = m(s_2 - s_1) = mc_{avg} \ln \frac{T_2}{T_1}$$

$$= (50 \text{ kg})(0.45 \text{ kJ/kg} \cdot \text{K}) \ln \frac{285 \text{ K}}{500 \text{ K}}$$

$$= -12.65 \text{ kJ/K}$$

(b) The temperature of the lake water remains constant during this process at 285 K. Also, the amount of heat transfer from the iron block to the lake is determined from an exergy balance on the iron block to be

$$\underbrace{E_{in} - E_{out}}_{\substack{\text{Net exergy transfer} \\ \text{by heat, work and mass}}} = \underbrace{\Delta E_{system}}_{\substack{\text{Change in internal, kinetic,} \\ \text{potential, etc., energies}}}$$

$$-Q_{out} = \Delta U = mc_{avg}(T_2 - T_1)$$

or

$$Q_{out} = mc_{avg}(T_1 - T_2) = (50 \text{ kg})(0.45 \text{ kJ/kg} \cdot \text{K})(500 - 285) \text{ K} = 4\,838 \text{ kJ}$$

Then the entropy change of the lake becomes

$$\Delta S_{lake} = \frac{Q_{lake}}{T_{lake}} = \frac{+4838 \text{ kJ}}{285 \text{ K}} = 16.97 \text{ kJ/K}$$

(c) The entropy generated during this process can be determined by applying an entropy balance on an *extended system* that includes the iron block and its immediate surroundings so that the boundary temperature of the extended system is at 285 K at all times:

$$\underbrace{S_{in} - S_{out}}_{\substack{\text{Net entropy transfer} \\ \text{by heat and mass}}} + \underbrace{S_{gen}}_{\substack{\text{Entropy} \\ \text{generation}}} = \underbrace{\Delta S_{system}}_{\substack{\text{Change} \\ \text{in entropy}}}$$

$$-\frac{Q_{out}}{T_b} + S_{gen} = \Delta S_{system}$$

or

$$S_{gen} = \Delta S_{system} + \frac{Q_{out}}{T_b} = \frac{4\,838 \text{ kJ}}{285 \text{ K}} + (-12.65 \text{ kJ/K}) = 4.32 \text{ kJ/K}$$

Discussion The entropy generated can also be determined by taking the iron block and the entire lake as the system, which is an isolated system, and applying an entropy balance. An isolated system involves no heat or entropy transfer, and thus the entropy generation in this case becomes equal to the total entropy change,

$$S_{gen} = \Delta S_{total} = \Delta S_{system} + \Delta S_{lake} = -12.65 + 16.97 = 4.32 \text{ kJ/K}$$

which is the same result obtained above.

4-4 A frictionless piston-cylinder device contains a saturated liquid-vapor mixture of water at 100 ℃. During a constant-pressure process, 600 kJ of heat is transferred to the surrounding air at 25 ℃. As a result, part of the water vapor contained in the cylinder condenses. Determine (a) the entropy change of the water and (b) the total entropy generation during this heat transfer process.

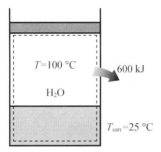

Fig. P4-4

Solution Saturated liquid – vapor mixture of water loses heat to its surroundings, and some of the vapor condenses. The entropy change of the water and the total entropy generation are to be determined.

Assumptions ① There are no irreversibilities involved within the system boundaries, and thus the process is internally reversible. ② The water temperature remains constant at 100 ℃ everywhere, including the boundaries.

Analysis We first take the *water in the cylinder* as the system (Fig. P4-4). This is a *closed system* since no mass crosses the system boundary during the process. We note that the pressure and thus the temperature of water in the cylinder remain constant during this process. Also, the entropy of the system decreases during the process because of heat loss.

(a) Noting that water undergoes an internally reversible isothermal process, its entropy change can be determined from

$$\Delta S_{system} = \frac{Q}{T_{system}} = \frac{-600 \text{ kJ}}{(100+273)\text{K}} = -1.61 \text{ kJ/K}$$

(b) To determine the total entropy generation during this process, we consider the *extended system*, which includes the water, the piston – cylinder device, and the region immediately outside the system that experiences a temperature change so that the entire boundary of the extended system is at the surrounding temperature of 25 ℃. The entropy balance for this *extended system* (system + immediate surroundings) yields

$$\underbrace{S_{in} - S_{out}}_{\substack{\text{Net entropy transfer} \\ \text{by heat and mass}}} + \underbrace{S_{gen}}_{\substack{\text{Entropy} \\ \text{generation}}} = \underbrace{\Delta S_{system}}_{\substack{\text{Change} \\ \text{in entropy}}}$$

$$-\frac{Q_{out}}{T_b} + S_{gen} = \Delta S_{system}$$

or

$$S_{gen} = \Delta S_{system} + \frac{Q_{out}}{T_b} = \frac{600 \text{ kJ}}{(25+273)\text{K}} + (-1.61 \text{ kJ/K}) = 0.4 \text{ kJ/K}$$

The entropy generation in this case is entirely due to irreversible heat transfer through a finite temperature difference. Note that the entropy change of this extended system is equivalent to the entropy change of water since the piston-cylinder device and the immediate surroundings do not experience any change of state at any point, and thus any change in any property, including entropy.

Discussion For the sake of argument, consider the reverse process (i.e., the transfer of 600 kJ of heat from the surrounding air at 25 ℃ to saturated water at 100 ℃) and see if the increase of entropy principle can detect the impossibility of this process. This time, heat transfer will be to the water(heat gain instead of heat loss), and thus the entropy change of water will be 1.61 kJ/K. Also, the entropy transfer at the boundary of the extended system will have the same magnitude but opposite direction. This will result in an entropy generation of 0.4 kJ/K. The negative sign for the entropy generation indicates that the reverse process is *impossible*.

To complete the discussion, let us consider the case where the surrounding air temperature is a differential amount below 100 ℃ (say 99.999 ⋯ 9 ℃) instead of being 25 ℃. This time, heat transfer from the saturated water to the surrounding air will take place through a differential temperature difference rendering this process *reversible*. It can be shown that $S_{gen} = 0$ for this process. Remember that reversible processes are idealized processes, and they can be approached but never reached in reality.

4-5 A heat engine receives heat from a source at 1 200 K at a rate of 500 kJ/s and rejects the waste heat to a medium at 300 K (Fig. P4-5). The power output of the heat engine is 180 kW. Determine the reversible power and the irreversibility rate for this process.

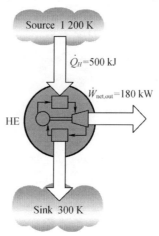

Fig. P4-5

Solution The operation of a heat engine is considered. The reversible power and the irreversibility rate associated with this operation are to be determined.

Analysis The reversible power for this process is the amount of power that a reversible heat engine, such as a Carnot heat engine, would produce when operating between the same temperature limits, and is determined to be:

$$\dot{W}_{rev} = \eta_{th,rev} \dot{Q}_{in} = \left(1 - \frac{T_{sink}}{T_{source}}\right) \dot{Q}_{in} = \left(1 - \frac{300 \text{ K}}{1\ 200 \text{ K}}\right)(500 \text{ kW}) = 375 \text{ kW}$$

This is the maximum power that can be produced by a heat engine operating between the specified temperature limits and receiving heat at the specified rate. This would also represent the *available power* if 300 K were the lowest temperature available for heat rejection.

The irreversibility rate is the difference between the reversible power (maximum power that could have been produced) and the useful power output:

$$\dot{I} = \dot{W}_{\text{rev,out}} - \dot{W}_{\text{u,out}} = 375 - 180 = 375 \text{ kW}$$

Discussion Note that 195 kW of power potential is wasted during this process as a result of irreversibilities. Also, the 500 − 375 = 125 kW of heat rejected to the sink is not available for converting to work and thus is not part of the irreversibility.

4 − 6 Refrigerant − 134a is to be compressed from 0.14 MPa and 10 ℃ to 0.8 MPa and 50 ℃ steadily by a compressor. Taking the environment conditions to be 20 ℃ and 95 kPa, determine the exergy change of the refrigerant during this process and the minimum work input that needs to be supplied to the compressor per unit mass of the refrigerant.

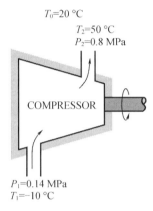

Fig. P4 − 6

Solution Refrigerant − 134a is being compressed from a specified inlet state to a specified exit state. The exergy change of the refrigerant and the minimum compression work per unit mass are to be determined.

Assumptions ①Steady operating conditions exist. ②The kinetic and potential energies are negligible.

Analysis We take the *compressor* as the system (Fig. P4 − 6). This is a *control volume* since mass crosses the system boundary during the process. Here the question is the exergy change of a fluid stream, which is the change in the flow exergy ψ.

The properties of the refrigerant at the inlet and the exit states are

Inlet state:

$$\left.\begin{array}{l} P_1 = 0.14 \text{ MPa} \\ T_1 = -10 \text{ ℃} \end{array}\right\} h_1 = 246.36 \text{ kJ/kg}, s_1 = 0.9724 \text{ kJ/kg} \cdot \text{K}$$

Exit state:

$$\left.\begin{array}{l} P_2 = 0.8 \text{ MPa} \\ T_2 = 50 \text{ ℃} \end{array}\right\} h_2 = 286.69 \text{ kJ/kg} \quad s_1 = 0.9802 \text{ kJ/kg} \cdot \text{K}$$

The exergy change of the refrigerant during this compression process is determined directly from $\Delta\psi = \psi_2 - \psi_1 = (h_2 - h_1) - T_0(s_2 - s_1) + \dfrac{V_2^2 - V_1^2}{2} + g(z_2 - z_1)$ to be

$$\begin{aligned}\Delta\psi &= \psi_2 - \psi_1 = (h_2 - h_1) - T_0(s_2 - s_1) + \dfrac{V_2^2 - V_1^2}{2}{\to}0 + g(z_2 - z_1){\to}0 \\ &= (h_2 - h_1) - T_0(s_2 - s_1) \\ &= (286.69 - 246.36)\,\text{kJ/kg} - (293\ \text{K})[(0.980\ 2 - 0.972\ 4)\,\text{kJ/kg}\cdot\text{K}] \\ &= 38\ \text{kJ/kg}\end{aligned}$$

Therefore, the exergy of the refrigerant increases during compression by 38.0 kJ/kg.

The exergy change of a system in a specified environment represents the reversible work in that environment, which is the minimum work input required for work-consuming devices such as compressors. Therefore, the increase in exergy of the refrigerant is equal to the minimum work that needs to be supplied to the compressor:

$$w_{\text{in, min}} = \psi_2 - \psi_1 = 38\ \text{kJ/kg}$$

Discussion Note that if the compressed refrigerant at 0.8 MPa and 50 ℃ were to be expanded to 0.14 MPa and 10 ℃ in a turbine in the same environment in a reversible manner, 38.0 kJ/kg of work would be produced.

Chapter 5

5 – 1 A mass of 200 g of saturated liquid water is completely vaporized at a constant pressure of 100 kPa. Determine (a) the volume change and (b) the amount of exergy transferred to the water.

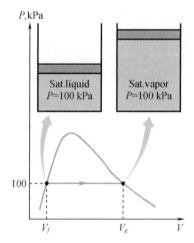

Fig. P5 – 1

Solution Saturated liquid water is vaporized at constant pressure. The volume change and the exergy transferred are to be determined.

Analysis (a) The process described is illustrated on a $P - v$ diagram in Fig. P5 – 1. The volume change per unit mass during a vaporization process is v_{fg}, which is the difference between v_g and v_f. Reading these values from Table A – 5 at 100 kPa and substituting yield

$$v_{fg} = v_g - v_f = 1.6941 - 0.001043 = 1.6931\ \text{m}^3/\text{kg}$$

ANSWERS

Thus,
$$\Delta V = m v_{fg} = (0.2 \text{ kg})(1.6931 \text{ m}^3/\text{kg}) = 0.3386 \text{ m}^3$$

(b) The amount of exergy needed to vaporize a unit mass of a substance at a given pressure is the enthalpy of vaporization at that pressure, which is $h_{fg} = 2257.5$ kJ/kg for water at 100 kPa. Thus, the amount of exergy transferred is

$$m h_{fg} = (0.2 \text{ kg})(2257.5 \text{ kJ/kg}) = 451.5 \text{ kJ}$$

Discussion Note that we have considered the first four decimal digits of v_{fg} and disregarded the rest. This is because vg has significant numbers to the first four decimal places only, and we do not know the numbers in the other decimal places. Copying all the digits from the calculator would mean that we are assuming $v_g = 1.694100$, which is not necessarily the case. It could very well be that $v_g = 1.694138$ since this number, too, would truncate to 1.6941. All the digits in our result (1.6931) are significant. But if we did not truncate the result, we would obtain $v_{fg} = 1.693057$, which falsely implies that our result is accurate to the sixth decimal place.

5 – 2 An 80 L vessel contains 4 kg of refrigerant – 134a at a pressure of 160 kPa. Determine (a) the temperature, (b) the quality, (c) the enthalpy of the refrigerant, and (d) the volume occupied by the vapor phase.

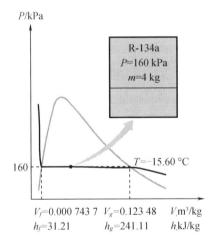

Fig. P5 – 2

Solution A vessel is filled with refrigerant – 134a. Some properties of the refrigerant are to be determined.

Analysis (a) The state of the saturated liquid – vapor mixture is shown in Fig. P5 – 2. At this point we do not know whether the refrigerant is in the compressed liquid, superheated vapor, or saturated mixture region. This can be determined by comparing a suitable property to the saturated liquid and saturated vapor values. From the information given, we can determine the specific volume:

$$v = \frac{V}{m} = \frac{0.08 \text{ m}^3}{4 \text{ kg}} = 0.02 \text{ m}^3/\text{kg}$$

At 160 kPa, we read

$$v_f = 0.0007437 \text{ m}^3/\text{kg}$$
$$v_g = 0.12348 \text{ m}^3/\text{kg} \quad (\text{Table A}-9)$$

Obviously, $v_f < v < v_g$, and, the refrigerant is in the saturated mixture region. Thus, the temperature must be the saturation temperature at the specified pressure:
$$T = T_{\text{sat @ 160kPa}} = -15.6 \text{ °C}$$

(b) Quality can be determined from
$$x = \frac{v - v_f}{v_{fg}} = \frac{0.02 - 0.0007437}{0.12348 - 0.0007437} = 0.157$$

(c) At 160 kPa, we also read from Table A – 12 that h_f = 31.21 kJ/kg and h_{fg} = 209.90 kJ/kg. Then,
$$\begin{aligned} h &= h_f + xh_{fg} \\ &= 31.21 \text{ kJ/kg} + (0.157)(209.9 \text{ kJ/kg}) \\ &= 64.2 \text{ kJ/kg} \end{aligned}$$

(d) The mass of the vapor is
$$m_g = xm_t = (0.157)(4 \text{ kg}) = 0.628 \text{ kg}$$
and the volume occupied by the vapor phase is
$$V_g = m_g v_g = (0.628 \text{ kg})(0.12348 \text{ m}^3/\text{kg}) = 0.0775 \text{ m}^3 (\text{or } 77.5 \text{ L})$$
The rest of the volume (2.5 L) is occupied by the liquid.

5 – 3 Determine the temperature of water at a state of $P = 0.5$ MPa and $h = 2890$ kJ/kg.

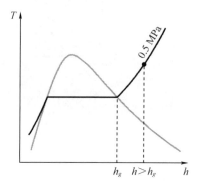

Fig. P5 – 3

Solution The temperature of water at a specified state is to be determined.

Analysis At 0.5 MPa, the enthalpy of saturated water vapor is $hg = 2748.1$ kJ/kg. Since $h > hg$, as shown in Fig. 3 – 41, we again have superheated vapor. Under 0.5 MPa in Table A – 6 we read

T/°C	h/(kJ/kg)
200	2 855.8
250	2 961.0

Obviously, the temperature is between 200 ℃ and 250 ℃. By linear interpolation it is determined to be

$$T = 216.3 \ ℃$$

5 – 4 A rigid container that is filled with R – 134a is heated. The initial pressure and final pressure are 300 kPa and 600 kPa, respectively. The temperature and total enthalpy are to be determined at the initial and final states.

 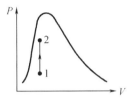

Fig. P5 – 4

Analysis This is a constant volume process. The specific volume is

$$v_1 = v_2 = \frac{V}{m} = \frac{0.014 \ \text{m}^3}{10 \ \text{kg}} = 0.001 \ 4 \ \text{m}^3/\text{kg}$$

The initial state is determined to be a mixture, and thus the temperature is the saturation temperature at the given pressure. From Table A – 12 by interpolation

$$T_1 = T_{\text{sat @ 300 kPa}} = 0.61 \ ℃$$

and

$$x_1 = \frac{v_1 - v_f}{v_{fg}} = \frac{(0.001 \ 4 - 0.000 \ 773 \ 6) \ \text{m}^3/\text{kg}}{(0.067 \ 978 - 0.000 \ 773 \ 6) \ \text{m}^3/\text{kg}} = 0.009 \ 321$$

$$h_1 = h_f + x_1 h_{fg} = 52.67 + (0.009 \ 321)(198.13) = 54.52 \ \text{kJ/kg}$$

The total enthalpy is then

$$H_1 = m h_1 = (10 \ \text{kg})(54.52 \ \text{kJ/kg}) = 545.2 \ \text{kJ}$$

The final state is also saturated mixture. Repeating the calculations at this state,

$$T_2 = T_{\text{sat @ 600 kPa}} = 21.55 \ ℃$$

$$x_2 = \frac{v_2 - v_f}{v_{fg}} = \frac{(0.001 \ 4 - 0.000 \ 819 \ 9) \ \text{m}^3/\text{kg}}{(0.034 \ 295 - 0.000 \ 819 \ 9) \ \text{m}^3/\text{kg}} = 0.017 \ 33$$

$$h_2 = h_f + x_2 h_{fg} = 81.51 + (0.017 \ 33)(180.90) = 84.64 \ \text{kJ/kg}$$

$$H_2 = m h_2 = (10 \ \text{kg})(84.64 \ \text{kJ/kg}) = 846.4 \ \text{kJ}$$

Chapter 6

6 – 1 A 5 m × 5 m × 3 m room shown in Fig. P6 – 1 contains air at 25 ℃ and 100 kPa at a relative humidity of 75 percent. Determine (a) the partial pressure of dry air, (b) the specific humidity, (c) the enthalpy per unit mass of the dry air, and (d) the masses of the dry air and water vapor in the room.

Solution The relative humidity of air in a room is given. The dry air pressure, specific humidity, enthalpy, and the masses of dry air and water vapor in the room are to be determined.

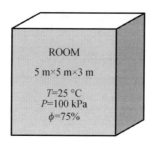

Fig. P6-1

Assumptions The dry air and the water vapor in the room are ideal gases.

Properties The constant-pressure specific heat of air at room temperature is $c_p = 1.005$ kJ/kg · K (Table A-10a). For water at 25 ℃, we have $T_{sat} = 3.1698$ kPa and $h_g = 2546.5$ kJ/kg (Table A-4).

Analysis (a) The partial pressure of dry air can be determined from $P = P_a + P_v$:

$$P_a = P - P_v$$

where

$$P_v = \varphi P_g = \varphi P_{sat\ @\ 25℃} = (0.75)(3.1698\ \text{kPa}) = 2.38\ \text{kPa}$$

Thus,

$$P_a = (100 - 2.38)\ \text{kPa} = 97.62\ \text{kPa}$$

(b) The specific humidity of air is determined:

$$\omega = 0.622 \frac{P_v}{P - P_v} = \frac{(0.622)(2.38\ \text{kPa})}{(100 - 2.38)\ \text{kPa}} = 0.0152\ \text{kg H}_2\text{O/kg dry air}$$

(c) The enthalpy of air per unit mass of dry air is determined from Eq. 14-12:

$$h = h_a + \omega h_v \cong c_p T + \omega h_v$$
$$= (1.005\ \text{kJ/kg} \cdot ℃)(25\ ℃) + (0.0152)(2546.5\ \text{kJ/kg})$$
$$= 63.8\ \text{kJ/kg dry air}$$

The enthalpy of water vapor (2546.5 kJ/kg) could also be determined from the approximation given by $h_g(T) \cong 2500.9 + 1.82T$:

$$h_{g\ @\ 25℃} \cong 2500.9 + 1.82(25) = 2546.4\ \text{kJ/kg}$$

which is almost identical to the value obtained from Table A-4.

(d) Both the dry air and the water vapor fill the entire room completely. Therefore, the volume of each gas is equal to the volume of the room:

$$V_a = V_v = V_{room} = (5\ \text{m})(5\ \text{m})(3\ \text{m}) = 75\ \text{m}^3$$

The masses of the dry air and the water vapor are determined from the ideal gas relation applied to each gas separately:

$$m_a = \frac{P_a V_a}{R_a T} = \frac{(97.62\ \text{kPa})(75\ \text{m}^3)}{(0.287\ \text{kPa} \cdot \text{m}^3/\text{kg} \cdot \text{K})(298\ \text{K})} = 85.61\ \text{kg}$$

$$m_v = \frac{P_v V_v}{R_v T} = \frac{(2.38\ \text{kPa})(75\ \text{m}^3)}{(0.4615\ \text{kPa} \cdot \text{m}^3/\text{kg} \cdot \text{K})(298\ \text{K})} = 1.3\ \text{kg}$$

The mass of the water vapor in the air could also be determined from $\omega = \dfrac{m_v}{m_a}$:

ANSWERS

$$m_v = \omega m_a = (0.015\ 2)(65.61\ \text{kg}) = 1.3\ \text{kg}$$

6 – 2 The dry-and the wet-bulb temperatures of atmospheric air at 1 atm (101.325 kPa) pressure are measured with a sling psychrometer and determined to be 25 ℃ and 15 ℃, respectively. Determine (a) the specific humidity, (b) the relative humidity, and (c) the enthalpy of the air.

Solution Dry-and wet-bulb temperatures are given. The specific humidity, relative humidity, and enthalpy are to be determined.

Properties The saturation pressure of water is 1.705 7 kPa at 15 ℃, and 3.169 8 kPa at 25 ℃ (Table A – 4). The constant-pressure specific heat of air at room temperature is c_p = 1.005 kJ/kg · K (Table A – 10a).

Analysis (a) The specific humidity ω_1 is determined to be,

$$\omega_1 = \frac{c_p(T_2 - T_1) + \omega_2 h_{fg2}}{h_{g1} - h_{f2}}$$

where T_2 is the wet-bulb temperature and ω_2 is

$$\omega_2 = \frac{0.622 P_{g2}}{P_2 - P_{g2}} = \frac{(0.622)(1.705\ 7\ \text{kPa})}{(101.325 - 1.705\ 7)\text{kPa}}$$
$$= 0.010\ 65\ \text{kg H}_2\text{O/kg dry air}$$

Thus,

$$\omega_1 = \frac{(1.005\ \text{kJ/kg} \cdot \text{℃})[(15-25)\text{℃}] + (0.010\ 65)(2\ 465.4\ \text{kJ/kg})}{(2\ 546.5 - 62.982)\text{kJ/kg}}$$
$$= 0.006\ 53\ \text{kg H}_2\text{O/kg dry air}$$

(b) The relative humidity φ_1 is determined to be

$$\varphi_1 = \frac{\omega_1 P_2}{(0.622 + \omega_1) P_{g1}} = \frac{(0.006\ 53)(101.325\ \text{kPa})}{(0.622 + 0.006\ 53)(3.169\ 8\ \text{kPa})} = 0.332\ \text{or}\ 33.2\%$$

(c) The enthalpy of air per unit mass of dry air is determined:

$$h_1 = h_{a_1} + \omega_1 h_{v_1} \cong c_p T_1 + \omega_1 h_{g1}$$
$$= (1.005\ \text{kJ/kg} \cdot \text{℃})(25\ \text{℃}) + (0.006\ 53)(2\ 546.5\ \text{kJ/kg})$$
$$= 41.8\ \text{kJ/kg dry air}$$

Discussion The previous property calculations can be performed easily using EES or other programs with built-in psychrometric functions.

6 – 3 An air-conditioning system is to take in outdoor air at 10 ℃ and 30 percent relative humidity at a steady rate of 45 m³/min and to condition it to 25 ℃ and 60 percent relative humidity. The outdoor air is first heated to 22 ℃ in the heating section and then humidified by the injection of hot steam in the humidifying section. Assuming the entire process takes place at a pressure of 100 kPa, determine (a) the rate of heat supply in the heating section and (b) the mass flow rate of the steam required in the humidifying section.

Solution Outdoor air is first heated and then humidified by steam injection. The rate of heat transfer and the mass flow rate of steam are to be determined.

Assumptions ①This is a steady-flow process and thus the mass flow rate of dry air remains constant during the entire process. ②Dry air and water vapor are ideal gases. ③The kinetic and potential exergy changes are negligible.

Properties The constant-pressure specific heat of air at room temperature is $c_p = 1.005$ kJ/kg · K, and its gas constant is $R_a = 0.287$ kJ/kg · K (Table A − 10a). The saturation pressure of water is 1.228 1 kPa at 10 ℃, and 3.169 8 kPa at 25 ℃. The enthalpy of saturated water vapor is 2 519.2 kJ/kg at 10 ℃, and 2 541.0 kJ/kg at 22 ℃ (Table A − 4).

Analysis We take the system to be the *heating* or the *humidifying section*, as appropriate. The schematic of the system and the psychrometric chart of the process are shown in Fig. P6 − 3. We note that the amount of water vapor in the air remains constant in the heating section ($\omega_1 = \omega_2$) but increases in the humidifying section ($\omega_3 = \omega_2$).

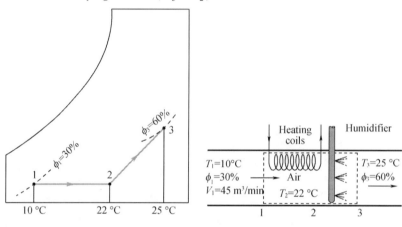

Fig. P6 − 3

(a) Applying the mass and exergy balances on the heating section gives

Dry air mass balance:
$$\dot{m}_{a_1} = \dot{m}_{a_2} = \dot{m}_a$$

Water mass balance:
$$\dot{m}_{a_1}\omega_1 = \dot{m}_{a_2}\omega_2 \rightarrow \omega_1 = \omega_2$$

Exergy balance:
$$\dot{Q}_{in} + \dot{m}_a h_1 = \dot{m}_a h_2 \rightarrow \dot{Q}_{in} = \dot{m}_a(h_2 - h_1)$$

The psychrometric chart offers great convenience in determining the properties of moist air. However, its use is limited to a specified pressure only, which is 1 atm (101.325 kPa) for the one given in the appendix. At pressures other than 1 atm, either other charts for that pressure or the relations developed earlier should be used. In our case, the choice is clear:

$$P_{v_1} = \varphi_1 P_{g_1} = \varphi P_{\text{sat @ 10 ℃}} = (0.3)(1.2281 \text{ kPa}) = 0.368 \text{ kPa}$$

$$P_{a_1} = P_1 - P_{v_1} = (100 - 0.368) \text{ kPa} = 99.632 \text{ kPa}$$

$$v_1 = \frac{R_a T_1}{P_a} = \frac{(0.287 \text{ kPa} \cdot \text{m}^3/\text{kg} \cdot \text{K})(283 \text{ K})}{99.632 \text{ kPa}} = 0.815 \text{ m}^3/\text{kg dry air}$$

$$\dot{m}_a = \frac{\dot{V}_1}{v_1} = \frac{45 \text{ m}^3/\text{min}}{0.815 \text{ m}^3/\text{kg}} = 55.2 \text{ kg/min}$$

ANSWERS

$$\omega_1 = \frac{0.622 P_{v_1}}{P_1 - P_{v_1}} = \frac{0.622(0.368 \text{ kPa})}{(100 - 0.368) \text{ kPa}} = 0.0023 \text{ kg H}_2\text{O/kg dry air}$$

$$h_1 = c_p T_1 + \omega_1 h_{g_1} = (1.005 \text{ kJ/kg} \cdot \text{°C})(10 \text{ °C}) + (0.0023)(2519.2 \text{ kJ/kg})$$
$$= 15.8 \text{ kJ/kg dry air}$$

$$h_2 = c_p T_2 + \omega_2 h_{g_2} = (1.005 \text{ kJ/kg} \cdot \text{°C})(22 \text{ °C}) + (0.0023)(2541 \text{ kJ/kg})$$
$$= 28 \text{ kJ/kg dry air}$$

since $\omega_1 = \omega_2$. Then the rate of heat transfer to air in the heating section becomes

$$\dot{Q}_{\text{in}} = \dot{m}_a(h_2 - h_1) = (55.2 \text{ kg/min})[(28 - 15.8) \text{ kJ/kg}] = 673 \text{ kJ/min}$$

(b) The mass balance for water in the humidifying section can be expressed as

$$\dot{m}_{a_2}\omega_2 + \dot{m}_w = \dot{m}_{a_3}\omega_3$$

or

$$\dot{m}_w = \dot{m}_{a_3}(\omega_3 - \omega_2)$$

where

$$\omega_3 = \frac{0.622 \varphi_3 P_{g_3}}{P_3 - \varphi_3 P_{g_3}} = \frac{0.622(0.6)(3.1698 \text{ kPa})}{[100 - (0.6)(3.1698)] \text{ kPa}} = 0.01206 \text{ kg H}_2\text{O/kg dry air}$$

Thus,

$$\dot{m}_w = (55.2 \text{ kg/min})(0.01206 - 0.0023) = 0.539 \text{ kg/min}$$

Discussion The result 0.539 kg/min corresponds to a water requirement of close to one ton a day, which is significant.

6-4 Air enters an evaporative (or swamp) cooler at 14.7 psi, 95 °F, and 20 percent relative humidity, and it exits at 80 percent relative humidity. Determine (a) the exit temperature of the air and (b) the lowest temperature to which the air can be cooled by this evaporative cooler.

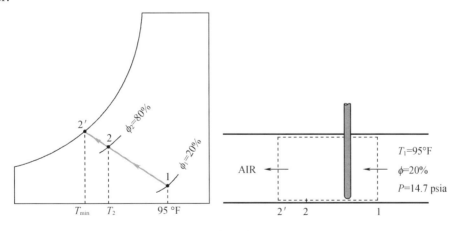

Fig. P6-4

Solution Air is cooled steadily by an evaporative cooler. The temperature of discharged air and the lowest temperature to which the air can be cooled are to be determined.

Analysis The schematic of the evaporative cooler and the psychrometric chart of the process are shown in Fig. P6-4.

(a) If we assume the liquid water is supplied at a temperature not much different from the exit temperature of the airstream, the evaporative cooling process follows a line of constant wet bulb temperature on the psychrometric chart. That is,

$$T_{wb} = const$$

The wet bulb temperature at 95 °F and 20% relative humidity is determined from the psychrometric chart to be 66.0 °F. The intersection point of the $T_{wb} = 66.0$ °F and the $\phi = 80\%$ lines is the exit state of the air. The temperature at this point is the exit temperature of the air, and it is determined from the psychrometric chart to be

$$T_2 = 70.4 \text{ °F}$$

(b) In the limiting case, air leaves the evaporative cooler saturated ($\varphi = 100\%$), and the exit state of the air in this case is the state where the $T_{wb} = 66.0$ °F line intersects the saturation line. For saturated air, the dry and the wet bulb temperatures are identical. Therefore, the lowest temperature to which air can be cooled is the wet bulb temperature, which is

$$T_{min} = T_{2'} = 66 \text{ °F}$$

Discussion Note that the temperature of air drops by as much as 30 °F in this case by evaporative cooling.

Chapter 7

7 – 1 What is the dew-point temperature?

Dew-point temperature is the temperature at which condensation begins when air is cooled at constant pressure.

7 – 2 In summer, the outer surface of a glass filled with iced water frequently "sweats." How can you explain this sweating?

The outsider surface temperature of the glass may drop below the dew-point temperature of the surrounding air, causing the moisture in the vicinity of the glass to condense. After a while, the condensate may start dripping down because of gravity.

7 – 3 What is the difference between the specific humidity and the relative humidity?

Specific humidity is the amount of water vapor present in a unit mass of dry air.

Relative humidity is the ratio of the actual amount of vapor in the air at a given temperature to the maximum amount of vapor air can hold at that temperature.

7 – 4 Carbon dioxide flows steadily through a converging – diverging nozzle, as shown in Fig. P7 – 4. Calculate the critical pressure and temperature of carbon dioxide.

Solution For the flow discussed in Example 17 – 3, the critical pressure and temperature are to be calculated.

Assumptions ①The flow is steady, adiabatic, and one-dimensional. ②Carbon dioxide is an ideal gas with constant specific heats.

Properties The specific heat ratio of carbon dioxide at room temperature is $k = 1.289$ (Table A – 10a).

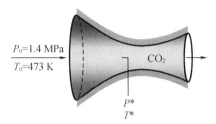

Fig. P7-4

Analysis The ratios of critical to stagnation temperature and pressure are determined to be

$$\frac{T^*}{T_0} = \frac{2}{k+1} = \frac{2}{1.289+1} = 0.8737$$

$$\frac{P^*}{P_0} = \left(\frac{2}{k+1}\right)^{k/(k-1)} = \left(\frac{2}{1.289+1}\right)^{1.289/(1.289-1)} = 0.5477$$

Noting that the stagnation temperature and pressure are, $T_0 = 473$ K and $P_0 = 1400$ kPa, we see that the critical temperature and pressure in this case are

$$T^* = 0.8737 T_0 = (0.8737)(473 \text{ K}) = 413 \text{ K}$$

$$P^* = 0.5477 T_0 = (0.5477)(1400 \text{ kPa}) = 767 \text{ kPa}$$

Discussion Note that these values agree with those listed in Table 7-1, as expected. Also, property values other than these at the throat would indicate that the flow is not critical, and the Mach number is not unity.

7-5 Nitrogen enters a duct with varying flow area at $T_1 = 400$ K, $P_1 = 100$ kPa, and $Ma_1 = 0.3$. Assuming steady isentropic flow, determine T_2, P_2, and Ma_2 at a location where the flow area has been reduced by 20 percent.

Fig. P7-5

Solution Nitrogen gas enters a converging nozzle. The properties at the nozzle exit are to be determined.

Assumptions ①Nitrogen is an ideal gas with $k = 1.4$, ②Flow through the nozzle is steady, one-dimensional, and isentropic.

Analysis The schematic of the duct is shown in Fig. P7-5. For isentropic flow through a duct, the area ratio A/A^* (the flow area over the area of the throat where Ma = 1) is also listed in Table A-13. At the initial Mach number of $Ma_1 = 0.3$, we read

$$\frac{A_1}{A^*} = 2.0351 \quad \frac{T_1}{T_0} = 0.9823 \quad \frac{P_1}{P_0} = 0.9395$$

With a 20 percent reduction in flow area, $A_2 = 0.8 A_1$,

$$\frac{A_2}{A^*} = \frac{A_2}{A_1}\frac{A_1}{A^*} = (0.8)(2.035\ 1) = 1.628\ 1$$

and For this value of A_2/A^* from Table A-32, we read

$$\frac{T_2}{T_0} = 0.970\ 1 \qquad \frac{P_2}{P_0} = 0.899\ 3 \qquad Ma = 0.391$$

Here we chose the subsonic Mach number for the calculated A_2/A^* instead of the supersonic one because the duct is converging in the flow direction and the initial flow is subsonic. Since the stagnation properties are constant for isentropic flow, we can write

$$\frac{T_2}{T_1} = \frac{T_2/T_0}{T_1/T_0} \rightarrow T_2 = T_1\left(\frac{T_2/T_0}{T_1/T_0}\right) = (400\ \text{K})\left(\frac{0.970\ 1}{0.982\ 3}\right) = 395\ \text{K}$$

$$\frac{P_2}{P_1} = \frac{P_2/P_0}{P_1/P_0} \rightarrow P_2 = P_1\left(\frac{P_2/P_0}{P_1/P_0}\right) = (100\ \text{kPa})\left(\frac{0.899\ 3}{0.939\ 5}\right) = 95.7\ \text{kPa}$$

which are the temperature and pressure at the desired location.

Discussion Note that the temperature and pressure drop as the fluid accelerates in a converging nozzle.

7-6 Air enters a converging-diverging nozzle, shown in Fig. P7-6, at 1.0 MPa and 800 K with a negligible velocity. The flow is steady, one-dimensional, and isentropic with $k = 1.4$. For an exit Mach number of $Ma = 2$ and a throat area of 20 cm^2, determine (a) the throat conditions, (b) the exit plane conditions, including the exit area, and (c) the mass flow rate through the nozzle.

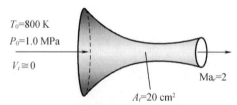

Fig. P7-6

Solution Air flows through a converging-diverging nozzle. The throat and the exit conditions and the mass flow rate are to be determined.

Assumptions ①Air is an ideal gas with constant specific heats at room temperature. ②Flow through the nozzle is steady, one-dimensional, and isentropic.

Properties The specific heat ratio of air is given to be $k = 1.4$. The gas constant of air is 0.287 kJ/kg · K.

Analysis ①The exit Mach number is given to be ②Therefore, the flow must be sonic at the throat and supersonic in the diverging section of the nozzle. Since the inlet velocity is negligible, the stagnation pressure and stagnation temperature are the same as the inlet temperature and pressure, $P_0 = 1.0$ MPa and $T_0 = 800$ K. The stagnation density is

$$\rho_0 = \frac{P_0}{RT_0} = \frac{1\ 000\ \text{kPa}}{(0.287\ \text{kPa} \cdot \text{m}^3/\text{kg} \cdot \text{K})(800\ \text{K})} = 4.355\ \text{kg/m}^3$$

(a) At the throat of the nozzle $Ma = 1$, and from Table A-32 we read

ANSWERS

$$\frac{P^*}{P_0}=0.528\ 3 \quad \frac{T^*}{T_0}=0.833\ 3 \quad \frac{\rho^*}{\rho_0}=0.633\ 9$$

Thus,
$$P^*=0.528\ 3P_0=(0.528\ 3)(1\ \text{MPa})=0.528\ 3$$
$$T^*=0.833\ 3T_0=(0.833\ 3)(800\ \text{K})=666.6\ \text{K}$$
$$\rho^*=0.633\ 9\rho_0=(0.633\ 9)(4.355\ \text{kg/m}^3)=2.761\ \text{kg/m}^3$$

Also,
$$V^*=c^*=\sqrt{kRT^*}=\sqrt{(1.4)(0.287\ \text{kJ/kg}\cdot\text{K})(666.6\ \text{K})\left(\frac{1\ 000\ \text{m}^2/\text{s}^2}{1\ \text{kJ/kg}}\right)}=517\ \text{m/s}$$

(b) Since the flow is isentropic, the properties at the exit plane can also be calculated by using data from Table A – 13. For $Ma=2$ we read

$$\frac{P_e}{P_0}=0.127\ 8 \quad \frac{T_e}{T_0}=0.555\ 6 \quad \frac{\rho_e}{\rho_0}=0.23 \quad M_{a_t}^*=1.633 \quad \frac{A_e}{A^*}=1.687\ 5$$

Thus,
$$P_e=0.127\ 8P_0=(0.127\ 8)(10\ \text{MPa})=0.127\ 8\ \text{MPa}$$
$$T_e=0.555\ 6T_0=(0.555\ 6)(800\ \text{K})=444.5\ \text{K}$$
$$\rho_e=0.23\rho_0=(0.23)(4.355\ \text{kg/m}^3)=1.002\ \text{kg/m}^3$$
$$A_e=1.687\ 5A^*=(1.687\ 5)(20\ \text{cm}^2)=33.75\ \text{cm}^2$$

and
$$V_e=Ma_e^*c^*=(1.633)(517.5\ \text{m/s})=845.1\ \text{m/s}$$

The nozzle exit velocity could also be determined from $V_e=Ma_e c_e$, where c_e is the speed of sound at the exit conditions:

$$V_e=Ma_e c_e=Ma_e\sqrt{kRT_e}=2\sqrt{(1.4)(0.287\ \text{kJ/kg}\cdot\text{K})(444.5\ \text{K})\left(\frac{1000\ \text{m}^2/\text{s}^2}{1\ \text{kJ/kg}}\right)}$$
$$=845.2\ \text{m/s}$$

(c) Since the flow is steady, the mass flow rate of the fluid is the same at all sections of the nozzle. Thus it may be calculated by using properties at any cross section of the nozzle. Using the properties at the throat, we find that the mass flow rate is

$$\dot{m}=\rho^*A^*V^*=(2.761\ \text{kg/m}^3)(20\times10^{-4}\ \text{m}^2)(517.5\ \text{m/s})=2.86\ \text{kg/s}$$

Discussion Note that this is the highest possible mass flow rate that can flow through this nozzle for the specified inlet conditions.

Chapter 8

8 – 1 What is the difference between air-standard assumptions and the cold-air-standard assumptions?

The cold air standard assumptions involves the additional assumption that air can be treated as an ideal gas with constant specific heats as room temperature.

8 – 2 What four processes make up the ideal Otto cycle?

The four processes that make up the Otto cycle are ①isentropic compression, ②v = constant heat addition, ③isentropic expansion, and ④v = constant heat rejection.

8 – 3 What four processes make up the simple ideal Rankine cycle?

The four processes that make up the simple ideal Rankine cycle are ①Isentropic compression in a pump, ②P = constant heat addition in a boiler, ③Isentropic expansion in a turbine, and ④ P = constant heat rejection in a condenser.

8 – 4 An ideal Otto cycle with air as the working fluid has a compression ratio of 9.5. The highest pressure and temperature in the cycle, the amount of heat transferred, the thermal efficiency, and the mean effective pressure are to be determined.

Assumptions ①The air-standard assumptions are applicable. ②Kinetic and potential exergy changes are negligible. ③Air is an ideal gas with constant specific heats.

Properties The properties of air at room temperature are c_p = 1.005 kJ/kg · K, c_v = 0.718 kJ/kg · K, R = 0.287 kJ/kg · K, and k = 1.4(Table A – 2).

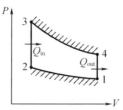

Fig. P8 – 4

Analysis (a) Process 1 – 2: isentropic compression.

$$T_2 = T_1 \left(\frac{v_1}{v_2}\right)^{k-1} = (308 \text{ K})(9.5)^{0.4} = 757.9 \text{ K}$$

$$\frac{P_2 v_2}{T_2} = \frac{P_1 v_1}{T_1} \longrightarrow P_2 = \frac{v_1}{v_2}\frac{T_2}{T_1}P_1 = (9.5)\left(\frac{757.9 \text{ K}}{308 \text{ K}}\right)(100 \text{ kPa}) = 2\,338 \text{ kPa}$$

Process 3 – 4: isentropic expansion.

$$T_3 = T_4 \left(\frac{v_4}{v_3}\right)^{k-1} = (800 \text{ K})(9.5)^{0.4} = 1\,969 \text{ K}$$

Process 2 – 3: v = constant heat addition.

$$\frac{P_3 v_3}{T_3} = \frac{P_2 v_2}{T_2} \longrightarrow P_3 = \frac{T_3}{T_2}P_2 = \left(\frac{1969 \text{ K}}{757.9 \text{ K}}\right)(2\,338 \text{ kPa}) = 6\,072 \text{ kPa}$$

(b) $$m = \frac{P_1 V_1}{RT_1} = \frac{(100 \text{ kPa})(0.000\,6 \text{ m}^3)}{(0.287 \text{ kPa} \cdot \text{m}^3/\text{kg} \cdot \text{K})(308 \text{ K})} = 6.788 \times 10^{-4} \text{ kg}$$

$$Q_{in} = m(u_3 - u_2) = mc_v(T_3 - T_2)$$
$$= (6.788 \times 10^{-4} \text{ kg})(0.718 \text{ kJ/kg} \cdot \text{K})(1\,969 - 757.9) \text{ K} = 0.590 \text{ kJ}$$

(c) Process 4 – 1: v = constant heat rejection.

$$Q_{out} = m(u_4 - u_1) = mc_v(T_4 - T_1)$$
$$= -(6.788 \times 10^{-4} \text{ kg})(0.718 \text{ kJ/kg} \cdot \text{K})(800 - 308) \text{ K} = 0.240 \text{ kJ}$$

$$W_{net} = Q_{in} - Q_{out} = 0.590 - 0.240 = 0.350 \text{ kJ}$$

$$\eta_{th} = \frac{W_{net,out}}{Q_{in}} = \frac{0.350 \text{ kJ}}{0.590 \text{ kJ}} = 59.4\%$$

(d) $$V_{min} = V_2 = \frac{V_{max}}{r}$$

ANSWERS

$$\text{MEP} = \frac{W_{net,out}}{V_1 - V_2} = \frac{W_{net,out}}{V_1(1 - 1/r)} = \frac{0.350 \text{ kJ}}{(0.0006 \text{ m}^3)(1 - 1/9.5)}\left(\frac{\text{kPa} \cdot \text{m}^3}{\text{kJ}}\right) = 652 \text{ kPa}$$

8 – 5 An ideal diesel cycle has a a cutoff ratio of 1.2. The power produced is to be determined.

Assumptions ①The air-standard assumptions are applicable. ②Kinetic and potential exergy changes are negligible. ③Air is an ideal gas with constant specific heats.

Properties The properties of air at room temperature are $c_p = 1.005$ kJ/kg · K, $c_v = 0.718$ kJ/kg · K, $R = 0.287$ kJ/kg · K, and $k = 1.4$(Table A – 10a).

Analysis The specific volume of the air at the start of the compression is

$$v_1 = \frac{RT_1}{P_1} = \frac{(0.287 \text{ kPa} \cdot \text{m}^3/\text{kg} \cdot \text{K})(288 \text{ K})}{95 \text{ kPa}} = 0.8701 \text{ m}^3/\text{kg}$$

The total air mass taken by all 8 cylinders when they are charged is

$$m = N_{cyl}\frac{\Delta V}{v_1} = N_{cyl}\frac{\pi B^2 S/4}{v_1} = (8)\frac{\pi(0.10 \text{ m})^2(0.12 \text{ m})/4}{0.8701 \text{ m}^3/\text{kg}} = 0.008665 \text{ kg}$$

The rate at which air is processed by the engine is determined from

$$\dot{m} = \frac{m\dot{n}}{N_{rev}} = \frac{(0.008665 \text{ kg/cycle})(1600/60 \text{ rev/s})}{2 \text{ rev/cycle}} = 0.1155 \text{ kg/s}$$

since there are two revolutions per cycle in a four-stroke engine. The compression ratio is

$$r = \frac{1}{0.05} = 20$$

At the end of the compression, the air temperature is

$$T_2 = T_1 r^{k-1} = (288 \text{ K})(20)^{1.4-1} = 954.6 \text{ K}$$

Application of the first law and work integral to the constant pressure heat addition gives

$$q_{in} = c_p(T_3 - T_2) = (1.005 \text{ kJ/kg} \cdot \text{K})(2273 - 954.6) \text{K} = 1325 \text{ kJ/kg}$$

while the thermal efficiency is

$$\eta_{th} = 1 - \frac{1}{r^{k-1}}\frac{r_c^k - 1}{k(r_c - 1)} = 1 - \frac{1}{20^{1.4-1}}\frac{1.2^{1.4} - 1}{1.4(1.2 - 1)} = 0.6867$$

The power produced by this engine is then

$$\dot{W}_{net} = \dot{m}w_{net} = \dot{m}\eta_{th}q_{in}$$
$$= (0.1155 \text{ kg/s})(0.6867)(1325 \text{ kJ/kg})$$
$$= 105.1 \text{ kW}$$

8 – 6 Consider a steam power plant operating on the ideal reheat Rankine cycle. Steam enters the high-pressure turbine at 15 MPa and 600 ℃ and is condensed in the condenser at a pressure of 10 kPa. If the moisture content of the steam at the exit of the low-pressure turbine is not to exceed 10.4 percent, determine (a) the pressure at which the steam should be reheated and (b) the thermal efficiency of the cycle. Assume the steam is reheated to the inlet temperature of the high – pressure turbine.

Solution A steam power plant operating on the ideal reheat Rankine cycle is considered. For a specified moisture content at the turbine exit, the reheat pressure and the thermal efficiency are to be determined.

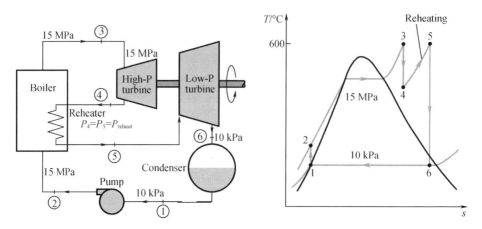

Fig. P8-6

Assumptions ①Steady operating conditions exist. ②Kinetic and potential exergy changes are negligible.

Analysis The schematic of the power plant and the $T-s$ diagram of the cycle are shown in Fig. P8-6. We note that the power plant operates on the ideal reheat Rankine cycle. Therefore, the pump and the turbines are isentropic, there are no pressure drops in the boiler and condenser, and steam leaves the condenser and enters the pump as saturated liquid at the condenser pressure.

(a) The reheat pressure is determined from the requirement that the entropies at states 5 and 6 be the same:

State 6:
$$P_6 = 10 \text{ kPa}$$
$$x_6 = 0.896 (\text{sat. mixture})$$
$$s_6 = s_f + x_6 s_{fg} = 0.649\ 2 + 0.896(7.499\ 6) = 7.368\ 8 \text{ kJ/kg} \cdot \text{K}$$

Also,
$$h_6 = h_f + x_6 h_{fg} = 191.81 + 0.896(2\ 392.1) = 2\ 335.1 \text{ kJ/kg}$$

Thus,
State 5:
$$\left.\begin{array}{l} T_5 = 600\ ℃ \\ s_5 = s_6 \end{array}\right\} \quad P_5 = 4 \text{ MPa} \quad h_5 = 3\ 674.9 \text{ kJ/kg}$$

Therefore, steam should be reheated at a pressure of 4 MPa or lower to prevent a moisture content above 10.4 percent.

(b) To determine the thermal efficiency, we need to know the enthalpies at all other states:

State 1:
$$\left.\begin{array}{l} P_1 = 10 \text{ kPa} \\ \text{Sat. liquid} \end{array}\right\} \quad h_1 = h_{f@10\text{ kPa}} = 191.81 \text{ kJ/kg} \quad v_1 = v_{f@10\text{ kPa}} = 0.001\ 01 \text{ m}^3/\text{kg}$$

State 2:
$$P_2 = 15 \text{ MPa}$$

ANSWERS

$$s_2 = s_1$$

$$w_{\text{pump,in}} = v_1(P_2 - P_1) = (0.001\ 01\ \text{m}^3/\text{kg})[(15\ 000 - 10)\text{kPa}]\left(\frac{1\ \text{kJ}}{1\ \text{kPa}\cdot\text{m}^3}\right)$$

$$= 15.14\ \text{kJ/kg}$$

$$h_2 = h_1 + w_{\text{pump,in}} = (191.81 + 15.14)\ \text{kJ/kg} = 206.95\ \text{kJ/kg}$$

State 3:

$$\left.\begin{array}{l} P_3 = 15\ \text{MPa} \\ T_3 = 600\ ^\circ\text{C} \end{array}\right\} \quad h_3 = 3\ 583.1\ \text{kJ/kg} \quad s_3 = 6.679\ 6\ \text{kJ/kg}\cdot\text{K}$$

State 4:

$$\left.\begin{array}{l} P_4 = 4\ \text{MPa} \\ s_4 = s_3 \end{array}\right\} \quad T_4 = 375.5\ ^\circ\text{C} \quad h_5 = 3\ 155\ \text{kJ/kg}$$

Thus

$$q_{\text{in}} = (h_3 - h_2) + (h_3 - h_2)$$
$$= (3\ 583.1 - 206.95) + (3\ 674.9 - 3\ 155)\ \text{kJ/kg}$$
$$= 3\ 896.1\ \text{kJ/kg}$$

$$q_{\text{out}} = h_6 - h_1 = (2\ 335.1 - 191.81)\ \text{kJ/kg} = 2\ 143.3\ \text{kJ/kg}$$

Chapter 9

9 –1 Define the concept of phase equilibrium.

There is no transformation from the liquid phase to the vapor phase, and the two phases are in phase equilibrium.

9 –2 What is the phase rule?

In general, the number of independent variables associated with a multicomponent, multiphase system is given by the Gibbs phase rule, expressed as $IV = C - PH + 2$.

where IV = the number of independent variables, C = the number of components, and PH the number of phases present in equilibrium.

9 –3 What is ideal solution?

An ideal solution or ideal mixture is a solution with thermodynamic properties analogous to those of a mixture of ideal gases and the vapor pressure of the solution obeys Raoult's law.

9 –4 In absorption refrigeration systems, a two-phase equilibrium mixture of liquid ammonia (NH_3) and water (H_2O) is frequently used. Consider one such mixture at 40 ℃, shown in Fig. P9 –4. If the composition of the liquid phase is 70 percent NH_3 and 30 percent H_2O by mole numbers, determine the composition of the vapor phase of this mixture.

Solution A two-phase mixture of ammonia and water at a specified temperature is considered. The composition of the liquid phase is given, and the composition of the vapor phase is to be determined.

Assumptions The mixture is ideal and thus Raoult's law is applicable.

Properties The saturation pressures of H_2O and NH_3 at 40 ℃ are $P_{H_2O,\text{sat}} = 7.385\ 1\ \text{kPa}$ and $P_{NH_3,\text{sat}} = 1\ 554.33\ \text{kPa}$.

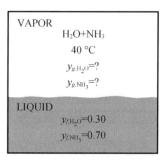

Fig. P9-4

Analysis The vapor pressures are determined from

$$P_{H_2O, \text{gas side}} = y_{H_2O, \text{liquid side}} P_{H_2O, \text{sat}}(T) = 0.3(7.3851 \text{ kPa}) = 2.22 \text{ kPa}$$

$$P_{NH_3, \text{gas side}} = y_{NH_3, \text{liquid side}} P_{NH_3, \text{sat}}(T) = 0.7(1554.33 \text{ kPa}) = 1088.03 \text{ kPa}$$

The total pressure of the mixture is

$$P_{\text{total}} = P_{H_2O} + P_{NH_3} = 2.22 + 1088.03 = 1090.25 \text{ kPa}$$

Then the mole fractions in the gas phase are

$$y_{H_2O, \text{gas side}} = \frac{P_{H_2O, \text{gas side}}}{P_{\text{total}}} = \frac{2.22 \text{ kPa}}{1090.25 \text{ kPa}} = 0.002$$

$$y_{NH_3, \text{gas side}} = \frac{P_{NH_3, \text{gas side}}}{P_{\text{total}}} = \frac{1088.03 \text{ kPa}}{1090.25 \text{ kPa}} = 0.998$$

Discussion Note that the gas phase consists almost entirely of ammonia, making this mixture very suitable for absorption refrigeration.

Chapter 10

10-1 Why is the reversed Carnot cycle executed within the saturation dome not a realistic model for refrigeration cycles?

Because the compression process involves the compression of a liquid-vapor mixture which requires a compressor that will handle two phases, and the expansion process involves the expansion of high moisture content refrigerant.

10-2 Does the ideal vapor-compression refrigeration cycle involve any internal irreversibilities?

Yes, the throttling process is an internally irreversible process.

10-3 An inventor claims to have developed a heat engine. The inventor reports temperature, heat transfer, and work output measurements. The claim is to be evaluated.

Analysis The highest thermal efficiency a heat engine operating between two specified temperature limits can have is the Carnot efficiency, which is determined from

$$\eta_{\text{th,max}} = \eta_{\text{th,C}} = 1 - \frac{T_L}{T_H} = 1 - \frac{290 \text{ K}}{500 \text{ K}} = 0.42 \text{ or } 42\%$$

The actual thermal efficiency of the heat engine in question is

$$\eta_{\text{th}} = \frac{W_{\text{net}}}{Q_H} = \frac{300 \text{ kJ}}{700 \text{ kJ}} = 0.429 \text{ or } 42.9\%$$

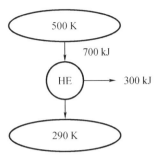

Fig. P10-3

which is greater than the maximum possible thermal efficiency. Therefore, this heat engine is a PMM2 and the claim is false.

10 – 4 A reversible heat pump with specified reservoir temperatures is considered. The entropy change of two reservoirs is to be calculated and it is to be determined if this heat pump satisfies the increase in entropy principle.

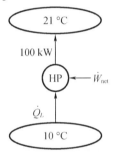

Fig. P10-4

Assumptions The heat pump operates steadily.

Analysis Since the heat pump is completely reversible, the combination of the coefficient of performance expression, first Law, and thermodynamic temperature scale gives

$$\text{COP}_{\text{HP,rev}} = \frac{1}{1 - T_L/T_H} = \frac{1}{1 - (283 \text{ K})/(294 \text{ K})} = 26.73$$

The power required to drive this heat pump, according to the coefficient of performance, is then

$$\dot{W}_{\text{net,in}} = \frac{\dot{Q}_H}{\text{COP}_{\text{HP,rev}}} = \frac{100 \text{ kW}}{26.73} = 3.741 \text{ kW}$$

According to the first law, the rate at which heat is removed from the low-temperature exergy reservoir is

$$\dot{Q}_L = \dot{Q}_H - \dot{W}_{\text{net,in}} = 100 \text{ kW} - 3.741 \text{ kW} = 96.26 \text{ kW}$$

The rate at which the entropy of the high temperature reservoir changes, according to the definition of the entropy, is

$$\Delta \dot{S}_H = \frac{\dot{Q}_H}{T_H} = \frac{100 \text{ kW}}{294 \text{ K}} = 0.340 \text{ kW/K}$$

and that of the low-temperature reservoir is

$$\Delta \dot{S}_L = \frac{\dot{Q}_L}{T_L} = \frac{-96.26 \text{ kW}}{283 \text{ K}} = -0.340 \text{ kW/K}$$

The net rate of entropy change of everything in this system is

$$\Delta \dot{S}_{total} = \Delta \dot{S}_H + \Delta \dot{S}_L = 0.340 - 0.340 = 0 \text{ kW/K}$$

as it must be since the heat pump is completely reversible.

APPENDIX

TABLE A – 1 Properties of air

Ideal-gas properties of air

T /K	h /(kJ/kg)	P_f	u /(kJ/kg)	v_f	s^0/(kJ/kg·K)	T /K	h /(kJ/kg)	P_f	u /(kJ/kg)	v_f	s^0/(kJ/kg·K)
200	199.97	0.336 3	142.56	1 707.0	1.295 59	420	421.26	4.522	300.69	266.6	2.041 42
210	209.97	0.398 7	149.69	1 512.0	1.344 44	430	431.43	4.915	307.99	251.1	2.065 33
220	219.97	0.469 0	156.82	1 346.0	1.391 05	440	441.61	5.332	315.30	236.8	2.088 70
230	230.02	0.547 7	164.00	1 205.0	1.435 57	450	451.80	5.775	322.62	223.6	2.111 61
240	240.02	0.635 5	171.13	1 084.0	1.478 24	460	462.02	6.245	329.97	211.4	2.134 07
250	250.05	0.732 9	178.28	979.0	1.519 17	470	472.24	6.742	337.32	200.1	2.156 04
260	260.09	0.840 5	185.45	887.8	1.558 48	480	482.49	7.268	344.70	189.5	2.177 60
270	270.11	0.959 0	192.60	808.0	1.596 34	490	492.74	7.824	352.08	179.7	2.198 76
280	280.13	1.088 9	199.75	738.0	1.632 79	500	503.02	8.411	359.49	170.6	2.219 52
285	285.14	1.158 4	203.33	706.1	1.650 55	510	513.32	9.031	366.92	162.1	2.239 93
290	290.16	1.231 1	206.91	676.1	1.668 02	520	523.63	9.684	374.36	154.1	2.259 97
295	295.17	1.306 8	210.49	647.9	1.685 15	530	533.98	10.37	381.84	146.7	2.279 67
298	298.18	1.354 3	212.64	631.9	1.695 28	540	544.35	11.10	389.34	139.7	2.299 06
300	300.19	1.386 0	214.07	621.2	1.702 03	550	555.74	11.86	396.86	133.1	2.318 09
305	305.22	1.468 6	217.67	596.0	1.718 65	560	565.17	12.66	404.42	127.0	2.336 85
310	310.24	1.554 6	221.25	572.3	1.734 98	570	575.59	13.50	411.97	121.2	2.355 31
315	315.27	1.644 2	224.85	549.8	1.751 06	580	586.04	14.38	419.55	115.7	2.373 48
320	320.29	1.737 5	228.42	528.6	1.766 90	590	596.52	15.31	427.15	110.6	2.391 40
325	325.31	1.834 5	232.02	508.4	1.782 49	600	607.02	16.28	434.78	105.8	2.409 02
330	330.34	1.935 2	235.61	489.4	1.797 83	610	617.53	17.30	442.42	101.2	2.426 44
340	340.42	2.149	242.82	454.1	1.827 90	620	628.07	18.36	450.09	96.92	2.443 55
350	350.49	2.379	250.02	422.2	1.857 08	630	638.63	19.84	457.78	92.84	2.460 48
360	360.58	2.626	257.24	393.4	1.885 43	640	649.22	20.64	465.50	88.99	2.477 16
370	370.67	2.892	264.46	367.2	1.913 13	650	659.84	21.86	473.25	85.34	2.493 64
380	380.77	3.176	271.69	343.4	1.940 01	660	670.47	23.13	481.01	81.89	2.509 85
390	390.88	3.481	278.93	321.5	1.966 33	670	681.14	24.46	488.81	78.61	2.525 89
400	400.98	3.806	286.16	301.6	1.991 94	680	691.82	25.85	496.62	75.50	2.541 75
410	411.12	4.153	293.43	283.3	2.016 99	690	702.52	27.29	504.45	72.56	2.557 31
700	713.27	28.80	512.33	69.76	2.572 77	1 300	1 395.97	330.9	1 022.82	11.275	3.273 45

TABLE A−1 (Continued)

T /K	h /(kJ/kg)	P_r	u /(kJ/kg)	v_r	s^0/(kJ/kg·K)	T /K	h /(kJ/kg)	P_r	u /(kJ/kg)	v_r	s^0/(kJ/kg·K)
710	724.04	30.38	520.23	67.07	2.588 10	1 320	1 419.76	352.5	1 040.88	10.747	3.291 60
720	734.82	32.02	528.14	64.53	2.603 19	1 340	1 443.60	375.3	1 058.94	10.247	3.309 59
730	745.62	33.72	536.07	62.13	2.618 03	1 360	1 467.49	399.1	1 077.10	9.780	3.327 24
740	756.44	35.50	544.02	59.82	2.632 80	1 380	1 491.44	424.2	1 095.26	9.337	3.344 74
750	767.29	37.35	551.99	57.63	2.647 37	1 400	1 515.42	450.5	1 113.52	8.919	3.362 00
760	778.18	39.27	560.01	55.54	2.661 76	1 420	1 539.44	478.0	1 131.77	8.526	3.379 01
780	800.03	43.35	576.12	51.64	2.690 13	1 440	1 563.51	506.9	1 150.13	8.153	3.395 86
800	821.95	47.75	592.30	48.08	2.717 87	1 460	1 587.63	537.1	1 168.49	7.801	3.412 47
820	843.98	52.59	608.59	44.84	2.745 04	1 480	1 611.79	568.8	1 186.95	7.468	3.428 92
840	866.08	57.60	624.95	41.85	2.771 70	1 500	1 635.97	601.9	1 205.41	7.152	3.445 16
860	888.27	63.09	641.40	39.12	2.797 83	1 520	1 660.23	636.5	1 223.87	6.854	3.461 20
880	910.56	68.98	657.95	36.61	2.823 44	1 540	1 684.51	672.8	1 242.43	6.569	3.477 12
900	932.93	75.29	674.58	34.31	2.848 56	1 560	1 708.82	710.5	1 260.99	6.301	3.492 76
920	955.38	82.05	691.28	32.18	2.873 24	1 580	1 733.17	750.0	1 279.65	6.046	3.508 29
940	977.92	89.28	708.08	30.22	2.897 48	1 600	1757.57	791.2	1298.30	5.804	3.523 64
960	1 000.55	97.00	725.02	28.40	2.921 28	1 620	1 782.00	834.1	1 316.96	5.574	3.538 79
980	1 023.25	105.2	741.98	26.73	2.944 68	1 640	1 806.46	878.9	1 335.72	5.355	3.553 81
1 000	1 046.04	114.0	758.94	25.17	2.967 70	1 660	1 830.96	925.6	1 354.48	5.147	3.568 67
1 020	1 068.89	123.4	776.10	23.72	2.990 34	1 680	1 855.50	974.2	1 373.24	4.949	3.583 35
1 040	1 091.85	133.3	793.36	23.29	3.012 60	1 700	1 880.1	1 025	1 392.7	4.761	3.597 9
1 060	1 114.86	143.9	810.62	21.14	3.034 49	1 750	1 941.6	1 161	1 439.8	4.328	3.633 6
1 080	1 137.89	155.2	827.88	19.98	3.056 08	1 800	2 003.3	1 310	1 487.2	3.994	3.668 4
1 100	1 161.07	167.1	845.33	18.896	3.077 32	1 850	2 065.3	1 475	1 534.9	3.601	3.702 3
1 120	1 184.28	179.7	862.79	17.886	3.098 25	1 900	2 127.4	1 655	1 582.6	6.295	3.735 4
1 140	1 207.57	193.1	880.35	16.946	3.118 83	1 950	2 189.7	1 852	1 630.6	3.022	3.767 7
1 160	1 230.92	207.2	897.91	16.064	3.139 16	2 000	2 252.1	2 068	1 678.7	2.776	3.799 4
1 180	1 254.34	222.2	915.57	15.241	3.159 16	2 050	2 314.6	2 303	1 726.8	2.555	3.830 3
1 200	1 277.79	238.0	933.33	14.470	3.178 88	2 100	2 377.7	2 559	1 775.3	2.356	3.860 5
1 220	1 301.31	254.7	951.09	13.747	3.198 34	2 150	2 440.3	2 837	1 823.8	2.175	3.890 1
1 240	1 324.93	272.3	968.95	13.069	3.217 51	2 200	2 503.2	3 138	1 872.4	2.012	3.919 1
1 260	1 348.55	290.8	986.90	12.435	3.236 38	2 250	2 566.4	3 464	1 921.3	1.864	3.947 4
1 280	1 372.24	310.4	1 004.76	11.835	3.255 10						

Note: The properties P_r (relative pressure) and V_r (relative specific volume) are dimensionless quantities used in the analysis of isentropic processes, and should not be confused with the properties pressure and specific volume.

Source: Kenneth Wark, Thermodymanics, 4th ed, (New York. McGraw−Hill, 1983), pp. 785−86, table A−5. Originaly published in J. H. Keenan and J. Kaye. Gas Tables (New York: John Wiley & Sons, 1948).

APPENDIX

TABLE A-2 Molar mass, gas constant, and critical-point properties

Substance	Formula	$M/(\text{g/mol})$	$R/[\text{J}/(\text{mol}\cdot\text{K})]$	T_c/K	p_c/MPa	Z_c
Ethyne	C_2H_2	26.04	319	309	6.28	0.274
Air	—	28.97	287	133	3.77	0.284
Ammonia	NH_3	17.04	488	406	11.28	0.242
Argon	Ar	39.94	208	151	4.86	0.290
Benzene	C_6H_6	78.11	106	563	4.93	0.274
n-Butane	C_4H_{10}	58.12	143	425	3.80	0.274
Carbon dioxide	CO_2	44.01	189	304	7.39	0.276
Carbon monoxide	CO	28.01	297	133	3.50	0.294
Ethane	C_2H_6	30.07	277	305	4.88	0.285
Ethyl alcohol	C_2H_5OH	46.07	180	516	6.38	0.249
Ethylene	C_2H_4	28.05	296	283	5.12	0.270
Helium	He	4.003	2 077	5.2	0.23	0.300
Hydrogen(normal)	H_2	2.018	4 124	33.2	1.30	0.304
Methane	CH_4	16.04	518	191	4.64	0.290
Methyl alcohol	CH_3OH	32.05	259	513	7.95	0.220
Nitrogen	N_2	28.01	297	126	3.39	0.291
n-Octane	C_8H_{18}	114.22	73	569	2.49	0.258
Oxygen	O_2	32.00	260	154	5.05	0.290
Propane	C_3H_8	44.09	189	370	4.27	0.276
Propylene	C_3H_6	42.08	198	365	4.62	0.276
Dichlorodifluoromethane(R-12)	CCl_2F_2	120.92	69	385	4.12	0.278
Chlorodifuoromethane(R-22)	$CHClF_2$	86.48	96	369	4.98	0.267
Tetrafluoroethane(R-134a)	CF_3CH_2F	102.03	81	374	4.07	0.260
Sulfur dioxide	SO_2	64.06	130	431	7.87	0.268
Water	H_2O	18.02	461	647.3	22.09	0.233

TABLE A−3 Properties of common liquids, solids, and foods

(a) Liquids

Substance	Boiling data at 1 atm		Freezing data		Liquid properties		
	Normal boiling point/℃	Latent heat of vaporization h_{fg}/(kJ/kg)	Freezing point/℃	Latent heat of fusion h_{if}/(kJ/kg)	Temperature /℃	Density ρ /(kg/m^3)	Specific heat c_p/ (kJ/kg·K)
Ammonia	−33.3	1357	−77.7	322.4	−33.3	682	4.43
					−20	665	4.52
					0	639	4.60
					25	602	4.80
Argon	−185.9	161.6	−189.3	28	−185.6	1394	1.14
Benzene	80.2	394	5.5	126	20	879	1.72
Brine (20% sodium chloride by mass)	103.9	—	−17.4	—	20	1150	3.11
n−Butane	−0.5	385.2	−138.5	80.3	−0.5	601	2.31
Carbon dioxide	−78.4*	230.5(at 0℃)	−56.6		0	298	0.59
Ethanol	78.2	838.3	−114.2	109	25	783	2.46
Ethyl alcohol	78.6	855	−156	108	20	789	2.84
Ethylene glycol	198.1	800.1	−10.8	181.1	20	1 109	2.84
Glycerine	179.9	974	18.9	200.6	20	1 261	2.32
Helium	−268.9	22.8	—	—	−268.9	146.2	22.8
Hydrogen	−252.8	445.7	−259.2	59.5	−252.8	70.7	10.0
Isobutane	−11.7	367.1	−160	105.7	−11.7	593.8	2.28
Kerosene	204～293	251	−24.9	—	20	820	2.00
Mercury	356.7	294.7	−38.9	11.4	25	13.560	0.139
Methane	−161.5	510.4	−182.2	58.4	−161.5	423	3.49
					−100	3.1	5.79
Methanol	64.5	1100	−97.7	99.2	25	787	2.55
Nitrogen	−195.8	198.6	−210	25.3	−195.8	809	2.06
					−160	596	2.97
Octane	124.8	306.3	−57.5	180.7	20	703	2.10
Oil (light)					25	910	1.80
Oxygen	−183	212.7	−218.8	13.7	−183	1 141	1.71
Petroleum	—	230～384			20	640	2.0
Propane	−42.1	427.8	−187.7	80.0	−42.1	581	2.25
					0	529	2.53
					50	449	3.13

TABLE A-3 (Continued)

(a) Liquids

Substance	Boiling data at 1 atm		Freezing data		Liquid properties		
	Normal boiling point/℃	Latent heat of vaporization h_{tg}/(kJ/kg)	Freezing point/℃	Latent heat of fusion h_d/(kJ/kg)	Temperature /℃	Density ρ /(kg/m³)	Specific heat c_p/ (kJ/kg·K)
Refrigerant-134a	-26.1	217.0	-96.6	—	-50	1 443	1.23
					-26.1	1 374	1.27
					0	1 295	1.34
					25	1 207	1.43
Water	100	2257	0.0	333.7	0	1 000	4.22
					25	997	4.18
					50	988	4.18
					75	975	4.19
					100	958	4.22

* Sublimation temperature. (At pressures below the triple-point pressure of 518 kPa, carbon dioxide exists as a solid or gas. Also, the freezing-point temperature of carbon dioxide is the triple-point temperature of -56.5 ℃.)

(b) Solids (Values are for room temperature unless indicated otherwise)

Substance	Density ρ/ (kg/m³)	Specific heat c_p/ (kJ/kg·K)	Substance	Density ρ/ (kg/m³)	Specific heat c_p/ (kJ/kg·K)
Metals			Nonmetals		
Aluminum			Asphalt	2 110	0.920
200 K		0.797	Brick, common	1 922	0.79
250 K		0.859	Brick, fireclay(500 ℃)	2 300	0.960
300 K	2 700	0.902	Concrete	2 300	0.653
350 K		0.929	Clay	1 000	0.920
400 K		0.949	Diamond	2 420	0.616
450 K		0.973	Glass, window	2 700	0.800
500 K		0.997	Glass, pyrex	2 230	0.840
Bronze (76% Cu, 2% Zn, 2% A)	8 280	0.400	Graphite	2 500	0.711
			Granite	2 700	1.017
Brass, yellow (65% Cu, 35% Zn)	8 310	0.400	Gypsum or plaster board	800	1.09
			ke		
Copper			200 K		1.56
-173 ℃		0.254	220 K		1.71
-100 ℃		0.342	240 K		1.86

TABLE A-3 (Continued)

Substance	Density ρ/ (kg/m³)	Specific heat c_p/ (kJ/kg·K)	Substance	Density ρ/ (kg/m³)	Specific heat c_p/ (kJ/kg·K)
Metals			Nonmetals		
−50 ℃		0.367	260 K		2.01
0 ℃		0.381	273 K	921	2.11
27 ℃	8 900	0.386	Limestone	1 650	0.909
100 ℃		0.393	Marble	2 600	0.880
200 ℃		0.403	Plywood (Douglas Fir)	545	1.21
Iron	7 840	0.45	Rubber (solt)	1 100	1.840
Lead	11 310	0.128	Rubber (hard)	1 150	2.009
Magnesium	1 730	1 000	Sand	1 520	0.800
Nickel	8 890	0.440	Stane	1 500	0.800
Silver	10 470	0.235	Woods, hard (maple, oak, etc.)	721	1.26
Steel, mild	7 830	0.500	Woods, soft (fir, pine, etc.)	513	1.38
Tungsten	19 400	0.130			

(c) Foods

Food	Water content, % (mass)	Freezing point/℃	Specific nest /(kJ/kg·K) Above freezing	Specific nest /(kJ/kg·K) Below freezing	Latent heat of fusion/ (kJ/kg)	Food	Water content, % (mass)	Freezing point/℃	Specific nest /(kJ/kg·K) Above freezing	Specific nest /(kJ/kg·K) Below freezing	Latent heat of fusion/ (kJ/kg)
Apples	84	−1.1	3.65	1.90	281	Lettuce	95	−0.2	4.02	2.04	317
Bananas	75	−0.8	3.35	1.78	251	Milk, whole	88	−0.6	3.79	1.95	294
Beef round	67	—	3.08	1.68	224	Oranges	87	−0.8	3.75	1.94	291
Broccoli	90	−0.6	3.86	1.97	301	Potatoes	78	−0.6	3.45	1.82	261
Butter	16	—	—	1.04	53	Salmon fish	64	−2.2	2.98	1.65	214
Cheese, swiss	39	−10.0	2.15	1.33	130	Shrimp	83	−2.2	3.62	1.89	277
Cherries	80	−1.8	3.52	1.85	267	Spinach	93	−0.3	3.96	2.01	311
Chicken	74	−2.8	3.32	1.77	247	Strawberries	90	−0.8	3.86	1.97	301
Corn, sweet	74	−0.6	3.32	1.77	247	Tomatoes, ripe	94	−0.5	3.99	2.02	314
Eggs, whole	74	−0.6	3.32	1.77	247	Turkey	64	—	2.98	1.65	214
Ice cream	63	−5.6	2.95	1.63	210	Watermelon	93	−0.4	3.96	2.01	311

Source: Values are obtained from various handbooks and other source or are calculated. Water content and freezing point data of foods are from ASHRAE, Handbook of Fundamentals, SI version (Atlanta. GA: American Society of Heating. Refrigerating and Air-Conditioning Engineers. ioc., 1993), Chapter 30, Table 1. Freezing point is that temperature at which freezing starts for fruits and vegetables, and the average freezing temperature for other foods.

APPENDIX

TABLE A-4 Saturated water—Temperature table

Temp $T/°C$	Sat. press P_{sat} /kPa	Specific volume /(m³/kg)		Internal exergy /(kJ/kg)			Enthalpy /(kJ/kg)			Entropy /(kJ/kg·K)		
		Sat. liqurd, v_l	Sat. vapot, v_g	Sat. liqurd, u_f	Evap. u_h	Sat. vapot, u_g	Sat. liquid, h_l	Evap. h_{fg}	Sat. vapot, b_g	Sat. liquid, s_l	Evap. s_{lg}	Sat. vapor. s_g
0.1	0.611 7	0.001 000	206.00	0.000	2 374.9	2 374.9	0.001	2 500.9	2 500.9	0.000 0	9.155 6	9.155 6
5	0.872 5	0.001 000	147.03	21.019	2 360.8	2 381.8	21.020	2 489.1	2 510.1	0.076 3	8.948 7	9.024 9
10	1.228 1	0.001 000	106.32	42.020	2 346.6	2 388.7	42.022	2 477.2	2 519.2	0.151 1	8.748 8	8.899 9
15	1.705 7	0.001 001	77.885	62.980	2 332.5	2 395.5	62.982	2 465.4	2 528.3	0.224 5	8.555 9	8.780 3
20	2.339 2	0.001 002	57.762	83.913	2 318.4	2 402.3	83.915	2 453.5	2 537.4	0.296 5	8.369 6	8.666 1
25	3.169 8	0.001 003	43.340	104.83	2 304.3	2 409.1	104.83	2 441.7	2 546.5	0.367 2	8.189 5	8.655 7
30	4.246 9	0.001 004	32.879	125.73	2 290.2	2 415.9	125.74	2 429.8	2 555.6	0.436 8	8.015 2	8.452 0
35	5.629 1	0.001 006	25.205	146.63	2 276.0	2 422.7	146.64	2 417.9	2 564.6	0.505 1	7.846 6	8.351 7
40	7.385 1	0.001 008	19.515	167.53	2 261.9	2 429.4	167.53	2 406.0	2 573.5	0.572 4	7.683 2	3.255 6
45	9.595 3	0.001 010	15.251	188.43	2 247.7	2 436.1	188.44	2 394.0	2 582.4	0.638 6	7.524 7	8.163 3
50	12.352	0.001 012	12.025	209.33	2 233.4	2 442.7	209.34	2 382.0	2 591.3	0.703 8	7.371 0	8.074 8
55	15.763	0.001 015	9.563 9	230.24	2 219.1	2 449.3	230.26	2 369.78	2 600.1	0.768 0	7.221 8	7.939 8
60	19.947	0.001 017	7.667 0	251.16	2 204.7	2 455.9	251.18	2 357.7	2 608.8	0.831 3	7.076 9	7.908 2
65	25.043	0.001 020	6.193 5	272.09	2 190.3	2 462.4	272.12	2 345.4	2 617.5	0.693 7	6.936 0	7.829 6
70	31.202	0.001 023	5.039 6	293.04	2 175.8	2 468.9	293.07	2 333.0	2 626.1	0.955 1	6.798 9	7.754 0
75	38.597	0.001 026	4.129 1	313.99	2 161.3	2 475.3	314.03	2 320.6	2 634.6	1.015 8	6.665 5	7.681 2
80	47.416	0.001 029	3.405 3	334.97	2 146.6	2 481.6	335.02	2 308.0	2 643.0	1.075 6	6.535 5	7.611 1
85	57.868	0.001 032	2.826 1	355.96	2 131.9	2 487.8	356.02	2 295.3	2 651.4	1.134 6	6.408 9	7.543 5
90	70.183	0.001 036	2.359 3	376.97	2 117.0	2 494.0	377.04	2 282.5	2 669.6	1.192 9	6.285 3	7.478 2
95	84.609	0.001 040	1.980 8	396.00	2 102.0	2 500.1	398.09	2 269.6	2 667.6	1.250 4	6.164 7	7.415 1
100	101.42	0.001 043	1.672 0	419.06	2 087.0	2 506.0	419.17	2 256.4	2 675.6	1.307 2	6.047 0	7.354 2
105	120.90	0.010 47	1.418 6	440.15	2 071.8	2 511.9	440.28	2 243.1	2 683.4	1.363 4	5.931 9	7.295 2
110	143.38	0.001 052	1.203 4	461.27	2 056.4	2 517.7	461.42	2 229.7	2 691.1	1.418 8	5.819 3	7.238 2
115	169.18	0.001 056	1.036 0	482.42	2 040.9	2 523.3	482.59	2 216.0	2 698.6	1.473 7	5.709 2	7.182 9
120	198.67	0.001 060	0.891 33	503.60	2 025.3	2 528.9	503.81	2 202.1	2 706.0	1.527 9	5.601 3	7.129 2
125	232.23	0.001 065	0.770 12	524.83	2 009.5	2 534.3	525.07	2 188.1	2 713.1	1.581 6	5.495 6	7.077 1
130	270.28	0.001 070	0.668 08	546.10	1 993.4	2 539.5	546.38	2 173.7	2 720.1	1.634 6	5.391 9	7.026 5
135	313.22	0.001 075	0.581 79	567.41	1 977.3	2 544.7	567.75	2 159.1	2 726.9	1.687 2	5.290 1	6.977 3
140	361.53	0.001 080	0.508 50	588.77	1 960.9	2 549.6	589.16	2 144.3	2 733.5	1.739 2	5.190 1	6.929 4
145	415.68	0.001 085	0.446 00	610.19	1 944.2	2 554.4	610.64	2 129.2	2 739.8	1.790 8	5.091 9	6.882 7
150	476.16	0.001 091	0.392 48	631.66	1 927.4	2 559.1	632.18	2 113.8	2 745.9	1.841 8	4.995 3	6.837 1

TABLE A−4 (Continued)

Temp $T/℃$	Sat. press P_{set} /kPa	Specific volume /(m³/kg)		Internal exergy /(kJ/kg)			Enthalpy /(kJ/kg)			Entropy /(kJ/kg·K)		
		Sat. liqurd, v_l	Sat. vapot, v_g	Sat. liqurd, u_f	Evap. u_h	Sat. vapot, u_g	Sat. liquid, h_l	Evap. h_{fg}	Sat. vapot, h_g	Sat. liquid, s_l	Evap. s_{lg}	Sat. vapor. s_g
155	543.49	0.001 096	0.346 48	653.19	1 910.3	2 563.6	653.79	2 098.0	2 751.8	1.892 4	4.900 2	6.792 7
160	618.23	0.001 102	0.306 80	674.79	1 893.0	2 567.8	675.47	2 082.0	2 757.5	1.942 6	4.806 6	6.749 2
165	700.93	0.001 108	0.272 44	696.46	1 875.4	2 571.9	697.24	2 065.6	2 762.8	1.992 3	4.714 3	6.706 7
170	792.18	0.001 114	0.242 60	718.20	1 857.5	2 575.7	719.08	2 048.8	2 767.9	2.041 7	4.623 3	6.665 0
175	892.60	0.001 121	0.216 59	740.02	1 839.4	2 579.4	741.02	2 031.7	2 772.7	2.090 6	4.533 5	6.624 2
180	1 002.8	0.001 127	0.193 84	761.92	1 820.9	2 582.8	763.05	2 014.2	2 777.2	2.139 2	4.444 8	6.584 1
185	1 123.5	0.001 134	0.173 90	783.91	1 802.1	2 586.0	785.19	1 996.2	2 781.4	2.187 5	4.357 2	6.544 7
190	1 255.2	0.001 141	0.156 36	806.00	1 783.0	2 589.0	807.43	1 977.9	2 785.3	2.235 5	4.270 5	6.505 9
195	1 398.8	0.001 149	0.140 89	828.18	1 763.6	2 591.7	829.78	1 959.0	2 788.8	2.283 1	4.184 7	5.467 8
200	1 554.9	0.001 157	0.127 21	850.46	1 743.7	2 594.2	852.26	1 939.8	2 792.0	2.330 5	4.099 7	6.430 2
205	1 724.3	0.001 164	0.115 08	872.86	1 723.5	2 596.4	874.87	1 920.0	2 794.8	2.377 6	4.015 4	6.393 0
210	1 907.7	0.001 173	0.104 29	895.38	1 702.9	2 598.3	897.61	1 899.7	2 797.3	2.424 5	3.931 8	6.356 3
215	2 105.9	0.001 181	0.094 680	918.02	1 681.9	2 599.9	920.50	1 878.8	2 799.3	2.471 2	3.848 9	6.320 0
220	2 319.6	0.001 190	0.086 094	940.79	1 660.5	2 601.3	943.65	1 857.4	2 801.0	2.517 5	3.766 4	6.284 0
225	2 549.7	0.001 199	0.078 405	963.70	1 638.6	2 602.3	966.76	1 835.4	2 802.2	2.563 9	3.684 4	6.248 3
230	2 797.1	0.001 209	0.071 505	986.75	1 616.1	2 602.9	990.14	1 812.8	2 802.9	2.610 0	3.602 8	6.212 8
235	3 062.6	0.001 219	0.065 300	1 010.0	1 593.2	2 603.2	1 013.7	1 789.5	2 803.2	2.656 0	3.521 6	6.177 5
240	3 347.0	0.001 229	0.059 707	1 033.4	1 569.8	2 603.1	1 037.5	1 765.5	2 803.0	2.701 8	3.440 5	6.142 4
245	3 651.2	0.001 240	0.054 656	1 056.9	1 545.7	2 602.7	1 061.5	1 740.8	2 802.2	2.747 6	3.359 5	6.107 2
250	3 976.2	0.001 252	0.050 085	1 080.7	1 521.1	2 601.8	1 085.7	1 715.3	2 801.0	2.793 3	3.278 8	6.072 1
255	4 322.9	0.001 263	0.045 941	1 104.7	1 495.8	2 600.5	1 110.1	1 689.0	2 799.1	2.839 0	3.197 9	6.036 9
260	4 692.3	0.001 276	0.042 175	1 128.8	1 469.9	2 598.7	1 134.8	1 661.8	2 796.6	2.884 7	3.116 9	6.001 7
265	5 085.3	0.001 289	0.038 748	1 153.3	1 443.2	2 596.5	1 159.8	1 633.7	2 793.5	2.930 4	3.035 8	5.966 2
270	5 503.0	0.001 303	0.035 622	1 177.9	1 415.7	2 593.7	1 185.1	1 604.6	2 789.7	2.976 2	2.954 2	5.930 5
275	5 946.4	0.001 317	0.032 767	1 202.9	1 387.4	2 590.3	1 210.7	1 574.5	2 785.2	3.022 1	2.872 3	5.894 4
280	6 416.6	0.001 333	0.030 153	1 228.2	1 358.2	2 586.4	1 236.7	1 543.2	2 779.9	3.068 1	2.789 8	5.857 9
285	6 914.6	0.001 349	0.027 756	1 253.7	1 328.1	2 581.8	1 263.1	1 510.7	2 773.7	3.114 4	2.706 6	5.821 0
290	7 441.8	0.001 366	0.025 554	1 279.7	1 296.9	2 576.5	1 289.8	1 476.9	2 766.7	3.160 8	2.622 5	5.783 4
295	7 999.0	0.001 384	0.023 528	1 306.0	1 264.5	2 570.5	1 317.1	1 441.6	2 758.7	3.207 6	2.537 4	5.745 0
300	8 587.9	0.001 404	0.021 659	1 332.7	1 230.9	2 563.6	1 344.8	1 404.8	2 749.6	3.254 8	2.451 1	5.705 9
305	9 209.4	0.001 425	0.019 932	1 360.0	1 195.9	2 555.8	1 373.1	1 366.3	2 739.4	3.302 4	2.363 3	5.665 7

APPENDIX

TABLE A-4 (Continued)

Temp $T/°C$	Sat. press P_{sat} /kPa	Specific volume /(m³/kg)		Internal exergy /(kJ/kg)			Enthalpy /(kJ/kg)			Entropy /(kJ/kg·K)		
		Sat. liquid, v_f	Sat. vapor, v_g	Sat. liquid, u_f	Evap. u_{fg}	Sat. vapor, u_g	Sat. liquid, h_f	Evap. h_{fg}	Sat. vapor, h_g	Sat. liquid, s_f	Evap. s_{fg}	Sat. vapor, s_g
310	9 865.0	0.001 447	0.018 333	1 387.7	1 159.3	2 547.1	1 402.0	1 325.9	2 727.9	3.350 6	2.273 7	5.624 3
315	10 556	0.001 472	0.016 849	1 416.1	1 121.1	2 537.2	1 431.6	1 283.4	2 715.0	3.399 4	2.182 1	5.581 6
320	11 284	0.001 499	0.015 470	1 445.1	1 080.9	2 526.0	1 462.0	1 238.5	2 700.6	3.449 1	2.088 1	5.537 2
325	12 051	0.001 528	0.014 183	1 475.0	1 038.5	2 513.4	1 493.4	1 191.0	2 684.3	3.499 8	1.991 1	5.490 8
330	12 858	0.001 560	0.012 979	1 505.7	993.5	2 499.2	1 525.8	1 140.3	2 666.0	3.551 6	1.890 6	5.442 2
335	13 707	0.001 597	0.011 848	1 537.5	945.5	2 483.0	1 559.4	1 086.0	2 645.4	3.605 0	1.785 7	5.390 7
340	14 601	0.001 638	0.010 783	1 570.7	893.8	2 464.5	1 594.6	1 027.4	2 622.0	3.660 2	1.675 6	5.335 8
345	15 541	0.001 685	0.009 772	1 605.5	837.7	2 443.2	1 631.7	963.4	2 595.1	3.717 9	1.558 5	5.276 5
350	16 529	0.001 741	0.008 806	1 642.4	775.9	2 418.3	1 671.2	892.7	2 563.9	3.778 8	1.432 6	5.211 4
355	17 570	0.001 808	0.007 872	1 682.2	706.4	2 388.6	1 714.0	812.9	2 526.9	8.844 2	1.294 2	5.138 4
360	18 666	0.001 895	0.006 950	1 726.2	625.7	2 351.9	1 761.5	720.1	2 481.6	3.916 5	1.137 3	5.053 7
365	19 822	0.002 015	0.006 009	1 777.2	526.4	2 303.6	1 817.2	605.5	2 422.7	4.000 4	0.948 9	4.949 3
370	21 044	0.002 217	0.004 953	1 844.5	385.6	2 230.1	1 891.2	443.1	2 334.3	4.111 9	0.689 0	4.800 9
373.95	22 064	0.003 106	0.003 106	2 015.7	0	2 015.7	2 084.3	0	2 084.3	4.407 1	0	4.407 0

TABLE A-5 Saturated water - Pressure table

Temp $T/°C$	Sat. press P_{sat} /kPa	Specific volume /(m³/kg)		Internal exergy /(kJ/kg)			Enthalpy /(kJ/kg)			Entropy /(kJ/kg·K)		
		Sat. liquid, v_f	Sat. vapor, v_g	Sat. liquid, u_f	Evap. u_{fg}	Sat. vapor, u_g	Sat. liquid, h_f	Evap. h_{fg}	Sat. vapor, h_g	Sat. liquid, s_f	Evap. s_{fg}	Sat. vapor, s_g
1.0	6.97	0.001 000	129.19	29.302	2 355.2	2 384.5	29.303	2 484.4	2 513.7	0.105 9	8.869 0	8.974 9
1.5	13.02	0.001 001	87.964	54.686	2 338.1	2 392.8	54.688	2 470.1	2 524.7	0.195 6	8.631 4	8.827 0
2.0	17.50	0.001 001	66.990	73.431	2 325.5	2 398.9	73.433	2 459.5	2 532.9	0.260 6	8.462 1	8.722 7
2.5	21.08	0.001 002	54.242	88.422	2 315.4	2 403.8	88.424	2 451.0	2 539.4	0.311 8	8.330 2	8.542 1
3.0	24.08	0.001 003	45.654	100.98	2 306.9	2 407.9	100.98	2 443.9	2 544.8	0.354 3	8.222 2	8.576 5
4.0	28.96	0.001 004	34.791	121.39	2 293.1	2 414.5	121.39	2 432.3	2 553.7	0.422 4	8.051 0	8.473 4
5.0	32.87	0.001 005	28.185	137.75	2 282.1	2 419.8	137.75	2 423.0	2 560.7	0.476 2	7.917 6	8.393 8
7.5	40.29	0.001 008	19.233	168.74	2 261.1	2 429.8	168.75	2 405.3	2 574.0	0.576 3	7.673 8	8.250 1
10	45.81	0.001 010	14.670	191.79	2 245.4	2 437.2	191.81	2 392.1	2 583.9	0.649 2	7.499 6	8.148 8
15	53.97	0.001 014	10.020	225.93	2 222.1	2 448.0	225.94	2 372.3	2 598.3	0.754 9	7.252 2	8.007 1

TABLE A-5 (Continued)

Temp $T/°C$	Sat. press P_{set} /kPa	Specific volume /(m³/kg)		Internal exergy /(kJ/kg)			Enthalpy /(kJ/kg)			Entropy /(kJ/kg·K)		
		Sat. liqurd, v_l	Sat. vapot, v_g	Sat. liqurd, u_f	Evap. u_h	Sat. vapot, u_g	Sat. liquid, h_l	Evap. h_{fg}	Sat. vapot, b_g	Sat. liquid, s_l	Evap. s_{lg}	Sat. vapor. s_g
20	60.06	0.001 017	7.648 1	251.40	2 204.6	2 456.0	251.42	2 357.5	2 608.9	0.832 0	7.075 2	7.907 3
25	64.96	0.001 020	6.203 4	271.93	2 190.4	2 462.4	271.96	2 345.5	2 617.5	0.893 2	6.937 0	7.830 2
30	69.09	0.001 022	5.228 7	289.24	2 178.5	2 467.7	289.27	2 335.3	2 624.6	0.944 1	6.823 4	7.767 5
40	75.86	0.001 026	3.993 3	317.58	2 158.8	2 476.3	317.62	2 318.4	2 636.1	1.026 1	6.643 0	7.669 1
50	81.32	0.001 030	3.240 3	340.49	2 142.7	2 483.2	340.54	2 304.7	2 645.2	1.091 2	6.501 9	7.593 1
75	91.76	0.001 037	2.217 2	384.36	2 111.8	2 496.1	384.44	2 278.0	2 662.4	1.213 2	6.242 6	7.455 8
100	99.61	0.001 043	1.694 2	417.40	2 088.2	2 505.6	417.51	2 257.5	2 675.0	1.302 8	6.056 2	7.358 9
101.325	99.97	0.001 043	1.673 4	418.95	2 087.0	2 506.0	419.06	2 256.5	2 675.6	1.306 9	6.047 6	7.554 5
125	105.97	0.001 048	1.375 0	444.23	2 068.8	2 513.0	444.36	2 240.6	2 684.9	1.374 1	5.910 0	7.284 1
150	111.35	0.001 053	1.159 4	466.97	2 052.3	2 519.2	467.13	2 226.0	2 693.1	1.433 7	5.789 4	7.223 1
175	116.04	0.001 057	1.003 7	486.82	2 037.7	2 524.5	487.01	2 213.1	2 700.2	1.485 0	5.686 5	7.171 6
200	120.21	0.001 061	0.885 78	504.50	2 024.6	2 529.1	504.71	2 201.6	2 706.3	1.530 2	5.596 8	7.127 0
225	123.97	0.001 064	0.793 29	520.47	2 012.7	2 533.2	520.71	2 191.0	2 711.7	1.570 6	5.517 1	7.087 7
250	127.41	0.001 067	0.718 73	535.08	2 001.8	2 536.8	535.35	2 181.2	2 716.5	1.607 2	5.445 3	7.052 5
275	130.58	0.001 070	0.657 32	548.57	1 991.6	2 540.1	548.86	2 172.0	2 720.9	1.640 8	5.380 0	7.020 7
300	133.52	0.001 073	0.605 82	561.11	1 982.1	2 543.2	561.43	2 163.5	2 724.9	1.671 7	5.320 0	6.991 7
325	136.27	0.001 076	0.561 99	572.84	1 973.1	2 545.9	573.19	2 155.4	2 728.6	1.700 5	5.264 5	6.965 0
350	138.865	0.001 079	0.524 22	583.89	1 964.6	2 548.5	584.26	2 147.7	2 732.0	1.727 4	5.212 8	6.940 2
375	141.30	0.001 081	0.491 33	594.32	1 956.6	2 550.9	594.73	2 140.4	2 735.1	1.752 6	5.164 5	6.917 1
400	143.61	0.001 084	0.462 42	604.22	1 948.9	2 553.1	604.66	2 133.4	2 738.1	1.776 5	5.119 1	6.895 5
450	147.90	0.001 088	0.413 92	622.65	1 934.5	2 557.1	623.14	2 120.3	2 743.4	1.820 5	5.035 6	6.856 1
500	151.83	0.001 093	0.374 83	639.54	1 921.2	2 560.7	640.09	2 108.0	2 748.1	1.860 4	4.960 3	6.820 7
550	155.46	0.001 097	0.342 61	655.16	1 908.8	2 563.9	655.77	2 096.6	2 752.4	1.897 0	4.891 6	6.788 6
600	158.83	0.001 101	0.315 60	669.72	1 897.1	2 566.8	670.38	2 085.8	2 756.2	1.930 8	4.828 5	6.759 3
650	161.98	0.001 104	0.292 60	683.37	1 886.1	2 569.4	684.08	2 075.5	2 759.6	1.962 3	4.769 9	6.732 2
700	164.95	0.001 108	0.272 78	696.23	1 875.6	2 571.8	697.00	2 065.8	2 762.8	1.991 8	4.715 3	6.707 1
750	167.75	0.001 111	0.255 52	708.40	1 865.6	2 574.0	709.24	2 056.4	2 765.7	2.019 5	4.664 2	6.683 7
800	170.41	0.001 115	0.240 35	719.97	1 856.1	2 576.0	720.87	2 047.5	5 768.3	2.045 7	4.616 0	5.661 6
850	172.94	0.001 118	0.226 90	731.00	1 846.9	2 577.9	731.95	2 038.8	2 770.8	2.070 5	4.570 5	6.640 9
900	175.35	0.001 121	0.214 89	741.55	1 838.1	2 579.6	742.56	2 030.5	2 773.0	2.094 1	4.527 3	6.621 3
950	177.66	0.001 124	0.204 11	751.67	1 829.6	2 581.3	752.74	2 022.4	2 775.2	2.116 6	4.486 2	6.602 7

APPENDIX

TABLE A−5 (Continued)

Temp $T/°C$	Sat. press P_{set} /kPa	Specific volume /(m³/kg)		Internal exergy /(kJ/kg)			Enthalpy /(kJ/kg)			Entropy /(kJ/kg·K)		
		Sat. liqurd, v_f	Sat. vapot, v_g	Sat. liqurd, u_f	Evap. u_h	Sat. vapot, u_g	Sat. liquid, h_f	Evap. h_{fg}	Sat. vapot, b_g	Sat. liquid, s_f	Evap. s_{fg}	Sat. vapor. s_g
1 000	179.88	0.001 127	0.194 36	761.39	1 821.4	2 582.8	762.61	2 014.5	2 777.1	2.138 1	4.447 0	6.585 0
1 100	184.06	0.001 133	0.177 45	779.78	1 805.7	2 585.5	781.03	1 999.6	2 780.7	2.178 5	4.373 45	6.552 0
1 200	187.96	0.001 138	0.163 26	796.96	1 790.9	2 587.8	798.33	1 985.4	2 783.8	2.215 9	4.305 8	6.521 7
1 300	191.60	0.001 144	0.151 19	813.10	1 776.8	2 589.9	814.59	1 971.9	2 786.5	2.250 8	4.242 8	6.493 6
1 400	195.04	0.001 149	0.140 78	828.35	1 763.4	2 591.8	829.96	1 958.9	2 788.9	2.283 5	4.184 0	6.467 5
1 500	198.29	0.001 154	0.131 71	842.82	1 750.6	2 593.4	844.55	1 946.4	2 791.0	2.314 3	4.128 7	6.443 0
1 750	205.72	0.001 166	0.113 44	876.12	1 720.6	2 596.7	878.16	1 917.1	2 795.2	2.384 4	4.003 3	6.337 7
2 000	212.38	0.001 177	0.099 587	906.12	1 693.0	2 599.1	908.47	1 889.8	2 798.3	2.446 7	3.892 3	6.339 0
2 250	218.41	0.001 187	0.088 717	933.54	1 667.3	2 600.9	936.21	1 864.3	2 800.5	2.502 9	3.792 6	6.295 4
2 500	223.95	0.001 197	0.079 952	958.87	1 643.2	2 602.1	961.87	1 840.1	2 801.9	2.554 2	3.701 6	6.255 8
3 000	233.85	0.001 217	0.066 667	1 004.6	1 598.5	2 603.2	1 008.3	1 794.9	2 803.2	2.645 4	3.540 2	6.185 6
3 500	242.56	0.001 235	0.057 061	1 045.4	1 557.6	2 603.0	1 049.7	1 753.0	2 802.7	2.725 3	3.399 1	6.124 4
4 000	250.35	0.002 52	0.049 779	10 082.4	1 519.3	2 601.7	1 087.4	1 713.5	2 800.8	2.796 56	3.273 1	6.069 6
5 000	263.94	0.001 286	0.039 448	1 148.1	1 448.9	2 597.0	1 154.5	1 639.7	2 794.2	2.920 7	3.053 0	5.973 7
6 000	275.59	0.001 319	0.032 449	1 205.8	1 384.1	2 589.9	1 213.8	1 570.9	2 784.6	3.027 5	2.862 7	5.890 2
7 000	285.83	0.001 352	0.027 378	1 258.0	1 323.0	2 581.0	1 267.5	1 505.2	2 772.6	3.122 0	2.692 7	5.814 8
8 000	295.01	0.001 384	0.035 25	1 306.0	1 264.5	2 570.5	1 317.1	1 441.6	2 758.7	3.207 7	2.537 3	5.745 0
9 000	303.35	0.001 418	0.020 480	1 350.5	1 207.6	2 558.5	1 363.7	1 379.3	2 742.9	3.286 6	2.392 5	5.679 1
10 000	311.00	0.001 452	0.018 028	1 393.3	1 151.8	2 545.2	1 407.8	1 317.6	2 725.5	3.360 3	2.255 6	5.615 9
11 000	318.08	0.001 488	0.015 988	1 433.9	1 096.6	2 530.4	1 450.2	1 256.1	2 706.3	3.429 9	2.124 5	5.554 4
12 000	324.68	0.001 526	0.014 264	1 473.0	1 041.3	2 514.3	1 491.3	1 194.1	2 685.4	3.496 4	1.997 5	5.493 9
13 000	330.85	0.001 566	0.012 781	1 511.0	985.5	2 496.6	1 531.4	1 131.3	2 662.7	3.560 6	1.873 0	5.433 6
14 000	336.67	0.001 610	0.012 487	1 548.4	928.7	2 477.1	1 571.0	1 067.0	2 637.9	3.623 2	1.749 7	5.372 8
15 000	342.16	0.001 657	0.010 341	1 585.5	870.3	2 455.7	1 610.3	1 000.5	2 610.8	3.684 8	1.626 1	5.310 8
16 000	347.36	0.001 710	0.009 312	1 622.6	809.4	2 432.0	1 649.9	931.1	2 581.0	3.746 1	1.500 5	5.246 6
17 000	352.29	0.001 776	0.008 374	1 660.2	745.1	2 405.4	1 690.3	857.4	2 547.7	3.808 2	1.370 9	5.179 1
18 000	356.99	0.001 840	0.007 504	1 699.1	675.9	2 375.0	1 732.2	77.8	2 510.0	3.872 0	1.234 3	5.106 54
19 000	361.47	0.001 926	0.006 677	1 740.3	598.9	2 339.2	1 776.8	689.2	2 466.0	3.939 6	1.086 0	5.025 6
20 000	365.75	0.002 038	0.005 862	1 785.8	509.0	2 294.8	1 826.6	585.5	2 412.1	4.014 6	0.916 4	4.931 0
21 000	369.83	0.002 207	0.004 994	1 841.6	391.9	2 233.5	1 888.0	450.4	2 338.4	4.107 1	0.700 5	4.807 6
22 000	373.71	0.002 703	0.003 644	1 951.7	140.8	2 092.4	2 011.1	161.5	2 172.6	4.294 2	0.249 6	4.543 9
22.064	373.95	0.003 106	0.003 106	2 015.7	0	2 015.7	2 084.3	0	2 084.3	4.407 0	0	4.407 0

TABLE A-6 Superheated water

T/°C	v/(m³/kg)	u/(kJ/kg)	h/(kJ/kg)	s/(kJ/kg·K)	v/(m³/kg)	u/(kJ/kg)	h/(kJ/kg)	s/(kJ/kg·K)	v/(m³/kg)	u/(kJ/kg)	h/(kJ/kg)	s/(kJ/kg·K)
	\multicolumn{4}{c}{P=0.01 MPa(45.81 °C)}					\multicolumn{4}{c}{P=0.05 MPa(81.32 °C)}	\multicolumn{4}{c}{P=0.10 MPa(99.61 °C)}
Sat.	14.670	2 437.2	2 583.9	8.148 8	3.240 3	2 483.2	2 645.2	7.593 1	1.694 1	2 505.6	2 675.0	7.358 9
50	14.867	2 443.3	2 592.0	8.174 1								
100	17.196	2 515.5	2 687.5	8.448 9	3.418 7	2 511.5	2 682.4	7.695 3	1.695 9	2 506.2	2 675.8	7.361 1
150	19.513	2 587.9	2 783.0	8.689 3	3.889 7	2 585.7	2 780.2	7.941 3	1.936 7	2 582.9	2 776.6	7.614 8
200	21.826	2 661.4	2 879.6	8.904 9	4.356 2	2 660.0	2 877.8	8.159 2	2.172 4	2 658.2	2 875.5	7.835 6
250	24.136	2 736.1	2 977.5	9.101 5	4.820 6	2 735.1	2 976.2	8.356 8	2.406 2	2 733.9	2 974.5	8.034 6
300	26.446	2 812.3	3 076.7	9.282 7	5.284 1	2 811.6	3 075.8	8.538 7	2.638 9	2 810.7	3 074.5	8.217 2
400	31.063	2 969.3	3 280.0	9.609 4	6.209 4	2 968.9	3 279.3	8.865 9	3.102 7	2 968.3	3 278.6	8.545 2
500	35.680	3 132.9	3 489.7	9.899 8	7.133 8	3 132.6	2 489.3	9.156 6	3.565 5	3 132.2	3 488.7	8.836 2
600	40.296	3 303.3	3 706.3	10.163 1	8.057 7	3 303.1	3 706.0	9.420 1	4.027 9	3 302.8	3 705.6	9.099 9
700	44.911	3 480.8	3 929.9	10.405 6	8.981 3	3 480.6	3 929.7	9.662 6	4.490 0	3 480.4	3 929.4	9.342 4
800	49.527	3 665.4	4 160.6	10.631 2	9.904 7	3 665.2	4 160.4	9.888 3	4.951 9	3 665.0	4 160.2	9.568 2
900	54.143	3 856.9	4 398.3	10.842 9	10.828 0	3 856.8	4 398.2	10.100 0	5.413 7	3 856.7	4 398.0	9.780 0
1 000	58.758	4 055.3	4 642.8	11.042 9	11.751 3	4 055.2	4 642.7	10.300 0	5.875 5	4 055.0	4 642.6	9.980 0
1 100	63.373	4 260.0	4 893.8	11.232 6	12.674 5	4 259.9	4 893.7	10.489 7	6.337 2	4 259.8	4 893.6	10.169 8
1 200	67.989	4 470.9	5 150.8	114 132	13.597 7	4 470.8	5 150.7	10.670 4	6.798 8	4 470.7	5 150.6	10.350 4
1 300	72.604	4 687.4	5 413.4	11.585 7	14.520 9	4 687.3	5 413.3	10.842 9	7.260 5	4 687.2	5 413.3	10.522 9
	\multicolumn{4}{c}{P=0.20 MPa(120.21 °C)}	\multicolumn{4}{c}{P=0.30 MPa(133.52 °C)}	\multicolumn{4}{c}{P=0.40 MPa(143.61 °C)}
Sat.	0.885 78	2 529.1	2 706.3	7.127 0	0.605 82	2 543.2	2 724.9	6.991 7	0.462 42	2 553.1	2 738.1	6.895 5
150	0.959 86	2 577.1	2 769.1	7.281 0	0.634 02	2 571.0	2 761.2	7.079 2	0.470 88	2 564.4	2 752.8	6.930 6
200	1.080 49	2 654.6	2 870.7	7.508 1	0.716 43	2 651.0	2 865.9	7.313 2	0.534 34	2 647.2	2 860.9	7.172 3
250	1.198 90	2 731.4	2 971.2	7.710 0	0.796 45	2 728.9	2 967.9	7.518 0	0.595 20	2 726.4	2 964.5	7.380 4
300	1.316 23	2 808.8	3 072.1	7.894 1	0.875 35	2 807.0	3 069.6	7.703 7	0.654 89	2 805.1	3 067.1	7.567 7
400	1.549 34	2 967.2	3 277.0	8.223 6	1.031 55	2 966.0	3 275.5	8.034 7	0.772 65	2 964.9	3 273.9	7.900 3
500	1.781 42	3 131.4	3 487.7	8.515 3	1.186 72	3 130.6	3 486.6	8.327 1	0.889 36	3 129.8	3 485.5	8.193 3
600	2.013 02	3 302.2	3 704.8	8.779 3	1.341 39	3 301.6	3 704.0	8.591 5	1.005 58	3 301.0	3 703.3	8.458 0
700	2.244 34	3 479.9	3 928.8	9.022 1	1.495 80	3 479.5	3 928.2	8.834 5	1.121 52	3 479.0	3 927.6	8.701 2
800	2.475 50	3 664.7	4 159.8	9.247 9	1.650 04	3 664.3	4 159.3	9.060 5	1.237 30	3 663.9	4 158.9	8.927 4
900	2.706 56	3 856.3	4 397.7	9.459 8	1.804 17	3 856.0	4 397.3	9.272 5	1.352 98	3 855.7	4 396.9	9.139 4
1 000	2.937 55	4 054.8	4 642.3	9.659 9	1.958 24	4.54.5	4 642.0	9.472 6	1.468 59	4 054.3	4 641.7	9.339 6
1 100	3.168 48	4 259.6	4 893.3	9.849 7	2.112 26	4 259.4	4 893.1	9.662 4	1.584 14	4 259.2	4 892.9	9.529 5
1 200	3.399 38	4 470.5	5 150.4	10.030 4	2.266 24	4 470.3	5 150.2	9.843 1	1.699 66	4 470.2	5 150.0	9.710 2
1 300	3.630 26	4 687.1	5 413.1	10.202 9	2.420 19	4 686.9	5 413.0	10.015 7	1.815 16	4 686.7	5 412.8	9.882 8

TABLE A−6 (Continued)

T /°C	v/ (m³/kg)	u/ (kJ/kg)	h/ (kJ/kg)	s/(kJ /kg·K)	v/ (m³/kg)	u/ (kJ/kg)	h/ (kJ/kg)	s/(kJ/ kg·K)	v/ (m³/kg)	u/ (kJ/kg)	h/ (kJ/kg)	s/(kJ /kg·K)
	P = 0.50 MPa (151.83 °C)				P = 0.60 MPa (158.83 °C)				P = 0.80 MPa (170.41 °C)			
Sat.	0.374 83	2 560.7	2 748.1	6.820 7	0.315 60	2 566.8	2 756.2	6.759 3	0.240 35	2 576.0	2 768.3	6.661 6
200	0.425 03	2 643.3	2 855.8	7.061 0	0.352 12	2 639.4	2 850.6	6.968 3	0.260 88	2 631.1	2 839.8	6.817 7
250	0.474 43	2 723.8	2 961.0	7.272 5	0.393 90	2 721.2	2 957.6	7.183 3	0.293 21	2 715.9	2 950.4	7.040 2
300	0.522 61	2 803.3	3 064.6	7.461 4	0.434 42	2 801.4	3 062.0	7.374 0	0.324 16	2 797.5	3 056.9	7.234 5
350	0.570 15	2 883.0	3 168.1	7.634 6	0.474 28	2 881.6	3 166.1	7.548 1	0.354 42	2 878.6	3 162.2	7.410 7
400	0.617 31	2 963.7	3 272.4	7.795 6	0.513 74	2 962.5	3 270.8	7.709 7	0.384 29	2 960.2	3 267.7	7.573 5
500	0.710 95	3 129.0	3 484.5	8.089 3	0.592 00	3 128.2	3 483.4	8.004 1	0.443 32	3 126.6	3 481.3	7.869 2
600	0.804 09	3 300.4	3 702.5	8.354 4	0.669 76	3 299.8	3 701.7	8.269 5	0.501 86	3 298.7	3 700.1	8.135 4
700	0.869 6	3 478.6	3 927.0	8.597 8	0.747 25	3 478.1	3 926.4	8.513 2	0.560 11	3 477.2	3 925.3	8.379 4
800	0.989 66	3 663.6	4 158.4	8.824 0	0.824 57	3 663.2	4 157.9	8.739 5	0.618 20	3 662.5	4 157.0	8.606 1
900	1.082 27	3 855.4	4 396.6	9.036 2	0.901 79	3 855.1	4 396.2	8.951 8	0.676 19	3 854.5	4 395.5	8.618 5
1 000	1.174 80	4 054.0	4 641.4	9.236 4	0.978 93	4 053.8	4 641.1	9.152 1	0.734 11	4 053.3	4 640.5	9.018 9
1 100	1.267 28	4 259.0	4 892.6	9.426 3	1.056 03	4 258.8	4 892.4	9.342 0	0.791 97	4 258.3	4 891.9	9.209 0
1 200	1.359 72	4 470.0	5 149.8	9.607 1	1.133 09	4 469.8	5 149.6	9.522 9	0.849 80	4 469.4	5 149.3	9.389 8
1 300	1.452 14	4 686.6	5 412.6	9.779 7	1.210 12	4 686.4	5 412.5	9.695 5	0.907 61	4 696.1	5 412.2	9.562 5
	P = 1.00 MPa (179.88 °C)				P = 1.20 MPa (187.96 °C)				P = 1.40 MPa (195.04 °C)			
Sat.	0.194 37	2 582.8	2 777.1	6.585 0	0.163 26	2 587.8	2 783.8	6.521 7	0.140 78	2 591.8	2 783.0	6.467 5
200	0.206 02	2 622.3	2 828.3	6.695 6	0.169 34	2 612.9	2 813.1	6.590 9	0.143 03	2 602.7	2 803.0	6.497 5
250	0.232 75	2 710.4	2 943.1	6.926 5	0.192 41	2 704.7	2 935.6	6.831 3	0.163 56	2 698.9	2 927.9	6.748 8
300	0.257 99	2 793.7	3 051.6	7.124 6	0.213 86	2 789.7	3 046.3	7.033 5	0.182 33	2 785.7	3 040.9	6.955 3
350	0.282 50	2 875.7	3 158.2	7.302 9	0.234 55	2 872.7	3 154.2	7.213 9	0.200 29	2 869.7	3 150.1	7.137 9
400	0.306 61	2 957.9	3 264.5	7.467 0	0.254 82	2 955.5	3 261.3	7.379 3	0.217 82	2 953.1	3 258.1	7.304 6
500	0.354 11	3 125.0	3 479.1	7.764 2	0.294 64	3 123.4	3 477.0	7.677 9	0.252 16	3 121.8	3 474.8	7.604 7
600	0.401 11	3 297.5	3 598.6	8.031 1	0.333 95	3 296.3	3 697.0	7.945 6	0.285 97	3 295.1	3 695.5	7.873 0
700	0.447 83	3 476.3	3 924.1	8.275 5	0.372 97	3 475.3	3 922.9	8.190 4	0.319 51	3 474.4	3 921.7	8.118 3
800	0.494 38	3 661.7	4 156.1	8.502 4	0.411 84	3 661.0	4 155.2	8.417 6	0.352 88	3 660.3	4 154.3	8.345 8
900	0.540 83	3 853.9	4 394.8	8.715 0	0.450 59	3 853.3	4 394.0	8.630 3	0.386 14	3 862.7	4 393.3	8.558 7
1 000	0.587 21	4 052.7	4 640.0	8.915 5	0.489 28	4 052.2	4 639.4	8.331 0	0.419 33	4 051.7	4 638.8	8.759 5
1 100	0.633 54	4 257.9	4 891.4	9.105 7	0.527 92	4 257.5	4 891.0	9.021 2	0.452 47	4 257.0	4 890.5	8.949 7
1 200	0.679 83	4 469.0	5 148.9	9.286 6	0.566 52	4 468.7	5 148.5	9.202 2	0.485 58	4 468.3	5 148.1	9.130 8
1 300	0.726 10	4 685.8	5 411.9	9.459 3	0.605 03	4 685.5	5 411.6	9.375 0	0.518 66	4 685.1	5 411.3	9.303 6

TABLE A – 6 (Continued)

T /°C	v/ (m³/kg)	u/ (kJ/kg)	h/ (kJ/kg)	s/ (kJ /kg·K)	v/ (m³/kg)	u/ (kJ/kg)	h/ (kJ/kg)	s/ (kJ/ kg·K)	v/ (m³/kg)	u/ (kJ/kg)	h/ (kJ/kg)	s/ (kJ /kg·K)
	P = 1.60 MPa (201.37 °C)				P = 1.80 MPa (207.11 °C)				P = 2.00 MPa (212.38 °C)			
Sat.	0.123 74	2 594.8	2 792.8	6.420 0	0.110 37	2 597.3	2 795.9	6.377 5	0.099 59	2 599.1	2 798.3	6.339 0
225	0.132 93	2 645.1	2 857.8	6.553 7	0.116 78	2 637.0	2 847.2	6.482 5	0.103 81	2 628.5	2 836.1	6.416 0
250	0.141 90	2 692.9	2 919.9	6.675 3	0.125 02	2 686.7	2 911.7	6.608 8	0.111 50	2 680.3	2 903.3	6.547 5
300	0.158 66	2 781.6	3 035.4	6.886 4	0.140 25	2 777.4	3 029.9	6.824 6	0.125 51	2 773.2	3 024.2	6.768 4
350	0.174 59	2 866.6	3 146.0	7.071 3	0.154 60	2 863.6	3 141.9	7.012 0	0.138 60	2 860.5	3 137.7	6.958 3
400	0.190 07	2 950.8	3 254.9	7.239 4	0.168 49	2 948.3	3 251.6	7.181 4	0.151 22	2 945.9	3 248.4	7.129 2
500	0.220 29	3 120.1	3 472.6	7.541 0	0.195 51	3 118.5	3 470.4	7.484 5	0.175 58	3 116.9	3 468.3	7.433 7
600	0.249 99	3 293.9	3 693.9	7.810 1	0.222 00	3 292.7	3 692.3	7.754 3	0.199 62	3 291.5	3 690.7	7.704 3
700	0.279 41	3 473.5	3 920.5	8.055 8	0.248 22	3 472.6	3 919.4	8.000 5	0.223 26	3 471.7	3 918.2	7.950 9
800	0.308 65	3 659.5	4 153.4	8.283 4	0.274 26	3 658.8	4 152.4	8.228 4	0.246 74	3 658.0	4 151.5	8.179 1
900	0.337 80	3 852.1	4 392.6	8.496 5	0.300 20	3 851.5	4 391.9	8.441 7	0.270 12	3 850.9	43 291.1	8.392 5
1 000	0.366 87	4 051.2	4 638.2	8.697 4	0.326 06	4 050.7	4 637.6	8.642 7	0.293 42	4 050.2	4 637.1	8.593 6
1 100	0.395 89	4 256.6	4 890.0	8.887 8	0.351 88	4 256.2	4 889.6	8.833 1	0.316 67	4 255.7	4 889.1	8.784 2
1 200	0.424 88	4 467.9	5 147.7	9.068 9	0.377 56	4 467.6	5 147.3	9.014 3	0.339 89	4 467.2	5 147.0	8.965 4
1 300	0.453 83	4 684.8	5 410.9	9.241 8	0.403 41	4 684.5	5 410.6	9.187 2	0.363 08	4 684.2	5 410.3	9.138 4
	P = 2.50 MPa (228.96 °C)				P = 3.00 MPa (233.85 °C)				P = 3.50 MPa (242.56 °C)			
Sat.	0.079 95	2 602.1	2 801.9	6.255 8	0.066 67	2 603.2	2 803.2	6.185 6	0.057 06	2 603.0	2 802.7	6.124 4
225	0.080 26	2 604.8	2 805.5	6.262 9								
250	0.087 05	2 663.3	2 880.9	6.410 7	0.070 63	2 644.7	2 856.5	6.289 3	0.058 76	2 624.0	2 829.7	6.176 4
300	0.098 94	2 762.2	3 009.6	6.645 9	0.081 18	2 750.8	2 994.3	6.541 2	0.068 45	2 738.8	2 978.4	6.448 4
350	0.109 79	2 852.5	3 127.0	6.842 4	0.090 56	2 844.4	3 116.1	6.745 0	0.076 80	2 836.0	3 104.9	6.660 1
400	0.120 12	2 939.8	3 240.1	7.017 0	0.099 38	2 933.6	3 231.7	6.923 5	0.084 56	2 927.2	3 223.2	6.842 8
450	0.130 15	3 026.2	3 351.6	7.176 8	0.107 89	3 021.2	3 344.9	7.085 6	0.091 98	3 016.1	3 338.1	7.007 4
500	0.139 99	3 112.8	3 462.8	7.325 4	0.116 20	3 108.6	3 457.2	7.235 9	0.099 19	31 304.5	3 451.7	7.159 3
600	0.159 31	3 288.5	3 686.8	7.597 9	0.132 45	3 285.5	3 682.8	7.510 3	0.113 25	3 282.5	3 678.9	7.435 7
700	0.178 35	3 469.3	3 915.2	7.845 5	0.148 41	3 467.0	3 912.2	7.759 0	0.127 02	3 464.7	3 909.3	7.685 5
800	0.197 22	3 656.2	4 149.2	8.074 4	0.164 20	3 654.3	4 146.9	7.988 5	0.140 61	3 652.5	4 144.6	7.915 6
900	0.215 97	3 849.4	4 389.3	8.288 2	0.149 88	3 847.9	4 387.5	8.202 8	0.154 10	3 846.4	4 385.7	8.130 4
1 000	0.234 66	4 049.0	4 635.6	8.489 7	0.195 49	4 047.7	4 634.2	8.404 5	0.167 51	4 046.4	4 632.7	8.332 4
1 100	0.253 30	4 254.7	4 887.9	8.680 4	0.211 05	4 253.6	4 886.7	8.595 5	0.180 87	4 252.5	4 885.6	8.523 6
1 200	0.271 90	4 466.3	5 146.0	8.861 8	0.226 58	4 465.3	5 145.1	8.777 1	0.194 20	4 464.4	5 144.1	8.705 3
1 300	0.290 48	4 683.4	5 409.5	9.034 9	0.242 07	4 682.6	5 408.8	8.950 2	0.207 50	4 681.8	5 408.0	8.878 6

TABLE A −6 (Continued)

T /°C	v/ (m³/kg)	u/ (kJ/kg)	h/ (kJ/kg)	s/(kJ /kg·K)	v/ (m³/kg)	u/ (kJ/kg)	h/ (kJ/kg)	s/(kJ/ kg·K)	v/ (m³/kg)	u/ (kJ/kg)	h/ (kJ/kg)	s/(kJ /kg·K)
	P = 4.0 MPa (350.35 °C)				P = 4.5 MPa (352.44 °C)				P = 5.0 MPa (263.94 °C)			
Sat.	0.049 78	2 601.7	2 800.8	6.069 6	0.044 06	2 599.7	2 798.0	6.019 8	0.039 45	2 597.0	2 794.2	5.973 7
275	0.054 61	2 668.9	2 887.3	6.231 2	0.047 33	2 651.4	2 864.4	6.142 9	0.041 44	2 632.3	2 839.5	6.057 1
300	0.058 87	2 726.2	2 961.7	6.363 9	0.051 38	2 713.0	2 944.2	6.285 4	0.045 35	2 699.0	2 925.7	6.211 1
350	0.066 47	2 827.4	3 093.3	6.584 3	0.058 42	2 818.6	3 081.5	6.515 3	0.051 97	2 809.5	3 069.3	6.451 6
400	0.073 43	2 920.8	3 214.5	6.771 4	0.064 77	2 9414.2	3 205.7	6.707 1	0.057 84	2 907.5	3 196.7	6.648 3
450	0.080 04	3 011.0	3 331.2	6.938 6	0.070 76	3 005.8	3 324.2	6.877 0	0.063 32	3 000.65	3 317.2	6.821 0
500	0.086 44	3 100.3	3 446.0	7.092 2	0.076 52	3 096.0	3 440.4	7.032 3	0.068 58	3 091.8	3 434.7	5.978 1
600	0.098 86	3 279.4	3 674.9	7.370 6	0.087 66	3 276.4	3 670.9	7.312 7	0.078 70	3 273.3	3 666.9	7.260 5
700	0.110 98	3 462.4	3 906.3	7.621 4	0.098 50	3 460.0	3 903.3	7.564 7	0.088 52	3 457.7	3 900.3	7.513 6
800	0.122 92	3 650.6	4 142.3	7.852 3	0.109 16	3 648.8	4 140.0	7.796 2	0.098 16	3 646.9	4 137.7	7.745 8
900	0.134 76	3 844.8	4 383.9	8.067 5	0.119 72	3 843.3	4 382.1	8.011 8	0.107 69	3 841.8	4 380.2	7.961 9
1 000	0.146 53	4 045.1	4 631.2	8.269 8	0.130 20	4 043.9	4 629.8	8.214 4	0.117 15	4 042.6	4 628.3	8.164 8
1 100	0.158 24	4 251.4	4 884.4	8.461 2	0.140 64	4 250.4	4 883.2	8.406 0	0.126 55	4 249.3	4 882.1	8.356 6
1 200	0.169 92	4 463.5	5 143.2	8.643 0	0.151 03	4 462.6	5 142.2	8.588 0	0.135 92	4 461.6	5 141.3	8.538 8
1 300	0.181 57	4 680.9	5 407.2	8.816 4	0.161 40	4 680.1	5 406.5	8.761 6	0.145 27	4 679.3	5 405.7	8.712 4
	P = 6.0 MPa (275.59 °C)				P = 7.0 MPa (285.83 °C)				P = 8.0 MPa (295.01 °C)			
Sat.	0.032 45	2 589.9	2 784.6	5.890 2	0.027 378	2 581.0	2 772.6	5.814 8	0.023 525	2 570.5	2 758.7	5.745 0
300	0.036 19	2 668.4	2 885.6	6.070 3	0.029 492	2 633.5	2 839.9	5.933 7	0.024 279	2 592.3	2 786.5	5.793 7
350	0.042 25	2 790.4	3 043.9	6.335 7	0.035 262	2 770.1	3 016.9	6.230 5	0.029 975	2 748.3	2 988.1	6.132 1
400	0.047 42	2 893.7	3 178.3	6.543 2	0.039 958	2 879.5	3 159.2	6.450 2	0.034 344	2 864.6	3 139.4	6.365 8
450	0.052 17	2 989.9	3 302.9	6.721 9	0.044 187	2 979.0	3 288.3	6.635 3	0.038 194	2 967.8	3 273.3	6.557 9
500	0.056 67	3 083.1	3 423.1	6.882 6	0.048 157	3 074.3	3 411.4	6.800 0	0.041 767	3 065.4	3 399.5	6.726 6
550	0.061 02	3 175.2	3 541.3	7.030 8	0.051 966	3 167.9	3 531.6	6.950 7	0.045 172	3 160.5	3 521.8	6.880 0
600	0.065 27	3 267.2	3 658.8	7.169 3	0.055 665	3 261.0	3 650.6	7.091 0	0.048 463	3 254.7	3 642.4	7.022 1
700	0.073 55	3 453.0	3 894.3	7.424 7	0.062 850	3 448.3	3 888.3	7.348 7	0.054 829	3 443.6	3 882.2	7.282 2
800	0.081 65	3 643.2	4 133.1	7.658 2	0.069 856	3 639.5	4 128.5	7.583 6	0.061 011	3 635.7	4 123.8	7.518 5
900	0.089 64	3 838.8	4 376.6	7.875 1	0.076 750	3 835.7	4 373.0	7.801 4	0.067 082	3 832.7	4 369.3	7.737 2
1 000	0.097 56	4 040.1	4 625.4	8.078 6	0.083 571	4 037.5	4 622.5	8.005 5	0.073 079	4 035.0	4 619.6	7.941 9
1 100	0.105 43	4 247.1	4 879.7	8.270 9	0.090 341	4 245.0	4 877.4	8.198 2	0.079 025	4 242.8	4 875.0	8.135 0
1 200	0.113 26	4 459.8	5 139.4	8.453 4	0.097 075	4 457.9	5 137.4	8.381 0	0.084 934	4 456.1	5 135.5	8.318 1
1 300	0.121 07	4 677.7	5 404.1	8.627 3	0.103 781	4 676.1	5 402.6	8.555 1	0.090 817	4 674.5	5 401.0	8.492 5

TABLE A−6 (Continued)

T /°C	v/ (m³/kg)	u/ (kJ/kg)	h/ (kJ/kg)	s/(kJ /kg·K)	v/ (m³/kg)	u/ (kJ/kg)	h/ (kJ/kg)	s/(kJ /kg·K)	v/ (m³/kg)	u/ (kJ/kg)	h/ (kJ/kg)	s/(kJ /kg·K)
	P = 9.0 MPa (303.35 °C)				P = 10.0 MPa (311.00 °C)				P = 12.5 MPa (327.81 °C)			
Sat.	0.020 489	2 558.5	2 742.9	5.679 1	0.018 028	2 545.2	2 725.5	5.615 9	0.013 496	2 505.6	2 674.3	5.463 8
325	0.023 284	2 647.6	2 857.1	5.873 8	0.019 877	2 611.6	2 810.3	5.759 6				
350	0.025 816	2 725.0	2 957.3	6.038 0	0.022 440	2 699.6	2 924.0	5.946 0	0.016 138	2 624.9	2 826.6	5.713 0
400	0.029 960	2 849.2	3 118.8	6.287 6	0.026 436	2 833.1	3 097.5	6.214 1	0.020 030	2 789.6	3 040.0	6.043 3
450	0.033 524	2 956.3	3 258.0	6.487 2	0.029 782	2 944.5	3 242.4	6.421 9	0.023 019	2 913.7	3 201.5	6.274 9
500	0.036 793	3 056.3	3 387.4	6.660 3	0.032 811	3 047.0	3 375.1	6.599 5	0.025 630	3 023.2	3 343.6	6.465 1
550	0.039 885	3 153.0	3 512.0	6.816 4	0.035 655	3 145.4	3 502.0	6.758 5	0.028 033	3 126.1	3 476.5	6.631 7
600	0.042 861	3 248.4	3 634.1	6.960 5	0.038 378	3 242.0	3 625.8	6.904 5	0.030 306	3 225.8	3 604.6	6.782 8
650	0.045 755	3 343.4	3 755.2	7.095 4	0.041 018	3 338.0	3 748.1	7.040 8	0.032 491	3 324.1	3 730.2	5.922 7
700	0.048 589	3 438.8	3 876.1	7.222 9	0.043 597	3 434.0	3 870.0	7.169 3	0.034 612	3 422.0	3 854.6	7.054 0
800	0.054 132	3 632.0	4 119.2	7.460 6	0.048 629	3 628.2	4 114.5	7.408 5	0.038 724	3 618.8	4 102.8	7.296 7
900	0.059 562	3 829.6	4 365.7	7.680 2	0.053 547	3 826.5	4 362.0	7.629 0	0.042 720	3 818.9	4 352.9	7.519 5
1 000	0.064 919	4 032.4	4 616.7	7.885 5	0.058 391	4 029.9	4 613.8	7.834 9	0.046 641	4 023.5	4 606.5	7.726 9
1 100	0.070 224	4 240.7	4 872.7	8.079 1	0.063 183	4 238.5	4 870.3	8.028 9	0.050 510	4 233.1	4 864.5	7.922 0
1 200	0.075 492	4 454.2	5 133.6	8.262 5	0.067 938	4 452.4	5 131.7	8.212 6	0.054 342	4 447.7	5 127.0	8.106 5
1 300	0.080 733	4 672.9	5 399.5	8.437 1	0.072 667	4 671.3	5 398.0	8.387 4	0.058 147	4 667.3	5 394.1	8.281 9
	P = 15.0 MPa (342.16 °C)				P = 17.5 MPa (354.67 °C)				P = 20.0 MPa (355.75 °C)			
Sat.	0.010 341	2 455.7	2 610.8	5.310 8	0.007 932	2 390.7	2 529.5	5.143 5	0.005 862	2 294.8	2 412.1	4.931 0
350	0.011 481	2 520.9	2 693.1	5.443 8								
400	0.015 671	2 740.6	2 975.7	5.881 9	0.012 463	2 684.3	2 902.4	5.721 1	0.009 950	2 617.9	2 816.9	5.552 6
450	0.018 477	2 880.8	3 157.9	6.143 4	0.015 204	2 845.4	3 111.4	6.021 2	0.012 721	2 807.3	3 061.7	5.904 3
500	0.020 828	2 998.4	3 310.8	6.348 0	0.017 385	2 972.4	3 276.7	6.242 4	0.014 793	2 945.3	3 241.2	6.144 6
550	0.022 945	3 106.2	3 450.4	6.523 0	0.019 305	3 085.8	3 423.6	6.426 6	0.016 571	3 064.7	3 396.2	6.339 0
600	0.024 921	3 209.3	3 583.1	6.679 6	0.021 073	3 192.5	3 561.3	6.589 0	0.018 185	3 175.3	3 539.0	6.507 5
650	0.026 804	3 310.1	3 712.1	6.823 3	0.022 742	3 295.8	3 693.8	6.736 6	0.019 695	3 281.4	3 675.3	6.659 3
700	0.028 621	3 409.8	3 839.1	6.957 3	0.024 342	3 397.5	3 823.5	6.873 5	0.021 134	3 385.1	3 807.8	6.799 1
800	0.032 121	3 609.3	4 091.1	7.203 7	0.027 405	3 599.7	4 079.3	7.123 7	0.023 870	3 590.1	4 067.5	7.053 1
900	0.035 503	3 811.2	4 343.7	7.428 8	0.030 348	3 803.5	4 334.6	7.351 1	0.026 484	3 795.7	4 325.4	7.282 9
1 000	0.038 808	4 017.1	4 599.2	7.637 8	0.033 215	4 010.7	4 592.0	7.561 6	0.029 020	4 004.3	4 584.7	7.495 0
1 100	0.042 062	4 227.7	4 858.6	7.833 9	0.036 029	4 222.3	4 852.8	7.758 8	0.031 504	4 216.9	4 847.0	7.693 3
1 200	0.045 279	4 443.1	5 122.3	8.019 2	0.038 806	4 438.5	5 117.6	7.944 9	0.033 952	4 433.8	5 112.9	7.880 2
1 300	0.048 469	4 663.3	5 390.3	8.195 2	0.041 556	4 659.2	5 386.5	8.121 5	0.036 371	4 655.2	5 382.7	8.057 4

TABLE A −6 (Continued)

T /°C	v/ (m³/kg)	u/ (kJ/kg)	h/ (kJ/kg)	s/(kJ /kg·K)	v/ (m³/kg)	u/ (kJ/kg)	h/ (kJ/kg)	s/(kJ /kg·K)	v/ (m³/kg)	u/ (kJ/kg)	h/ (kJ/kg)	s/(kJ /kg·K)
	P = 25.0 MPa				P = 30.0 MPa				P = 35.0 MPa			
375	0.001 978	1 799.9	1 849.4	4.034 5	0.001 792	1 738.1	1 791.9	3.931 3	0.001 701	1 702.8	1 762.4	3.872 4
400	0.006 005	2 428.5	2 578.7	5.140 0	0.002 798	2 068.9	2 152.8	4.475 8	0.002 105	1 914.9	1 988.6	4.214 4
425	0.007 886	2 607.8	2 805.0	5.470 8	0.005 299	2 452.9	2 611.8	5.147 3	0.003 434	2 253.3	2 373.5	4.775 1
450	0.009 176	2 721.2	2 950.6	5.675 9	0.006 737	2 618.9	2 821.0	5.442 2	0.004 957	2 497.5	2 671.0	5.194 6
500	0.011 143	2 887.3	3 165.9	5.964 3	0.008 691	2 824.0	3 084.8	5.795 6	0.006 933	2 755.3	2 997.9	5.633 1
550	0.012 736	3 020.8	3 339.2	6.181 6	0.010 175	2 974.5	3 279.7	6.040 3	0.008 348	2 925.8	3 218.0	5.909 3
600	0.014 140	3 140.0	3 493.5	6.363 7	0.011 445	3 103.4	3 446.8	6.237 3	0.009 523	3 065.6	3 399.0	6.122 9
650	0.015 430	3 251.9	3 637.7	6.524 3	0.012 590	3 221.7	3 599.4	6.407 4	0.010 565	3 190.9	3 560.7	6.303 0
700	0.016 643	3 359.9	3 776.0	6.670 2	0.013 654	3 334.3	3 743.9	6.559 9	0.011 523	3 308.3	3 711.6	6.462 3
800	0.018 922	3 570.7	4 043.8	6.932 2	0.015 628	3 551.2	4 020.0	6.830 1	0.013 278	3 531.6	3 996.3	6.740 9
900	0.021 075	3 780.2	4 307.1	7.166 8	0.017 473	3 764.6	4 288.8	7.069 5	0.014 904	3 749.0	4 270.6	6.985 3
1 000	0.023 150	3 991.5	4 570.2	7.382 1	0.019 240	3 978.6	4 555.8	7.288 0	0.016 450	3 955.8	4 541.5	7.206 9
1 100	0.025 172	4 206.1	4 835.4	7.582 5	0.020 954	4 195.2	4 823.9	7.490 6	0.017 942	4 184.4	4 812.4	7.411 8
1 200	0.027 157	4 424.6	5 103.5	7.771 0	0.022 630	4 415.3	5 094.2	7.680 7	0.019 398	4 406.1	5 085.0	7.603 4
1 300	0.029 115	4 647.2	5 375.1	7.949 4	0.024 279	4 639.2	5 367.6	7.860 2	0.020 827	4 631.2	5 360.2	7.784 1
	P = 40.0 MPa				P = 50.0 MPa				P = 60.0 MPa			
375	0.001 641	1 677.0	1 742.6	3.829 0	0.001 560	1 638.6	1 716.6	3.764 2	0.001 503	1 609.7	1 699.9	3.714 9
400	0.001 911	1 855.0	1 931.4	4.114 5	0.001 731	1 787.8	1 874.4	4.002 9	0.001 633	1 745.2	1 843.2	3.931 7
425	0.002 538	2 097.5	2 199.0	4.504 4	0.002 009	1 960.3	2 060.7	4.274 6	0.001 816	1 892.9	2 001.8	4.163 0
450	0.003 692	2 364.2	2 511.8	4.944 9	0.002 487	2 160.3	2 284.7	4.589 6	0.002 086	2 055.1	2 180.2	4.414 0
500	0.005 623	2 681.6	2 906.5	5.474 4	0.003 890	2 528.1	2 722.6	5.176 2	0.002 952	2 393.2	2 570.3	4.935 6
550	0.006 985	2 875.1	3 154.4	5.785 7	0.005 118	2 769.5	3 025.4	5.556 3	0.003 955	2 664.6	2 901.9	5.351 7
600	0.008 089	3 026.8	3 350.4	6.017 0	0.006 108	2 947.1	3 252.6	5.824 5	0.004 833	2 866.8	3 156.8	5.652 7
650	0.009 053	3 159.5	3 521.6	5.207 8	0.006 957	3 095.6	3 443.5	6.037 3	0.005 591	3 031.3	3 366.8	5.886 7
700	0.009 930	3 282.0	3 679.2	6.374 0	0.007 717	3 228.7	3 614.6	6.217 9	0.006 265	3 175.4	3 551.3	6.081 4
800	0.011 521	3 511.8	3 972.6	6.661 3	0.009 073	3 472.2	3 925.8	6.522 5	0.007 456	3 432.6	3 880.0	6.403 3
900	0.012 980	3 733.3	4 252.5	6.910 7	0.010 296	3 702.0	4 216.8	6.781 9	0.008 519	3 670.9	4 182.1	6.672 5
1 000	0.014 360	3 952.9	4 527.3	7.135 5	0.011 441	3 927.4	4 499.4	7.013 1	0.009 504	3 902.0	4 472.2	6.909 9
1 100	0.015 686	4 173.7	4 801.1	7.342 5	0.012 534	4 152.2	4 778.9	7.224 4	0.010 439	4 130.9	4 757.3	7.125 5
1 200	0.016 976	4 396.9	5 075.9	7.535 7	0.013 590	4 378.6	5 058.1	7.420 7	0.011 339	4 360.5	5 040.8	7.324 8
1 300	0.018 239	4 623.3	5 352.8	7.717 5	0.014 620	4 607.5	5 338.5	7.604 8	0.012 213	4 591.8	5 324.5	7.511 1

* The temperature in parentheses is the saturation temperature at the specified pressure.

properties of saturated vapor at the specified pressure.

TABLE A-7 Compressed liquid water

T /°C	v/ (m³/kg)	u/ (kJ/kg)	h/ (kJ/kg)	s/(kJ /kg·K)	v/ (m³/kg)	u/ (kJ/kg)	h/ (kJ/kg)	s/(kJ/ kg·K)	v/ (m³/kg)	u/ (kJ/kg)	h/ (kJ/kg)	s/(kJ /kg·K)
	P=5 MPa(263.94 °C)				P=10 MPa(311.00 °C)				P=15 MPa(342.16 °C)			
Sat.	0.001 286 2	1 148.1	1 154.5	2.920 7	0.001 452 2	1 393.3	1 407.9	3.360 3	0.001 657 2	1 585.5	1 610.3	3.684 8
0	0.000 997 7	0.04	5.03	0.000 1	0.000 995 2	0.12	10.07	0.000 3	0.000 992 8	0.18	15.07	0.000 4
20	0.000 999 6	83.61	88.61	0.295 4	0.000 997 3	83.31	93.28	0.294 3	0.000 995 1	83.01	97.93	0.293 2
40	0.001 005 7	166.92	171.95	0.570 5	0.001 003 5	166.33	176.37	0.568 5	0.001 001 3	165.75	180.77	0.556 6
60	0.001 014 9	250.29	255.36	0.828 7	0.001 012 7	249.43	259.55	0.826 0	0.001 010 5	248.58	263.74	0.823 4
80	0.001 026 7	333.82	338.96	1.072 3	0.001 024 4	332.69	342.94	1.069 1	0.001 022 1	331.59	346.92	1.065 9
100	0.001 041 0	417.65	422.85	1.303 4	0.001 038 5	416.23	426.62	1.299 6	0.001 036 1	414.85	430.39	1.295 8
120	0.001 057 6	501.91	507.19	1.523 6	0.001 054 9	500.18	510.73	1.519 1	0.001 052 2	498.50	514.28	1.514 8
140	0.001 076 9	856.80	592.18	1.734 4	0.001 073 8	584.72	595.45	1.729 3	0.001 070 8	582.69	598.75	1.724 3
160	0.001 098 8	672.65	678.04	1.937 4	0.001 095 4	670.06	681.01	1.931 6	0.001 092 0	667.63	684.01	1.925 9
180	0.001 124 0	759.47	765.09	2.133 8	0.001 120 0	756.48	767.68	2.127 1	0.001 116 0	753.58	770.32	2.120 6
	P=5 MPa(263.94 °C)				P=10 MPa(311.00 °C)				P=15 MPa(342.16 °C)			
200	0.001 153 1	847.92	853.68	2.325 1	0.001 148 2	844.32	855.80	2.317 4	0.001 143 5	840.84	858.00	2.310 0
220	0.001 186 8	938.39	944.32	2.512 7	0.001 180 9	934.01	945.82	2.503 7	0.001 175 2	929.81	947.43	2.495 1
240	0.001 226 8	1 031.6	1 037.7	2.698 3	0.001 219 2	1 026.2	1 038.3	2.687 6	0.001 212 1	1 021.0	1 039.2	2.677 4
260	0.001 275 5	1 128.5	1 134.9	2.884 1	0.001 265 3	1 121.6	1 134.3	2.871 0	0.001 256 0	1 115.1	1 134.0	2.858 6
280					0.001 322 6	1 221.8	1 235.0	3.056 5	0.001 309 6	1 213.4	1 233.0	3.041 0
300					0.001 398 0	1 329.4	1 343.3	3.248 8	0.001 378 3	1 317.6	1 338.3	3.227 9
320									0.001 473 3	1 431.9	1 454.0	3.426 3
340									0.001 631 1	1 567.9	1 592.4	3.655 5
	P=20 MPa(365.75 °C)				P=30 MPa				P=50 MPa			
Sat.	0.002 037 8	1 785.8	1 826.6	4.014 6								
0	0.000 990 4	0.23	20.03	0.000 5	0.000 985 7	0.29	29.86	0.000 3	0.000 976 7	0.29	49.13	-0.001 0
20	0.000 992 9	82.71	102.57	0.292 1	0.000 988 6	82.11	111.77	0.289 7	0.000 980 5	80.93	129.95	0.284 5
40	0.000 999 2	165.17	185.16	0.564 6	0.000 995 1	164.05	193.90	0.560 7	0.000 987 2	161.90	211.25	0.552 8
60	0.001 008 4	247.75	267.92	0.820 8	0.001 004 2	246.14	276.26	0.815 6	0.000 996 2	243.08	292.88	0.805 5
80	0.001 019 9	330.50	350.90	1.062 7	0.001 015 5	328.40	358.86	1.056 4	0.001 007 2	324.42	374.78	1.044 2
100	0.001 033 7	413.50	434.17	1.292 0	0.001 029 0	410.87	441.74	1.284 7	0.001 020 1	405.94	456.94	1.270 5
120	0.001 049 6	496.85	517.84	1.510 5	0.001 044 5	493.66	525.00	1.502 0	0.001 034 9	487.69	539.43	1.485 9
140	0.001 067 9	580.71	602.07	1.719 4	0.001 062 3	576.90	608.76	1.709 8	0.001 051 7	569.77	622.36	1.691 6
160	0.001 088 6	665.28	687.05	1.920 3	0.001 082 3	660.74	693.21	1.909 4	0.001 070 4	652.33	705.85	1.888 9
180	0.001 112 2	750.78	773.02	2.114 3	0.001 104 9	745.40	778.55	2.102 0	0.001 091 4	735.49	790.06	2.079 0

TABLE A −7 (Continued)

T /°C	v/ (m³/kg)	u/ (kJ/kg)	h/ (kJ/kg)	s/(kJ /kg·K)	v/ (m³/kg)	u/ (kJ/kg)	h/ (kJ/kg)	s/(kJ /kg·K)	v/ (m³/kg)	u/ (kJ/kg)	h/ (kJ/kg)	s/(kJ /kg·K)
	P = 20 MPa (365.75 °C)				P = 30 MPa				P = 50 MPa			
200	0.001 139 0	837.49	860.27	2.302 7	0.001 130 4	831.11	865.02	2.288 8	0.001 114 9	819.45	875.19	2.262 8
220	0.001 169 7	925.77	949.16	2.486 7	0.001 159 5	918.15	952.93	2.470 7	0.001 141 2	904.39	961.45	2.441 4
240	0.001 205 3	1 016.1	1 040.2	2.667 6	0.001 192 7	1 006.9	1 042.7	2.649 1	0.001 170 8	990.55	1 049.1	2.615 6
260	0.001 247 2	1 109.0	1 134.0	2.846 9	0.001 231 4	1 097.8	1 134.7	2.825 0	0.001 204 4	1 078.2	1 138.4	2.786 4
280	0.001 297 8	1 205.6	1 231.5	3.026 5	0.001 277 0	1 191.5	1 229.8	3.000 1	0.001 243 0	1 167.7	1 229.9	2.954 7
300	0.001 361 1	1 307.2	1 334.4	3.209 1	0.001 332 2	1 288.9	1 328.9	3.176 1	0.001 287 9	1 259.6	1 324.0	3.121 8
320	0.001 445 0	1 416.6	1 445.5	3.399 6	0.001 401 4	1 391.7	1 433.7	3.355 8	0.001 340 9	1 354.3	1 421.4	3.288 8
340	0.001 569 3	1 540.2	1 571.6	3.608 6	0.001 493 2	1 502.4	1 547.1	3.543 8	0.001 404 9	1 452.9	1 523.1	3.457 5
360	0.001 824 8	1 703.6	1 740.1	3.878 7	0.001 627 6	1 626.8	1 675.6	3.749 9	0.001 484 8	1 556.5	1 630.7	3.630 1
380					0.001 872 9	1 782.0	1 838.0	4.002 6	0.001 588 4	1 667.1	1 746.5	3.810 2

TABLE A −8 Saturated ice-water vapor

Temp T/°C	Sat. press P_{set} /kPa	Specific volume /(m³/kg)		Internal exergy /(kJ/kg)			Enthalpy /(kJ/kg)			Entropy /(kJ/kg·K)		
		Sat. liqurd, v_t	Sat. vapot, v_g	Sat. liqurd, u_f	Evap. u_h	Sat. vapot, u_g	Sat. liquid, h_t	Evap. h_{fg}	Sat. vapot, b_g	Sat. liquid, s_t	Evap. s_{tg}	Sat. vapor. s_g
0.01	0.611 69	0.001 091	205.99	−333.40	2 707.9	2 374.5	−333.40	2 833.9	2 500.5	−1.220 2	10.374	9.154
0	0.611 15	0.001 091	206.17	−333.43	2 707.9	2 374.5	−333.43	2 833.9	2 500.5	−1.220 4	10.375	9.154
−2	0.517 72	0.001 091	241.62	−337.63	2 709.4	2 371.8	−337.63	2 834.5	2 496.8	−1.235 8	10.453	9.218
−4	0.437 48	0.001 090	283.84	−341.80	2 710.8	2 369.0	−341.80	2 835.0	2 493.2	−1.251 3	10.533	9.282
−6	0.368 73	0.001 090	334.27	−345.94	2 712.2	2 366.2	−345.93	2 835.4	2 489.5	−1.266 7	10.613	9.347
−8	0.309 98	0.001 090	394.66	−350.04	2 713.5	2 363.5	−350.04	2 835.8	2 485.8	−1.282 1	10.695	9.413
−10	0.259 90	0.001 089	467.17	−354.12	2 714.8	2 360.7	−354.12	2 836.2	2 482.1	−1.297 6	10.778	9.480
−12	0.217 32	0.001 089	554.47	−358.17	2 716.1	2 357.9	−358.17	2 836.6	2 478.4	−1.313 0	10.862	9.549
−14	0.181 21	0.001 088	659.88	−362.18	2 717.3	2 355.2	−362.18	2 836.9	2 474.7	−1.328 4	10.947	9.618
−16	0.150 68	0.001 088	787.51	−366.17	2 718.6	2 352.4	−366.17	2 837.2	2 471.0	−1.343 9	11.033	9.689
−18	0.124 92	0.001 088	942.51	−370.13	2 719.7	2 349.6	−370.13	2 837.5	2 467.3	−1.359 3	11.121	9.761
−20	0.103 26	0.001 087	1 131.3	−374.06	2 720.9	2 346.8	−374.06	2 837.7	2 463.6	−1.374 8	11.209	9.835
−22	0.085 10	0.001 087	1 362.0	−377.95	2 722.0	2 344.1	−377.95	2 837.9	2 459.9	−1.390 3	11.300	9.909
−24	0.069 91	0.001 087	1 644.7	−381.82	2 723.1	2 341.3	−381.82	2 838.1	2 456.2	−1.405 7	11.391	9.985
−26	0.057 25	0.001 087	1 992.2	−385.66	2 724.2	2 338.5	−385.66	2 838.2	2 452.6	−1.421 2	11.484	10.063
−28	0.046 73	0.001 086	2 421.0	−389.47	2 725.2	2 335.7	−389.47	2 838.3	2 448.8	−1.436 7	11.578	10.141

TABLE A-8 (Continued)

Temp T/°C	Sat. press P_{set} /kPa	Specific volume /(m³/kg)		Internal exergy /(kJ/kg)			Enthalpy /(kJ/kg)			Entropy /(kJ/kg·K)		
		Sat. liqurd, v_l	Sat. vapot, v_g	Sat. liqurd, u_f	Evap. u_h	Sat. vapot, u_g	Sat. liquid, h_l	Evap. h_{fg}	Sat. vapot, b_g	Sat. liquid, s_l	Evap. s_{lg}	Sat. vapor. s_g
-30	0.038 02	0.001 086	2 951.7	-393.25	2 726.2	2 332.9	-393.25	2 838.4	2 445.1	-1.452 1	11.673	10.221
-32	0.030 82	0.001 086	3 610.9	-397.00	2 727.2	2 330.2	-397.00	2 838.4	2 441.4	-1.467 6	11.770	10.303
-34	0.024 90	0.001 085	4 432.4	-400.72	2 728.1	2 327.4	-400.72	2 838.5	2 437.7	-1.483 1	11.869	10.386
-36	0.020 04	0.001 085	5 460.1	-404.40	2 729.0	2 324.6	-404.40	2 838.4	2 434.0	-1.498 6	11.969	10.470
-38	0.016 08	0.001 085	6 750.5	-408.07	2 729.9	2 321.8	-408.07	2 838.4	2 430.3	-1.514 1	12.071	10.557
-40	0.012 85	0.001 084	8 376.7	-411.70	2 730.7	2 319.0	-411.70	2 838.3	2 426.6	-1.529 6	12.174	10.644

TABLE A-9 Saturated refrigerant-134a—Pressure table

Press T /kPa	Sat. temp. T_{set}/°C	Specific volume /(m³/kg)		Internal exergy /(kJ/kg)			Enthalpy /(kJ/kg)			Entropy /(kJ/kg·K)		
		Sat. liqurd, v_l	Sat. vapot, v_g	Sat. liqurd, u_f	Evap. u_{lg}	Sat. vapot, u_g	Sat. liquid, h_l	Evap. h_{fg}	Sat. vapot, h_g	Sat. liquid, s_l	Evap. s_{lg}	Sat. vapor. s_g
60	-36.95	0.000 709 8	0.311 21	3.798	205.32	209.12	3.841	223.95	227.79	0.016 34	0.948 07	0.964 41
70	-33.87	0.000 714 4	0.269 29	7.680	203.20	210.88	7.730	222.00	229.73	0.032 67	0.927 75	0.960 42
80	-31.13	0.000 718 5	0.237 53	11.15	201.30	212.46	11.21	220.25	231.46	0.047 11	0.909 99	0.957 10
90	-28.65	0.000 722 3	0.212 63	14.31	199.57	213.88	14.37	218.65	233.02	0.060 08	0.894 19	0.954 27
100	-26.37	0.000 725 9	0.192 54	17.21	4 197.98	215.19	17.28	217.16	234.44	0.071 88	0.879 95	0.951 83
120	-22.32	0.000 732 4	0.162 12	22.40	195.11	217.51	22.49	214.48	236.97	0.092 75	0.855 03	0.947 79
140	-18.77	0.000 738 3	0.140 14	26.98	192.57	219.54	27.08	212.08	239.16	0.110 87	0.833 68	0.944 56
160	-15.60	0.000 743 7	0.123 48	31.09	190.27	221.35	31.21	209.90	241.11	0.126 93	0.814 96	0.941 90
180	-12.73	0.000 748 7	0.110 41	34.83	188.16	222.99	34.97	207.90	242.86	0.141 39	0.798 26	0.939 65
200	-10.09	0.000 753 3	0.099 867	38.28	186.21	224.48	38.43	206.03	244.46	0.154 57	0.783 16	0.937 73
240	-5.38	0.000 762 0	0.083 897	44.48	182.67	227.14	44.66	202.62	247.28	0.177 94	0.756 64	0.934 58
280	-1.25	0.000 769 9	0.072 352	49.97	179.50	229.46	50.18	199.54	249.72	0.198 29	0.733 81	0.932 10
320	2.46	0.000 777 2	0.063 604	54.92	176.61	231.52	55.16	196.71	251.88	0.216 37	0.713 69	0.930 06
360	5.82	0.000 784 1	0.056 738	59.44	173.94	233.38	59.72	194.08	253.81	0.232 70	0.695 66	0.928 35
400	8.91	0.000 790 7	0.051 201	63.62	171.45	235.07	63.94	191.62	255.55	0.247 61	0.679 29	0.926 91
450	12.46	0.000 798 5	0.045 619	68.45	168.54	237.00	68.81	188.71	257.53	0.264 65	0.660 69	0.925 35
500	15.71	0.000 805 9	0.041 118	72.93	165.82	238.75	73.33	185.98	259.30	0.280 23	0.643 77	0.924 00
550	18.73	0.000 813 0	0.037 408	77.10	163.25	240.35	77.54	183.38	260.92	0.294 61	0.628 21	0.922 82
600	21.55	0.000 819 9	0.034 295	81.02	160.81	241.83	81.51	180.90	262.40	0.307 99	0.613 78	0.921 77
650	24.20	0.000 826 6	0.031 646	84.72	158.48	243.20	85.26	178.51	263.77	0.320 51	0.600 30	0.920 81

APPENDIX

TABLE A-9 (Continued)

Press T /kPa	Sat. temp. T_{sat}/°C	Specding volome /(m³/kg)		Interent engero /(kJ/kg)			Enthalpy /(kJ/kg)			Entrooy /(kJ/kg·K)		
		Sat. liqurd, v_l	Sat. vapot, v_g	Sat. liqurd, u_f	Evap. u_{lg}	Sat. vapot, u_g	Sat. liquid, h_l	Evap. h_{fg}	Sat. vapot, h_g	Sat. liquid, s_l	Evap. s_{lg}	Sat. vapor. s_g
700	26.69	0.000 833 1	0.029 361	88.24	156.24	244.48	88.82	176.21	265.03	0.332 30	0.587 63	0.919 94
750	29.06	0.000 839 5	0.027 371	91.59	154.08	245.67	92.22	173.98	266.20	0.343 45	0.575 67	0.919 12
800	31.31	0.000 845 8	0.025 621	94.79	152.00	246.79	95.47	171.82	267.29	0.354 04	0.564 31	0.918 35
850	33.45	0.000 852 0	0.024 069	97.87	149.98	247.85	98.60	169.71	268.31	0.364 13	0.553 49	0.917 62
900	35.51	0.000 858 0	0.022 683	100.83	148.01	248.85	101.61	167.66	259.26	0.373 77	0.543 15	0.916 92
950	37.48	0.000 864 1	0.021 438	103.69	146.10	249.79	104.51	165.64	270.15	0.383 01	0.533 23	0.916 24
1 000	39.37	0.000 870 0	0.020 313	106.45	144.23	250.68	107.32	163.67	270.99	0.391 89	0.523 68	0.915 58
1 200	46.29	0.000 893 4	0.016 715	116.70	137.11	253.81	117.77	156.10	273.87	0.424 41	0.488 63	0.913 03
1 400	52.40	0.000 916 6	0.014 107	125.94	130.43	256.37	127.22	148.90	276.12	0.453 15	0.457 34	0.910 50
1 600	57.88	0.000 940 0	0.012 123	134.43	124.04	258.47	135.93	141.93	277.86	0.479 11	0.428 73	0.907 84
1 800	62.87	0.000 963 9	0.010 559	142.33	117.83	260.17	144.07	135.11	279.17	0.502 94	0.402 04	0.904 98
2 000	67.45	0.000 988 6	0.009 288	149.78	111.73	261.51	151.76	128.33	280.09	0.525 09	0.376 75	0.901 84
2 500	77.54	0.001 056 6	0.006 936	166.99	96.47	263.45	169.63	111.16	280.79	0.575 31	0.316 95	0.892 26
3 000	86.16	0.001 140 6	0.005 275	183.04	80.22	263.26	186.46	92.63	279.09	0.621 18	0.257 76	0.878 94

TABLE A-10 Ideal-gas specific heats of various common gases

(a) At 300 K

Gas	Formula	Gas constant, R/(kJ/kg·K)	C_p/(kJ/kg·K)	c_v/(kJ/kg·K)	k
Air	—	0.287 0	1.005	0.718	1.400
Argon	Ar	0.208 1	0.520 3	0.312 2	1.667
Butane	C_4H_{10}	0.143 3	1.716 4	1.573 4	1.091
Carbon dioxide	CO_2	0.188 9	0.846	0.657	1.289
Carbon monoxide	CO	0.296 8	1.040	0.744	1.400
Ethane	C_2H_6	0.276 5	1.766 2	1.489 7	1.186
Ethylene	C_2H_4	0.296 4	1.548 2	1.251 8	1.237
Helium	He	2.076 9	5.192 6	3.115 6	1.667
Hydrogen	H_2	4.124 0	14.307	10.183	1.405
Methane	CH_4	0.518 2	2.253 7	1.735 4	1.299
Neon	Ne	0.411 9	1.029 9	0.617 9	1.667
Nitrogen	N_2	0.296 8	1.039	0.743	1.400
Octane	C_8H_{18}	0.072 9	1.711 3	1.638 5	1.044
Oxygen	O_2	0.259 8	0.918	0.658	1.395

Thermodynamics

TABLE A – 10 (Continued)

Gas	Formula	Gas constant, R/(kJ/kg·K)	C_p/(kJ/kg·K)	c_v/(kJ/kg·K)	k
Propane	C_3H_8	0.188 5	1.679 4	1.490 9	1.126
Steam	H_2O	0.461 5	1.872 3	1.410 8	1.327

Note: The unit kJ/kg · K is equivalent to kJ/kg · °C.

Source: Chemical and Process Thermodynamics 3/E by Kyle, B. G., © 2000. Adapted by permission of Person Education, Inc., Upper Saddle River. NJ.

(b) At various temperatures

Temperature /K	C_p/(kJ /kg·K)	C_v/(kJ /kg·K)	K	C_p/(kJ /kg·K)	C_v/(kJ /kg·K)	K	C_p/(kJ /kg·K)	C_v/(kJ /kg·K)	K
	Air			Carbon dioxide, CO_2			Carbon monoxide, CO		
250	1.003	0.716	1.401	0.791	0.602	1.314	1.039	0.743	1.400
300	1.005	0.718	1.400	0.846	0.657	1.288	1.040	0.744	0.399
350	1.008	0.721	1.398	0.895	0.706	1.268	1.043	0.746	1.398
400	1.013	0.726	1.395	0.939	0.750	1.252	1.047	0.751	1.395
450	1.020	0.733	1.391	0.978	0.790	1.239	1.054	0.757	1.392
500	1.029	0.742	1.387	1.014	0.825	1.229	1.063	0.767	1.387
550	1.040	0.753	1.381	1.046	0.857	1.220	1.075	0.778	1.382
600	1.051	0.764	1.376	1.075	0.886	1.213	1.087	0.790	1.376
650	1.063	0.776	1.370	1.102	0.913	1.207	1.100	0.803	1.370
700	1.075	0.788	1.364	1.126	0.937	1.202	1.113	0.816	1.364
750	1.087	0.800	1.359	1.148	0.959	1.197	1.126	0.829	1.358
800	1.099	0.812	1.354	1.169	0.980	1.193	1.139	0.842	1353
900	1.121	0.834	1.344	1.204	1.015	1.186	1.163	0.866	1.343
1000	1.142	0.855	1.336	1.234	1.045	1.181	1.185	0.888	1.335
	Hydrogen, H_2			Nitrogen, N_2			Oxygen, O_2		
250	14.051	9.927	1.416	1.039	0.742	1.400	0.913	0.653	1.398
300	14.307	10.183	1.405	1.039	0.743	1.400	0.918	0.658	1.395
350	14.427	10.302	1.400	1.041	0.744	1.399	0.928	0.668	1.389
400	14.476	10.352	1.398	1.044	0.747	1.397	0.941	0.681	1.382
450	14.501	10.377	1.398	1.049	0.752	1.395	0.956	0.696	1.373
500	14.513	10.389	1.397	1.056	0.759	1.391	0.972	0.712	1.365
550	14.530	10.405	1.396	1.065	0.768	1.387	0.988	0.728	1.358
600	14.546	10.422	1.396	1.075	0.778	1.382	1.003	0.743	1.350
650	14.571	10.447	1.395	1.086	0.789	1.376	1.017	0.758	1.343
700	14.604	10.480	1.394	1.098	0.801	1.371	1.031	0.771	1.337
750	14.645	10.521	1.392	1.110	0.813	1.365	1.043	0.783	1.332
800	14.695	10.570	1.390	1.121	0.825	1.360	1.054	0.794	1.327

TABLE A − 10 (Continued)

Temperature /K	C_p/(kJ /kg·K)	C_v/(kJ /kg·K)	K	C_p/(kJ /kg·K)	C_v/(kJ /kg·K)	K	C_p/(kJ /kg·K)	C_v/(kJ /kg·K)	K
	Hydrogen, H_2			Nitrogen, N_2			Oxygen, O_2		
900	14.822	10.698	1.385	1.145	0.849	1.349	1.074	0.814	1.319
1000	14.983	10.859	1.380	1.167	0.870	1.341	1.090	0.830	1.313

Source: Kenneth Wark, Themodynamics, 4th ed. (New York: McGraw − Hill, 1983), p. 783, Table A − 4M, Originally published in Tables or Thermaj Properties of Gasses, NBS Circular 564, 1955.

(c) At a function of temperature

Substance	Formula	a	b	c	d	Temperature range/K	Se error Max.	Se error Avg.
Nitrogen	N_2	28.90	-0.1571×10^{-2}	0.8081×10^{-5}	-2.873×10^{-9}	273 ~ 1 800	0.59	0.34
Oxygen	O_2	25.48	1.520×10^{-2}	-0.7155×10^{-5}	1.312×10^{-9}	273 ~ 1 800	1.19	0.28
Air	—	28.11	0.1967×10^{-2}	10.4802×10^{-5}	-1.966×10^{-9}	273 ~ 1 800	0.72	0.33
Hydrogen	H_2	29.11	-0.1916×10^{-2}	0.4003×10^{-5}	-0.8704×10^{-9}	273 ~ 1 800	1.01	0.26
Carbon monoxide	CO	28.16	0.1675×10^{-2}	0.5372×10^{-5}	-2.222×10^{-9}	273 ~ 1 800	0.89	0.37
Carbon dioxide	CO_2	22.26	5.981×10^{-2}	-3.501×10^{-5}	7.469×10^{-9}	273 ~ 1 800	0.67	0.22
Water vapor	H_2O	32.24	0.1923×10^{-2}	1.055×10^{-5}	-3.595×10^{-9}	273 ~ 1 800	0.53	0.24
Nitric oxide	NO	29.34	-0.09395×10^{-2}	0.9747×10^{-5}	-4.187×10^{-9}	273 ~ 1 500	0.97	0.36
Nitrous oxide	N_2O	24.11	5.8632×10^{-2}	-3.562×10^{-5}	10.58×10^{-9}	273 ~ 1 500	0.59	0.26
Nitrogen dioxide	NO_2	22.9	5.715×10^{-2}	-3.52×10^{-5}	7.87×10^{-9}	273 ~ 1 500	0.46	0.18
Ammonia	NH_3	27.568	2.5630×10^{-2}	0.99072×10^{-5}	-6.6909×10^{-9}	273 ~ 1 500	0.91	0.36
Sulfur	S_2	27.21	2.218×10^{-2}	-1.628×10^{-5}	3.986×10^{-9}	273 ~ 1 800	0.99	0.38
Sulfur dioxide	SO_2	25.78	5.795×10^{-2}	-3.812×10^{-5}	8.612×10^{-9}	273 ~ 1 800	0.45	0.24
Sulfur trioxide	SO_3	16.40	14.58×10^{-2}	-11.20×10^{-5}	32.42×10^{-9}	273 ~ 1 300	0.29	0.13
Acetylene	C_2H_2	21.8	9.2143×10^{-2}	-6.527×10^{-5}	18.21×10^{-9}	273 ~ 1 500	1.46	0.59
Benzene	C_6H_6	−36.22	48.475×10^{-2}	-31.57×10^{-5}	77.62×10^{-9}	273 ~ 1 500	0.34	0.20
Methanol	CH_4O	19.0	9.152×10^{-2}	-1.22×10^{-5}	-8.039×10^{-9}	273 ~ 1 000	0.18	0.08
Ethanol	C_2H_6O	19.9	20.96×10^{-2}	-10.38×10^{-5}	20.05×10^{-9}	273 ~ 1 500	0.40	0.22
Hydrogen chloride	HCl	30.33	-0.7620×10^{-2}	1.327×10^{-5}	-4.338×10^{-9}	273 ~ 1 500	0.22	0.08
Methane	CH_4	19.89	5.024×10^{-2}	1.269×10^{-5}	-11.01×10^{-9}	273 ~ 1 500	1.33	0.57
Ethane	C_2H_6	6.900	17.27×10^{-2}	-6.406×10^{-5}	7.285×10^{-9}	273 ~ 1 500	0.83	0.28
Propane	C_2H_8	−4.04	30.48×10^{-2}	-15.72×10^{-5}	31.74×10^{-9}	273 ~ 1 500	0.40	0.12
n − Butane	C_4H_{10}	3.96	37.15×10^{-2}	-18.34×10^{-5}	35.00×10^{-9}	273 ~ 1 500	0.54	0.24
i − Butane	C_4H_{10}	−7.913	41.60×10^{-2}	-23.01×10^{-5}	49.91×10^{-9}	273 ~ 1 500	0.25	0.13
n − Pentane	C_5H_{12}	6.774	45.43×10^{-2}	22.46×10^{-5}	42.29×10^{-9}	273 ~ 1 500	0.56	0.21
n − Hexane	C_6H_{14}	6.938	55.22×10^{-2}	28.65×10^{-5}	57.69×10^{-9}	273 ~ 1 500	0.72	0.20

TABLE A – 10 (Continued)

Substance	Formula	a	b	c	d	Temperature range/K	Se error Max.	Se error Avg.
Ethylene	C_2H_4	3.95	15.64×10^{-2}	-8.344×10^{-5}	17.67×10^{-9}	273 ~ 1 500	0.54	0.13
Propylene	C_3H_6	3.15	23.83×10^{-2}	-12.18×10^{-5}	24.62×10^{-9}	273 ~ 1 500	0.73	0.17

Source: B. G. Kyle, Chemical and Process Thermodynamics (Englewood Cliffs, NJ: Prentice – Hall, 1984), Used with permission.

TABLE A – 11 One-dimensional isentropic compressible flow functions for an ideal gas with $k = 1.4$

Ma	Ma^*	A/A^*	P/P_0	ρ/ρ_0	T/T_0
0	0	∞	1.000 0	1.000 0	1.000 0
0.1	0.109 4	5.821 8	0.993 0	0.995 0	0.998 0
0.2	0.218 2	2.963 5	0.972 5	0.980 3	0.992 1
0.3	0.325 7	2.035 1	0.939 5	0.956 4	0.982 3
0.4	0.431 3	1.590 1	0.895 6	0.924 3	0.969 0
0.5	0.534 5	1.339 8	0.843 0	0.885 2	0.952 4
0.6	0.634 8	1.188 2	0.784 0	0.840 5	0.932 8
0.7	0.731 8	1.094 4	0.720 9	0.791 6	0.910 7
0.8	0.825 1	1.038 2	0.656 0	0.740 0	0.886 5
0.9	0.914 6	1.008 9	0.591 3	0.687 0	0.860 6
1.0	1.000 0	1.000 0	0.528 3	0.633 9	0.833 3
1.2	1.158 3	1.030 4	0.412 4	0.531 1	0.776 4
1.4	1.299 9	1.114 9	0.314 2	0.437 4	0.718 4
1.6	1.425 4	1.250 2	0.235 3	0.355 7	0.661 4
1.8	1.536 0	1.439 0	0.174 0	0.286 8	0.606 8
2.0	1.633 0	1.687 5	0.127 8	0.230 0	0.555 6
2.2	1.717 9	2.005 0	0.093 5	0.184 1	0.508 1
2.4	1.792 2	2.403 1	0.068 4	0.147 2	0.464 7
2.6	1.857 1	2.896 0	0.050 1	0.117 9	0.425 2
2.8	1.914 0	3.500 1	0.036 8	0.094 6	0.389 4
3.0	1.964 0	4.234 6	0.027 2	0.076 0	0.357 1
5.0	2.236 1	25.000	0.001 6	0.011 3	0.166 7
∞	2.249 5	∞	0	0	0

$$Ma^* = Ma\sqrt{\frac{k+1}{2(k-1)M^2a}}$$

$$\frac{A}{A^*} = \frac{1}{Ma}\left[\left(\frac{2}{k+1}\right)\left(1 + \frac{k-1}{2}Ma^2\right)\right]^{0.5(k+1)/(k-1)}$$

$$\frac{P}{P_0} = \left(1 + \frac{k-1}{2}Ma^2\right)^{-k/(k-1)}$$

$$\frac{\rho}{\rho_0} = \left(1 + \frac{k-1}{2}Ma^2\right)^{-1/(k-1)}$$

APPENDIX

$$\frac{T}{T_0} = \left(1 + \frac{k-1}{2}Ma^2\right)^{-1}$$

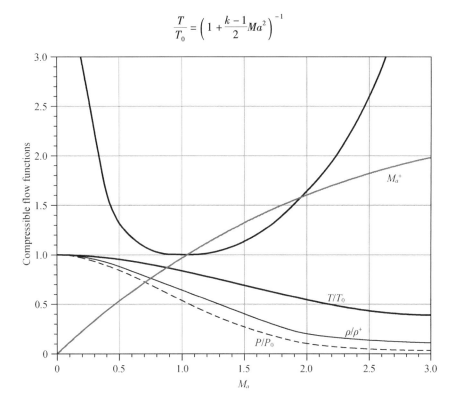

TABLE A – 12 Superheated water

T /°F	v/(ft³/lbm)	u/(Btu/lbm)	h/(Btu/lbm)	s/(Btu/lbm·R)	v/(ft³/lbm)	u/(Btu/lbm)	h/(Btu/lbm)	s/(Btu/lbm·R)	v/(ft³/lbm)	u/(Btu/lbm)	h/(Btu/lbm)	s/(Btu/lbm·R)
	$P=1.0$ psla(101.69 °F)				$P=5.0$ psla(162.18 °F)				$P=10$ psla(193.15 °F)			
Sat.	333.49	1 043.7	1 105.4	1.977 6	73.525	1 062.7	1 130.7	1.843 8	38.425	1 072.0	1 143.1	1.787 5
200	392.53	1 077.5	1 150.1	2.050 9	78.153	1 076.2	1 148.5	1.871 6	38.849	1 074.5	1 146.4	1.792 6
240	416.44	1 091.2	1 168.3	2.077 7	83.009	1 090.3	1 167.1	1.898 9	41.326	1 089.1	1 165.5	1.820 7
280	440.33	1 105.0	1 186.5	2.103 0	87.838	1 104.3	1 185.6	1.924 6	43.774	1 103.4	1 184.4	1.846 9
320	464.20	1 118.9	1 204.8	2.127 1	92.650	1 118.4	1 204.1	1.949 0	46.205	1 117.6	1 203.1	1.871 6
360	488.07	1 132.9	1 223.3	2.150 2	97.452	1 132.5	1 222.6	1.972 2	48.624	1 131.9	1 221.8	1.895 0
400	511.92	1 147.1	1 241.8	2.172 2	102.25	1 146.7	1 241.3	1.994 4	51.035	1 146.2	1 240.6	1.917 4
440	535.77	1 161.3	1 260.4	2.193 4	107.03	1 160.9	1 260.0	2.015 6	53.441	1 160.5	1 259.4	1.938 8
500	571.54	1 182.8	1 288.6	2.223 7	114.21	1 182.6	1 288.2	2.046 1	57.041	1 182.2	1 287.8	1.969 3
600	631.14	1 219.4	1 336.2	2.270 9	126.15	1 219.2	1 335.9	2.093 3	63.029	1 219.0	1 335.6	2.016 7
700	690.73	1 256.8	1 384.6	2.314 6	138.09	1 256.7	1 384.4	2.137 1	69.007	1 256.5	1 384.2	2.060 5
800	750.31	1 295.1	1 433.9	2.355 3	150.02	1 294.9	1 433.7	2.177 8	74.980	1 294.8	1 433.5	2.101 3
1 000	869.47	1 374.2	1 535.1	2.429 9	173.86	1 374.2	1 535.0	2.252 4	86.913	1 374.1	1 534.9	2.176 0
1 200	988.62	1 457.1	1 640.0	2.497 2	197.70	1 457.0	1 640.0	2.319 8	98.840	1 457.0	1 639.9	2.243 3
1 400	1 107.8	1 543.7	1 748.7	2.559 0	221.54	1 543.7	1 748.7	2.381 6	110.762	1 543.6	1 748.6	2.305 2

TABLE A – 12 (Continued)

T/°F	v/(ft³/lbm)	u/(Btu/lbm)	h/(Btu/lbm)	s/(Btu/lbm·R)	v/(ft³/lbm)	u/(Btu/lbm)	h/(Btu/lbm)	s/(Btu/lbm·R)	v/(ft³/lbm)	u/(Btu/lbm)	h/(Btu/lbm)	s/(Btu/lbm·R)
	\multicolumn{4}{c}{$P = 15$ psla (212.99 °F)}	\multicolumn{4}{c}{$P = 20$ psla (227.92 °F)}	\multicolumn{4}{c}{$P = 40$ psla (267.22 °F)}
Sat.	26.297	1 077.7	1 150.7	1.754 9	20.093	1 081.8	1 156.2	1.731 9	10.501	1 092.1	1 169.8	1.676 6
240	27.429	1 087.8	1 163.9	1.774 2	20.478	1 086.5	1 162.3	1.740 6				
280	29.085	1 102.4	1 183.2	1.801 0	21.739	1 101.4	1 181.9	1.767 9	10.713	1 097.3	1 176.6	1.685 8
320	30.722	1 116.9	1 202.2	1.826 0	22.980	1 116.1	1 201.2	1.793 3	11.363	1 112.9	1 197.1	1.712 8
360	32.348	1 131.3	1 221.1	1.849 6	24.209	1 130.7	1 220.2	1.817 1	11.999	1 128.1	1 216.9	1.737 6
400	33.965	1 145.7	1 239.9	1.872 1	25.429	1 145.1	1 239.3	1.839 8	12.625	1 143.1	1 236.5	1.761 0
440	35.576	1 160.1	1 258.8	1.893 6	26.644	1 159.7	1 258.3	1.861 4	13.244	1 157.9	1 256.0	1.783 1
500	37.986	1 181.9	1 287.3	1.924 3	28.458	1 181.6	1 286.9	1.892 2	14.165	1 180.2	1 285.0	1.814 3
600	41.988	1 218.7	1 335.3	1.971 8	31.467	1 218.5	1 334.9	1.939 8	15.686	1 217.5	1 333.6	1.862 5
700	45.981	1 256.3	1 383.9	2.015 6	34.467	1 256.1	1 383.7	1.983 7	17.197	1 255.3	1 382.6	1.906 7
800	49.967	1 294.6	1 433.3	2.056 5	37.461	1 294.5	1 433.1	2.024 7	18.702	1 293.9	1 432.3	1.947 8
1 000	57.930	1 374.0	1 534.8	2.131 2	43.438	1 373.8	1 534.6	2.099 4	21.700	1 373.4	1 534.1	2.022 7
1 200	65.885	1 456.9	1 639.8	2.198 6	49.407	1 456.8	1 639.7	2.166 8	24.691	1 456.5	1 639.3	2.090 2
1 400	73.836	1 543.6	1 748.5	2.260 4	55.373	1 543.5	1 748.4	2.228 7	27.678	1 543.3	1 748.1	2.152 2
1 600	81.784	1 634.0	1 861.0	2.317 8	61.335	1 633.9	1 860.9	2.286 1	30.662	1 633.7	1 860.7	2.209 6
	\multicolumn{4}{c}{$P = 60$ psla (292.69 °F)}	\multicolumn{4}{c}{$P = 80$ psla (312.02 °F)}	\multicolumn{4}{c}{$P = 100$ psla (327.81 °F)}
Sat.	7.175 6	1 098.1	1 177.8	1.644 2	5.473 3	1 102.3	1 183.4	1.621 2	4.432 7	1 105.5	1 187.5	1.603 2
320	7.486 3	1 109.6	1 192.7	1.663 6	5.544 0	1 105.9	1 187.9	1.627 1				
360	7.925 9	1 125.5	1 213.5	1.689 7	5.887 6	1 122.7	1 209.9	1.654 5	4.662 8	1 119.8	1 206.1	1.626 3
400	8.354 8	1 140.9	1 233.7	1.713 8	6.218 7	1 138.7	1 230.8	1.679 4	4.935 9	1 136.4	1 227.8	1.652 1
440	8.776 6	1 156.1	1 253.6	1.736 4	6.542 0	1 154.3	1 251.2	1.702 6	5.200 6	1 152.4	1 248.7	1.675 9
500	9.400 5	1 178.8	1 283.1	1.768 2	7.017 7	1 177.3	1 281.2	1.735 0	5.587 6	1 175.9	1 279.3	1.708 8
600	10.425 6	1 216.5	1 332.2	1.816 8	7.795 1	1 215.4	1 330.8	1.784 1	6.216 7	1 214.4	1 329.4	1.758 6
700	11.440 1	1 254.5	1 381.6	1.861 3	8.561 6	1 253.8	1 380.5	1.828 9	6.834 4	1 253.0	1 379.5	1.803 7
800	12.448 4	1 293.3	1 431.5	1.902 6	9.321 8	1 292.6	1 430.6	1.870 4	7.445 7	1 292.0	1 429.8	1.845 3
1 000	14.454 3	1 373.0	1 533.5	1.977 7	10.831 3	1 372.6	1 532.9	1.945 7	8.657 5	1 372.2	1 532.4	1.920 8
1 200	16.452 5	1 456.2	1 638.9	2.045 4	12.333 1	1 455.9	1 638.5	2.013 5	9.861 5	1 455.6	1 638.1	1.988 7
1 400	18.446 4	1 543.0	1 747.8	2.107 3	13.830 6	1 542.8	1 747.5	2.075 5	11.061 2	1 542.6	1 747.2	2.050 8
1 600	20.438	1 633.5	1 860.5	2.164 8	15.325 7	1 633.3	1 860.2	2.133 0	12.258 4	1 633.2	1 860.0	2.108 3
1 800	22.428	1 727.6	1 976.6	2.218 7	16.819 2	1 727.5	1 976.5	2.186 9	13.454 1	1 727.3	1 976.3	2.162 2
2 000	24.417	1 825.2	2 096.3	2.269 4	18.311 7	1 825.0	2 096.1	2.237 6	14.648 7	1 824.9	2 096.0	2.213 0

TABLE A – 12 (Continued)

T /°F	v/(ft³/lbm)	u/(Btu/lbm)	h/(Btu/lbm)	s/(Btu/lbm·R)	v/(ft³/lbm)	u/(Btu/lbm)	h/(Btu/lbm)	s/(Btu/lbm·R)	v/(ft³/lbm)	u/(Btu/lbm)	h/(Btu/lbm)	s/(Btu/lbm·R)
	\multicolumn{4}{c}{P = 120 psla (341.25 °F)}									
Sat.	3.728 9	1 107.9	1 190.8	1.588 3	3.220 2	1 109.9	1 193.4	1.575 7	2.834 7	1 111.6	1 195.5	1.564 7
350	3.844 6	1 116.7	1 202.1	1.602 3	3.258 4	1 113.4	1 197.8	1.581 1				
400	4.079 9	1 134.0	1 224.6	1.629 2	3.467 6	1 131.5	1 221.4	1.609 2	3.007 6	1 129.0	1 218.0	1.591 4
450	4.361 3	1 154.5	1 251.4	1.659 4	3.714 7	1 152.6	1 248.9	1.640 3	3.229 3	1 150.7	1 246.3	1.623 4
500	4.634 0	1 174.4	1 277.3	1.687 2	3.952 5	1 172.9	1 275.3	1.668 6	3.441 2	1 171.4	1 273.2	1.652 2
550	4.901 0	1 153.9	1 302.8	1.713 1	4.184 5	1 192.7	1 301.1	1.694 8	3.646 9	1 191.4	1 299.4	1.678 8
600	5.164 2	1 213.4	1 328.0	1.737 5	4.412 4	1 212.3	1 326.6	1.719 5	3.848 4	1 211.3	1 325.2	1.703 7
700	5.682 9	1 252.2	1 378.4	1.782 9	4.860 4	1 251.4	1 377.3	1.765 2	4.243 4	1 250.6	1 376.3	1.749 8
800	6.195 0	1 291.4	1 429.0	1.824 7	5.301 7	1 290.8	1 428.1	1.807 2	4.631 6	1 290.2	1 427.3	1.792 0
1 000	7.208 3	1 371.7	1 531.8	1.900 5	6.173 2	1 371.3	1 531.3	1.883 2	5.396 8	1 370.9	1 530.7	1.868 2
1 200	8.213 7	1 455.3	1 637.7	1.968 4	7.036 7	1 455.0	1 637.3	1.951 2	6.154 0	1 454.7	1 636.9	1.936 3
1 400	9.214 9	1 542.3	1 746.9	2.030 5	7.896 1	1 542.1	1 746.6	2.013 4	6.907 0	1 541.8	1 746.3	1.998 6
1 600	10.213 5	1 633.0	1 859.8	2.088 1	8.752 9	1 632.8	1 859.5	2.071 1	7.657 4	1 632.6	1 859.3	2.056 3
1 800	11.210 6	1 727.2	1 976.1	2.142 0	9.608 2	1 727.0	1 975.9	2.125 0	8.406 3	1 726.9	1 975.7	2.110 2
2 000	12.206 7	1 824.8	2 095.8	2.192 8	10.462 4	1 824.6	2 095.7	2.175 8	9.154 2	1 824.5	2 095.5	2.161 0
	\multicolumn{4}{c}{P = 180 psla (373.07 °F)}									
Sat.	2.532 2	1 113.0	1 197.3	1.554 8	2.288 2	1 114.1	1 198.8	1.546 0	2.042 3	1 115.3	1 200.3	1.536 0
400	2.649 0	1 126.3	1 214.5	1.575 2	2.361 5	1 123.5	1 210.9	1.560 2	2.072 8	1 119.7	1 206.0	1.542 7
450	2.851 4	1 148.7	1 243.7	1.608 2	2.548 8	1 146.7	1 241.0	1.594 3	2.245 7	1 144.1	1 237.6	1.578 3
500	3.043 3	1 169.8	1 271.2	1.637 6	2.724 7	1 168.2	1 269.0	1.624 3	2.405 9	1 166.2	1 266.3	1.609 1
550	3.228 6	1 190.2	1 297.7	1.664 6	2.893 9	1 188.9	1 296.0	1.651 6	2.559 0	1 187.2	1 293.8	1.637 0
600	3.409 7	1 210.2	1 323.8	1.689 7	3.058 6	1 209.1	1 322.3	1.677 1	2.707 5	1 207.7	1 320.5	1.662 8
700	3.763 5	1 249.8	1 375.2	1.736 1	3.379 6	1 249.0	1 374.1	1.723 8	2.995 6	1 248.0	1 372.7	1.709 9
800	4.110 4	1 289.5	1 426.5	1.778 5	3.693 4	1 288.9	1 425.6	1.766 4	3.276 5	1 288.1	1 424.5	1.752 8
900	4.453 1	1 329.7	1 478.0	1.817 9	4.003 1	1 329.2	1 477.3	1.805 9	3.553 0	1 328.5	1 476.5	1.792 5
1 000	4.792 9	1 370.5	1 530.1	1.854 9	4.309 9	1 370.1	1 529.6	1.843 0	3.826 8	1 369.5	1 528.9	1.829 6
1 200	5.467 4	1 454.3	1 636.5	1.923 1	4.918 2	1 454.0	1 636.1	1.911 3	4.368 9	1 453.6	1 635.6	1.898 1
1 400	6.137 7	1 541.6	1 746.0	1.985 5	5.522 2	1 541.4	1 745.7	1.973 7	4.906 8	1 541.1	1 745.4	1.960 6
1 600	6.805 4	1 632.4	1 859.1	2.043 2	6.123 8	1 632.2	1 858.8	2.031 5	5.442 2	1 632.0	1 858.6	2.018 4
1 800	7.471 6	1 726.7	1 975.6	2.097 1	6.723 8	1 726.5	1 975.4	2.085 5	5.976 0	1 726.4	1 975.2	2.072 4
2 000	8.136 7	1 824.4	2 095.4	2.147 9	7.322 7	1 824.3	2 095.3	2.136 3	6.508 7	1 824.1	2 095.1	2.123 2

Note: Column groupings — first block: P = 120 psla (341.25 °F), P = 140 psla (353.03 °F), P = 160 psla (363.64 °F); second block: P = 180 psla (373.07 °F), P = 200 psla (381.80 °F), P = 225 psla (391.80 °F).

TABLE A – 12 (Continued)

T /°F	v/(ft³/lbm)	u/(Btu/lbm)	h/(Btu/lbm)	s/(Btu/lbm·R)	v/(ft³/lbm)	u/(Btu/lbm)	h/(Btu/lbm)	s/(Btu/lbm·R)	v/(ft³/lbm)	u/(Btu/lbm)	h/(Btu/lbm)	s/(Btu/lbm·R)
	P = 120 psia (341.25 °F)				P = 140 psia (353.03 °F)				P = 160 psia (363.64 °F)			
Sat.	1.844 0	1 116.3	1 201.6	1.527 0	1.680 6	1 117.0	1 202.6	1.518 7	1.543 5	1 117.7	1 203.3	1.511 1
450	2.002 7	1 141.3	1 234.0	1.563 6	1.803 4	1 138.5	1 230.3	1.549 9	1.636 9	1 135.6	1 226.4	1.536 9
500	2.150 6	1 164.1	1 263.6	1.595 3	1.941 5	1 162.0	1 260.8	1.582 5	1.767 0	1 159.8	1 257.9	1.570 6
550	2.291 0	1 185.6	1 291.5	1.623 7	2.071 5	1 183.9	1 289.3	1.611 5	1.888 5	1 182.1	1 287.0	1.600 1
600	2.426 4	1 206.3	1 318.6	1.649 9	2.196 4	1 204.9	1 316.7	1.638 0	2.004 6	1 203.5	1 314.8	1.627 0
650	2.558 6	1 226.8	1 345.1	1.674 3	2.317 9	1 225.6	1 343.5	1.662 7	2.117 2	1 224.4	1 341.9	1.652 0
700	2.688 3	1 247.0	1 371.4	1.697 4	2.436 9	1 246.0	1 370.0	1.686 0	2.227 3	1 244.9	1 368.6	1.675 5
800	2.942 9	1 287.3	1 423.5	1.740 6	2.669 9	1 286.5	1 422.4	1.729 4	2.442 4	1 285.7	1 421.3	1.719 2
900	3.193 0	1 327.9	1 475.6	1.780 4	2.898 4	1 327.3	1 474.8	1.769 4	2.652 9	1 326.6	1 473.9	1.759 3
1 000	3.440 3	1 369.0	1 528.2	1.817 7	3.124 1	1 368.5	1 527.4	1.805 8	2.860 5	1 367.9	1 526.7	1.796 8
1 200	3.929 5	1 453.3	1 635.0	1.886 3	3.570 0	1 452.9	1 634.5	1.875 5	3.270 4	1 452.5	1 634.0	1.865 2
1 400	4.414 4	1 540.8	1 745.0	1.948 8	4.011 6	1 540.5	1 744.6	1.938 1	3.675 9	1 540.2	1 744.2	1.928 4
1 600	4.896 9	1 631.7	1 858.3	2.006 6	4.450 7	1 631.5	1 858.0	1.996 0	4.078 9	1 631.3	1 857.7	1.986 3
1 800	5.377 7	1 726.2	1 974.9	2.060 7	4.888 2	1 726.0	1 974.7	2.050 1	4.480 3	1 725.8	1 974.5	2.040 4
2 000	5.857 5	1 823.9	2 094.9	2.111 6	5.324 7	1 823.8	2 094.7	2.101 0	4.880 7	1 823.6	2 094.6	2.091 3
	P = 350 psia (431.74 °F)				P = 400 psia (444.62 °F)				P = 450 psia (456.31 °F)			
Sat.	1.326 3	1 118.5	1 204.4	1.497 3	1.161 7	1 119.0	1 205.0	1.485 2	1.032 4	1 119.2	1 205.2	1.474 2
450	1.373 9	1 129.3	1 218.3	1.512 8	1.174 7	1 122.5	1 209.4	1.490 1				
500	1.492 1	1 155.2	1 251.9	1.548 7	1.285 1	1 150.4	1 245.6	1.528 8	1.123 3	1 145.4	1 238.9	1.510 3
550	1.600 4	1 178.6	1 282.2	1.579 5	1.384 0	1 174.9	1 277.3	1.561 0	1.215 2	1 171.1	1 272.3	1.544 1
600	1.703 0	1 200.6	1 310.9	1.607 3	1.476 5	1 197.6	1 306.9	1.589 7	1.300 1	1 194.6	1 302.8	1.573 7
650	1.801 8	1 221.9	1 338.6	1.632 8	1.565 0	1 219.4	1 335.3	1.615 8	1.380 7	1 216.9	1 331.9	1.600 5
700	1.897 9	1 242.8	1 365.8	1.656 7	1.650 7	1 240.7	1 362.9	1.640 1	1.458 4	1 238.5	1 360.0	1.625 3
800	2.084 8	1 284.1	1 419.1	1.700 9	1.816 6	1 282.5	1 417.0	1.684 9	1.608 0	1 280.8	1 414.7	1.670 6
900	2.267 1	1 325.3	1 472.2	1.741 4	1.977 7	1 324.0	1 470.4	1.725 7	1.752 6	1 322.7	1 468.6	1.711 7
1 000	2.446 4	1 366.9	1 525.3	1.779 7	2.135 8	1 365.8	1 523.9	1.763 6	1.894 2	1 364.7	1 522.4	1.749 9
1 200	2.799 6	1 451.7	1 633.0	1.848 3	2.446 5	1 450.9	1 632.0	1.833 1	2.171 8	1 450.1	1 631.0	1.819 6
1 400	3.148 4	1 539.6	1 743.5	1.911 1	2.752 7	1 539.0	1 742.7	1.896 0	2.445 0	1 538.4	1 742.0	1.882 7
1 600	3.494 7	1 630.8	1 857.1	1.969 1	3.056 5	1 630.3	1 856.5	1.954 1	2.715 7	1 629.8	1 856.10	1.940 9
1 800	3.839 4	1 725.4	1 974.0	2.023 3	3.358 6	1 725.0	1 973.6	2.008 4	2.984 7	1 724.6	1 973.2	1.995 2
2 000	4.183 20	1 823.3	2 094.2	2.074 2	3.659 7	1 823.0	2 093.9	2.059 4	3.252 7	1 822.6	2 093.5	2.046 2

TABLE A – 12 (Continued)

T /°F	v/(ft³/lbm)	u/(Btu/lbm)	h/(Btu/lbm)	s/(Btu/lbm·R)	v/(ft³/lbm)	u/(Btu/lbm)	h/(Btu/lbm)	s/(Btu/lbm·R)	v/(ft³/lbm)	u/(Btu/lbm)	h/(Btu/lbm)	s/(Btu/lbm·R)
	P = 500 psla (467.04 °F)				P = 600 psla (486.24 °F)				P = 700 psla (503.13 °F)			
Sat.	0.928 15	1 119.1	1 205.0	1.464 2	0.770 20	1 118.3	1 203.9	1.446 3	0.655 89	1 116.9	1 201.9	1.430 5
500	0.993 04	1 140.1	1 231.9	1.492 8	0.795 26	1 128.2	1 216.5	1.459 6				
550	1.079 74	1 167.1	1 267.0	1.528 4	0.875 42	1 158.7	1 255.9	1.499 6	0.727 99	1 149.5	1 243.8	1.473 0
600	1.158 76	1 191.4	1 298.6	1.559 0	0.946 05	1 184.9	1 289.9	1.532 5	0.793 32	1 177.9	1 280.7	1.508 7
650	1.233 12	1 214.3	1 328.4	1.586 5	1.011 33	1 209.0	1 321.3	1.561 4	0.852 42	1 203.4	1 313.8	1.539 3
700	1.304 40	1 236.4	1 357.0	1.611 7	1.073 16	1 231.9	1 351.0	1.587 7	0.907 69	1 227.2	1 344.8	1.566 6
800	1.440 97	1 279.2	1 412.5	1.657 6	1.190 38	1 275.8	1 408.0	1.634 8	1.011 25	1 272.4	1 403.4	1.615 0
900	1.572 52	1 321.4	1 466.9	1.699 2	1.302 30	1 318.7	1 463.3	1.677 1	1.109 21	1 316.0	1 459.7	1.658 1
1 000	1.700 94	1 363.6	1 521.0	1.737 6	1.410 97	1 361.4	1 518.1	1.716 0	1.203 81	1 359.2	1 515.2	1.697 4
1 100	1.827 26	1 406.2	1 575.3	1.773 5	1.517 49	1 404.4	1 572.9	1.752 2	1.296 21	1 402.5	1 570.4	1.734 1
1 200	1.952 11	1 449.4	1 630.0	1.807 5	1.622 52	1 447.8	1 627.9	1.786 5	1.387 09	1 446.2	1 625.9	1.768 5
1 400	2.198 8	1 537.8	1 741.2	1.870 8	1.829 57	1 536.6	1 739.7	1.85 01	1.565 80	1 535.4	1 738.2	1.832 4
1 600	2.443 0	1 629.4	1 855.4	1.929 1	2.034 0	1 628.4	1 854.2	1.908 5	1.741 92	1 627.5	1 853.1	1.891 1
1 800	2.685 6	1 724.2	1 972.7	1.983 4	2.236 9	1 723.4	1 971.8	1.963 0	1.916 43	1 722.7	1 970.9	1.945 7
2 000	2.927 1	1 822.3	2 093.1	2.034 5	2.438 7	1 821.7	2 092.4	2.014 1	2.089 87	1 821.0	2 091.7	1.996 9
	P = 800 psla (518.27 °F)				P = 1 000 psla (544.65 °F)				P = 1 250 psla (572.45 °F)			
Sat.	0.569 20	1 115.0	1 199.3	1.416 2	0.446 04	1 110.1	1 192.6	1.390 6	0.345 49	1 102.0	1 181.9	1.362 3
550	0.615 86	1 139.4	1 230.5	1.447 6	0.453 75	1 115.2	1 199.2	1.397 2				
600	0.677 99	1 170.5	1 270.9	1.486 6	0.514 31	1 154.1	1 249.3	1.445 7	0.378 94	1 129.5	1 217.2	1.396 1
650	0.732 79	1 197.6	1 306.0	1.519 1	0.564 11	1 185.1	1 289.5	1.482 7	0.427 03	1 167.5	1 266.3	1.441 4
700	0.783 30	1 222.4	1 338.4	1.547 6	0.608 44	1 212.4	1 325.0	1.514 0	0.467 35	1 198.7	1 306.8	1.477 1
750	0.831 02	1 246.0	1 369.1	1.573 5	0.649 44	1 237.6	1 357.8	1.541 8	0.503 44	1 226.4	1 342.9	1.507 6
800	0.876 78	1 268.9	1 398.7	1.597 5	0.688 21	1 261.7	1 389.0	1.567 0	0.536 87	1 252.2	1 376.4	1.534 7
900	0.964 34	1 313.3	1 456.0	1.641 3	0.761 36	1 307.7	1 448.6	1.612 6	0.598 76	1 300.5	1 439.0	1.582 6
1 000	1.048 41	1 357.0	1 512.2	1.681 2	0.830 78	1 352.5	1 506.2	1.653 5	0.656 56	1 346.7	1 498.6	1.624 7
1 100	1.130 24	1 400.7	1 568.0	1.718 1	0.897 83	1 396.9	1 563.1	1.691 1	0.711 84	1 392.2	1 556.8	16 635
1 200	1.210 51	1 444.6	1 623.8	1.752 8	0.963 27	1 441.4	1 619.7	1.726 3	0.765 45	1 437.4	1 614.5	1.699 3
1 400	1.367 97	1 534.2	1 736.7	1.817 0	1.091 01	1 531.8	1 733.7	1.791 1	0.869 44	1 528.7	1 729.8	1.764 9
1 600	1.522 83	1 626.5	1 851.9	1.875 9	1.216 10	1 624.6	1 849.6	1.850 4	0.970 72	1 622.2	1 846.7	1.824 6
1 800	1.676 06	1 721.9	1 970.0	1.930 6	1.339 56	1 720.3	1 968.2	1.905 3	1.070 36	1 718.4	1 966.0	1.879 9
2 000	1.828 23	1 820.4	2 091.0	1.981 9	1.461 94	1 819.1	2 089.6	1.956 8	1.168 92	1 817.5	2 087.9	1.931 5

TABLE A – 12 (Continued)

T /°F	v/(ft³/lbm)	u/(Btu/lbm)	h/(Btu/lbm)	s/(Btu/lbm·R)	v/(ft³/lbm)	u/(Btu/lbm)	h/(Btu/lbm)	s/(Btu/lbm·R)	v/(ft³/lbm)	u/(Btu/lbm)	h/(Btu/lbm)	s/(Btu/lbm·R)
	$P = 1\,500$ psla (596.26 °F)				$P = 1\,750$ psla (617.17 °F)				$P = 2\,000$ psla (536.85 °F)			
Sat.	0.276 95	1 092.1	1 169.1	1.336 2	0.226 81	1 080.5	1 153.9	1.311 2	0.188 15	1 066.8	1 136.4	1.286 3
600	0.281 89	1 097.2	1 175.4	1.342 3								
650	0.333 10	1 147.2	1 239.7	1.401 6	0.262 92	1 122.8	1 207.9	1.360 7	0.205 86	1 091.4	1 167.6	1.314 6
700	0.371 98	1 183.6	1 286.9	1.443 3	0.302 52	1 166.8	1 264.7	1.410 8	0.248 94	1 147.6	1 239.8	1.378 3
750	0.405 35	1 214.4	1 326.9	1.477 1	0.334 56	1 201.5	1 309.8	1.448 9	0.280 74	1 187.4	1 291.3	1.421 8
800	0.435 50	1 242.2	1 363.1	1.506 4	0.362 66	1 231.7	1 349.1	1.480 7	0.307 63	1 220.5	1 334.3	1.456 7
850	0.463 56	1 268.2	1 396.9	1.532 8	0.388 35	1 259.3	1 385.1	1.508 8	0.331 69	1 250.0	1 372.8	1.486 7
900	0.490 15	1 293.1	1 429.2	1.556 9	0.412 38	1 285.4	1 419.0	1.534 1	0.353 90	1 277.5	1 408.5	1.513 4
1 000	0.540 31	1 340.9	1 490.8	1.600 7	0.457 19	1 334.9	1 482.9	1.579 5	0.394 79	1 328.7	1 474.9	1.560 6
1 100	0.587 81	1 387.3	1 550.5	1.640 2	0.499 17	1 382.4	1 544.1	1.620 1	0.432 66	1 377.5	1 537.6	1.602 1
1 200	0.633 55	1 433.3	1 609.2	1.676 7	0.539 32	1 429.2	1 603.9	1.657 2	0.468 64	1 425.1	1 598.5	1.640 0
1 400	0.721 72	1 525.7	1 726.0	1.743 2	0.616 21	1 522.6	1 722.1	1.724 5	0.537 08	1 519.5	1 718.3	1.708 1
1 600	0.807 14	1 619.8	1 843.8	1.803 3	0.690 31	1 617.4	1 840.9	1.785 2	0.602 69	1 615.0	1 838.0	1.769 3
1 800	0.890 90	1 716.4	1 963.7	1.858 9	0.762 73	1 714.5	1 961.5	1.841 0	0.666 60	1 712.5	1 959.2	1.825 5
2 000	0.973 58	1 815.9	2 086.1	1.910 8	0.834 06	1 814.2	2 084.3	1.893 1	0.729 42	1 812.6	2 082.6	1.877 8
	$P = 2\,500$ psla (668.17 °F)				$P = 3\,000$ psla (695.41 °F)				$P = 3\,500$ psla			
Sat.	0.130 76	1 031.2	1 091.7	1.233 0	0.084 60	969.8	1 016.8	1.158 7				
650									0.024 92	663.7	679.9	0.863 2
700	0.168 49	1 098.4	1 176.3	1.307 2	0.098 38	1 005.3	1 059.9	1.196 0	0.030 65	760.0	779.9	0.951 1
750	0.203 27	1 154.9	1 249.0	1.368 6	0.148 40	1 114.1	1 196.5	1.311 8	0.104 60	1 057.6	1 125.4	1.243 4
800	0.229 49	1 195.9	1 302.0	1.411 6	0.176 01	1 167.5	1 265.3	1.367 6	0.136 39	1 134.3	1 222.6	13 224
850	0.251 74	1 230.1	1 346.6	1.446 3	0.197 71	1 208.2	1 317.9	1.408 6	0.158 47	1 183.8	1 286.5	1.372 1
900	0.271 65	1 260.7	1 386.4	1.476 1	0.216 40	1 242.8	1 362.9	1.442 3	0.176 59	1 223.4	1 337.8	1.410 6
950	0.290 01	1 289.1	1 423.3	1.502 8	0.233 21	1 273.9	1 403.3	1.471 6	0.192 45	1 257.8	1 382.4	1.442 8
1 000	0.307 26	1 316.1	1 458.2	1.527 1	0.248 76	1 302.8	1 440.9	1.497 8	0.206 87	1 289.0	1 423.0	1.471 1
1 100	0.339 49	1 367.3	1 524.4	1.571 0	0.277 32	1 356.8	1 510.8	1.544 1	0.232 89	1 346.1	1 496.9	1.520 1
1 200	0.369 66	1 416.6	1 587.6	1.610 3	0.303 67	1 408.0	1 576.6	1.585 0	0.256 54	1 399.3	1 565.4	1.562 7
1 400	0.426 31	1 513.3	1 710.5	1.680 2	0.352 49	1 507.0	1 702.7	1.656 7	0.299 78	1 500.7	1 694.8	1.636 4
1 600	0.480 04	1 610.1	1 832.2	1.742 4	0.398 30	1 605.3	1 826.4	1.719 9	0.339 94	1 600.4	1 820.5	1.700 6
1 800	0.532 05	1 708.6	1 954.8	1.799 1	0.442 37	1 704.7	1 950.3	1.777 3	0.378 33	1 700.8	1 945.8	1.758 6
2 000	0.582 95	1 809.4	2 079.1	1.851 8	0.485 32	1 806.1	2 075.6	1.830 4	0.415 61	1 802.9	2 072.1	1.812 1

TABLE A-12 (Continued)

T /°F	v/(ft³/lbm)	u/(Btu/lbm)	h/(Btu/lbm)	s/(Btu/lbm·R)	v/(ft³/lbm)	u/(Btu/lbm)	h/(Btu/lbm)	s/(Btu/lbm·R)	v/(ft³/lbm)	u/(Btu/lbm)	h/(Btu/lbm)	s/(Btu/lbm·R)
	P = 4 000 psla				P = 5 000 psla				P = 6 000 psla			
650	0.024 48	657.9	676.1	0.857 7	0.023 79	648.3	670.3	0.848 5	0.023 25	640.3	666.1	0.840 8
700	0.028 71	742.3	763.6	0.934 7	0.026 78	721.8	746.6	0.915 6	0.025 64	708.1	736.5	0.902 8
750	0.063 70	962.1	1 009.2	1.141 0	0.033 73	821.8	853.0	1.005 4	0.029 81	788.7	821.8	0.974 7
800	0.105 20	1 094.2	1 172.1	1.273 4	0.059 37	986.9	1 041.8	1.158 1	0.039 49	897.1	941.0	1.071 1
850	0.128 48	1 156.7	1 251.8	1.335 5	0.085 51	1 092.4	1 171.5	1.259 3	0.058 15	1 018.6	1 083.1	1.181 9
900	0.146 47	1 202.5	1 310.9	1.379 9	0.103 90	1 155.9	1 252.1	1.319 8	0.075 84	1 103.5	1 187.7	1.260 3
950	0.161 76	1 240.7	1 360.5	1.415 7	0.118 63	1 203.9	1 313.6	1.364 3	0.090 10	1 163.7	1 263.7	1.315 3
1 000	0.175 38	1 274.6	1 404.4	1.446 3	0.131 28	1 244.0	1 365.5	1.400 4	0.102 08	1 211.4	1 324.7	1.357 8
1 100	0.199 57	1 335.1	1 482.8	1.498 3	0.152 98	1 312.2	1 453.8	1.459 0	0.122 11	1 288.4	1 424.0	1.423 7
1 200	0.221 21	1 390.3	1 554.1	1.542 6	0.171 85	1 372.1	1 531.1	1.507 0	0.139 11	1 353.4	1 507.8	1.475 8
1 300	0.241 28	1 443.0	1 621.6	1.582 1	0.189 02	1 427.8	1 602.7	1.549 0	0.154 34	1 412.5	1 583.8	1.520 3
1 400	0.260 28	1 494.3	1 687.0	1.618 2	0.205 08	1 481.4	1 671.1	1.585 8	0.168 41	1 468.4	1 655.4	1.559 8
1 600	0.296 20	1 595.5	1 814.7	1.683 5	0.235 05	1 585.6	1 803.1	1.654 2	0.194 38	1 575.7	1 791.5	1.629 4
1 800	0.330 33	1 696.8	1 941.4	1.742 2	0.263 20	1 689.0	1 932.5	1.714 2	0.218 53	1 681.1	1 923.7	1.690 7
2 000	0.363 35	1 799.7	2 068.6	1.796 1	0.290 23	1 793.2	2 061.7	1.768 9	0.241 55	1 786.7	2 054.9	1.746 3

The temperature in parentheses is the saturation temperature at the specified pressure.

Properties of saturated vapor at the specified pressure.

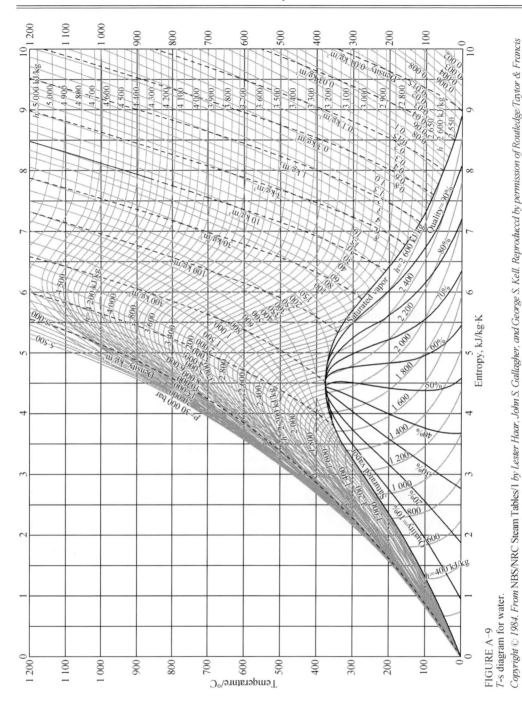

FIGURE A-9
T-s diagram for water.
Copyright © 1984. From NBS/NRC Steam Tables/1 by Lester Haar, John S. Gallagher, and George S. Kell. Reproduced by permission of Routledge/Taylor & Francis Books, Inc.

Fig. B-1 T-s diagram for water

FIGURE B−1 $T - s$ diagram for water

APPENDIX

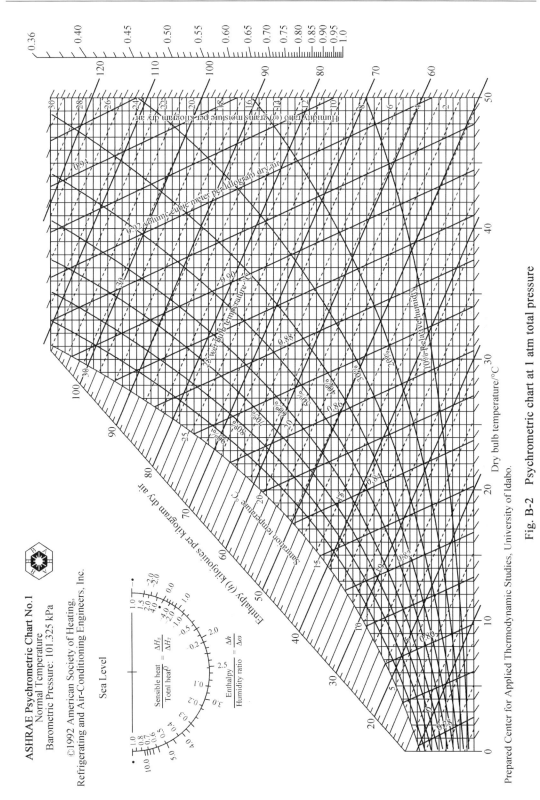

FIGURE B-2 Psychrometric chart at 1 atm total pressure

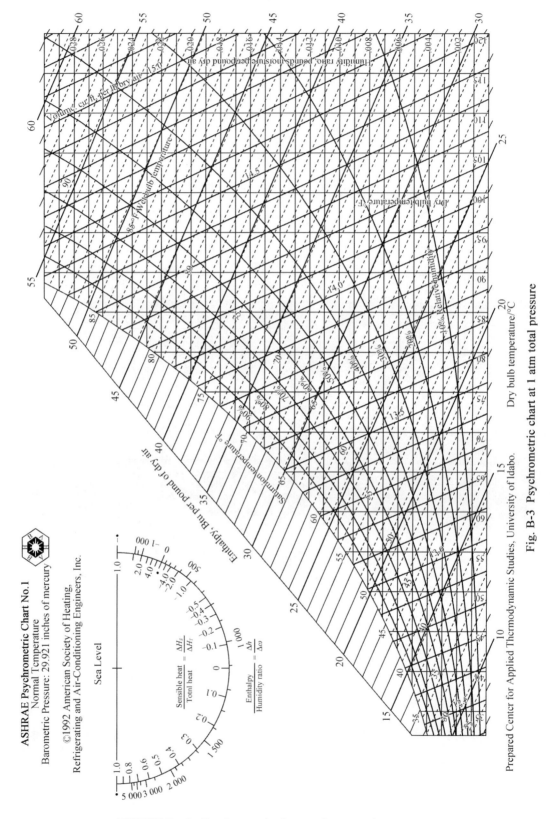

FIGURE B – 3 Psychrometric chart at 1 atm total pressure

APPENDIX

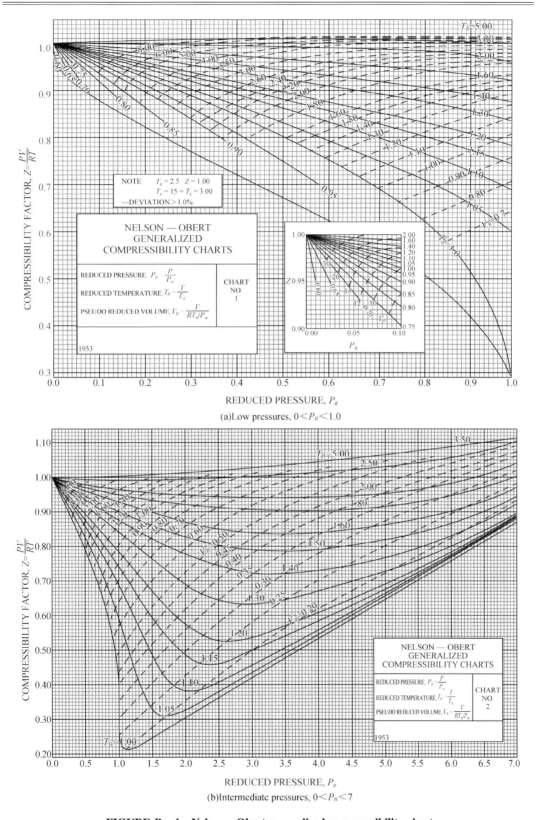

FIGURE B – 4 Nelson – Obert generalized compressibility chart

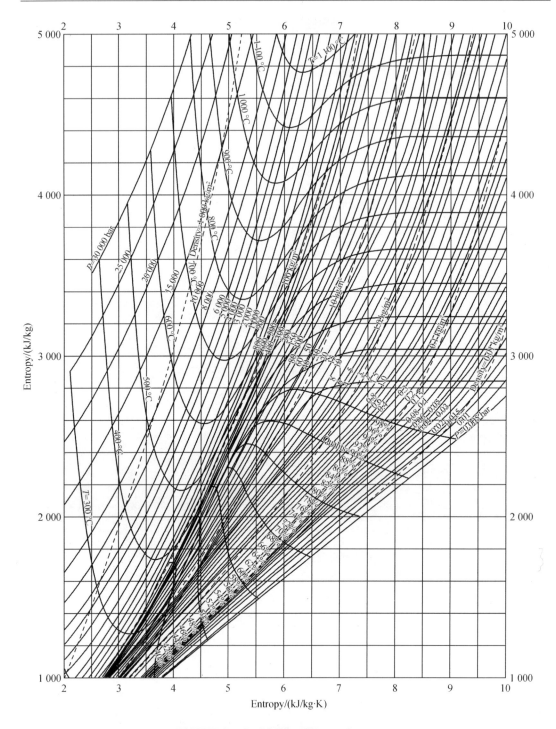

FIGURE B-5 Mollier diagram for water